T0345327

Handbook of Artificial Intelligence and Wearables

The ever-changing world of wearable technologies makes it difficult for experts and practitioners to keep up with the most recent developments. This handbook provides a solid understanding of the significant role that AI plays in the design and development of wearable technologies along with applications and case studies.

Handbook of Artificial Intelligence and Wearables: Applications and Case Studies presents a deep understanding of AI and its involvement in wearable technologies. The book discusses the key role that AI plays and goes on to discuss the challenges and possible solutions. It highlights the more recent advances along with real-world approaches for the design and development of the most popular AI-enabled wearable devices such as smart fitness trackers, AI-enabled glasses, sports wearables, disease diagnostic devices, and more, complete with case studies.

This book will be a valuable source for researchers, academics, technologists, industrialists, practitioners, and all people who wish to explore the applications of AI and the part it plays in wearable technologies.

Handbook of Artificial Intelligence and Wearables

Applications and Case Studies

Edited by
Hemachandran K, Manjeet Rege,
Zita Zoltay Paprika, K. V. Rajesh Kumar, and
Shahid Mohammad Ganie

CRC Press
Taylor & Francis Group
Boca Raton London New York

CRC Press is an imprint of the
Taylor & Francis Group, an **informa** business

Designed cover image: iStock

MATLAB® and Simulink® are trademarks of The MathWorks, Inc. and are used with permission. The MathWorks does not warrant the accuracy of the text or exercises in this book. This book's use or discussion of MATLAB® or Simulink® software or related products does not constitute endorsement or sponsorship by The MathWorks of a particular pedagogical approach or particular use of the MATLAB® and Simulink® software.

First edition published 2024

by CRC Press
2385 NW Executive Center Drive, Suite 320, Boca Raton FL 33431

and by CRC Press
4 Park Square, Milton Park, Abingdon, Oxon, OX14 4RN

CRC Press is an imprint of Taylor & Francis Group, LLC

© 2024 selection and editorial matter, Hemachandran K, Manjeet Rege, Zita Zoltay Paprika, K. V. Rajesh Kumar, Shahid Mohammad Ganie; individual chapters, the contributors

Reasonable efforts have been made to publish reliable data and information, but the author and publisher cannot assume responsibility for the validity of all materials or the consequences of their use. The authors and publishers have attempted to trace the copyright holders of all material reproduced in this publication and apologize to copyright holders if permission to publish in this form has not been obtained. If any copyright material has not been acknowledged please write and let us know so we may rectify in any future reprint.
Except as permitted under U.S. Copyright Law, no part of this book may be reprinted, reproduced, transmitted, or utilized in any form by any electronic, mechanical, or other means, now known or hereafter invented, including photocopying, microfilming, and recording, or in any information storage or retrieval system, without written permission from the publishers.

For permission to photocopy or use material electronically from this work, access www.copyright.com or contact the Copyright Clearance Center, Inc. (CCC), 222 Rosewood Drive, Danvers, MA 01923, 978-750-8400. For works that are not available on CCC please contact mpkbookspermissions@tandf.co.uk

Trademark notice: Product or corporate names may be trademarks or registered trademarks and are used only for identification and explanation without intent to infringe.

ISBN: 978-1-032-68493-2 (hbk)
ISBN: 978-1-032-68670-7 (pbk)
ISBN: 978-1-032-68671-4 (ebk)

DOI: 10.1201/9781032686714

Typeset in Times New Roman
by Deanta Global Publishing Services, Chennai, India

Contents

About the Editors

Hemachandran K is the Director of the AI Research Centre and Area Chair of the Analytics Department at Woxsen University, India. He is an ambassador of the AI Accelerator Institute and an Advisory Board member in many international and national companies such as AptAI Labs, USA, Agzitence Pvt, Ltd, and many more. He served as an effective resource person at various national and international scientific conferences and also gave lectures on topics related to artificial intelligence. He is currently serving as an Expert at UNESCO and ATL Mentor of Change. He has edited seven books and authored three books. He was bestowed a Best University Faculty Award at Woxsen University for two consecutive years (2022–2023 and 2021–2022) and also at Ashoka Institute of Engineering and Technology in 2019–2020. He has expertise in natural language processing, computer vision, building video recommendation systems, and autonomous robots. He is working on various real-time use cases and projects in collaboration with industries across the globe.

Manjeet Rege is a Professor and Chair of the Department of Software Engineering and Data Science; and Director of the Center for Applied Artificial Intelligence at the University of St. Thomas, USA. He is an author, mentor, thought leader, and frequent public speaker on big data, machine learning, and artificial intelligence technologies. He is also the co-host of the "All Things Data" podcast that brings together leading data scientists, technologists, business model experts, and futurists to discuss strategies to utilize, harness, and deploy data science, data-driven strategies, and enable digital transformation. Apart from being engaged in research, he regularly consults with various organizations to provide expert guidance for building Big Data and AI practice and applying innovative data science approaches. He has published in various peer-reviewed reputed publications such as IEEE Transactions on Knowledge and Data Engineering, *Data Mining & Knowledge Discovery Journal*, IEEE International Conference on Data Mining, and the World Wide Web Conference. He is on the editorial review board of the *Journal of Computer Information Systems* and regularly serves on the program committees of various international conferences.

Zita Zoltay Paprika graduated from the Karl Marx University of Budapest, Hungary in 1981 and received her university doctorate degree from the same institution in 1983. She worked at the Bureau for Systems Analysis of the State Committee for Technical Development as an analyst till 1987. She joined the Business Economics Department of the Budapest University of Economic Sciences, Hungary in 1988. She got her PhD in Business Administration in 1999. In 2000 she was appointed as the Director for International Affairs at the Budapest University of Economic Sciences and Public Administration, Hungary. Her main fields of interest are Decision Sciences and Business Economics. She was involved in different international and national projects during which activity she was project manager on several occasions. She was also acting as the USAID auditor in project evaluations. In 2005–2006 she got a Fulbright Research Scholarship and spent an academic year at California State University Sacramento, USA. She was the Chair of the Department of Decision

Sciences from 2005 to 2020 and was appointed as the Program Director of the International Study Programs at the Faculty of Business Administration, Corvinus University of Budapest from 2007 to 2014. From January 1st, 2014 to January 31st, 2020 she was the Dean of the Corvinus Business School at Corvinus University of Budapest. During this period Corvinus Business School achieved EQUIS and AMBA accreditations. From February 1st, 2020 until 15 October she was appointed as Vice President of International Relations and Accreditations at Corvinus University of Budapest. She served on the EFMD Programme Accreditation Board from 2016 to 2022. She is a member of the International Advisory Board of ESSCA School of Management and Burgundy School of Business in France. Since 2021 she has been a member of the Chinese European International Business School (CEIBS) International Advisory Board. Dr. Zoltay Paprika has been involved in several EQUIS and EFMD Programme accreditations as a Chair or PRT member. She is an editorial board member of the *Central European Business Review* and the *International Journal of Business Research and Management*. In January 2023 she was appointed as an Associate Director of EFMDa.

K. V. Rajesh Kumar is currently working as an Assistant Professor at the School of Business in the business analytics area at Woxsen University, Hyderabad, India. He is a holding Chair Professor of France Belanger of Information Systems. He is serving as a Co-Chairperson of the Centre of Excellence – Sports Analytics and Management and also he is Advertflair Chair Professor in Marketing and Advertising (AI Projects). He earned his MTech in Embedded Systems from JNTU, Kakinada, and AMIE in ECE from the Institution of Engineers, Kolkata. He did research and his PhD at VIT University, Chennai, India. He was earlier associated with DRDO – DIPAS Govt. of India as a Junior Research Fellow and upGrad Educational Private Limited where he guided students in selecting research problems and in writing a thesis, to deal with AI/ML models, for those who are pursuing M.Sc. in Data Science/AI ML courses from Liverpool John Moores University, UK, through upGrad. His research specialization is Human Computer Interface based signal Data Analytics for Bio-Medical Applications, Machine, and Deep Learning, and Business Analytics. He is a reviewer for the *American Journal of Artificial Intelligence*. Recently he presented a talk on sports analytics at the Data Con LA+ IM Data 2022 conference, California, US.

Shahid Mohammad Ganie is an Assistant Professor at the AI Research Centre in the School of Business at Woxsen University, India. He is a holding Chair Professor of Manjeet Rege Professor of Data Science and AI. He is serving as a Co-Chairperson of the Centre of Excellence – SCIN Data. He obtained his PhD (Machine Learning in Healthcare Informatics) and a master's degree in Computer Applications (MCA), where he was a Gold Medalist, from the School of Mathematical and Computer Sciences, BGSB University, Jammu and Kashmir, India. He also holds a BSc in Mathematics from the University of Kashmir. His research interests include Data Analytics, Machine Learning, Deep Learning, Healthcare Informatics, Smart Healthcare Systems, and Medical Image Computing. He has published various research articles in refereed journals, edited books, and reputed international conferences. In addition, he serves as a reviewer for international/national reputed journals such as the *Journal of Big Data, Applied Artificial Intelligence, Frontiers in Physiology, Connection Science,* and *Healthcare Analytics.* Furthermore, he is a member of the AI Research Centre at Woxsen University, India. He holds a research affiliation with the Woxsen Future Lab (Interdisciplinary/Multidisciplinary Research Lab) at Woxsen University. Also, he serves as a member of the Technology Advisory Board, AI Research Centre, Woxsen University.

List of Contributors

Sami Alshmrany
Faculty of Computer and Information Systems, Islamic University of Madinah, Saudi Arabia

M. Anand
DR. M.G.R Educational and Research Institute, Chennai, Tamil Nadu-641014, India

Harishchander Anandaram
Centre for Excellence in Computer Engineering and Networking, Amrita School of Artificial Intelligence, Coimbatore, Tamil Nadu, India

Abdul Basit Andrabi
Assistant Professor Department of Computer Engineering, Islamic University of Science and Technology (IUST) Awantipora, J&K, India

Syed Immamul Ansarullah
Department of Computer Application, GDC Sumbal J&K, India

Pisipati Sai Anurag
Woxsen University, Hyderabad, India

Ezendu Ariwa
Professor, Warwick University, United Kingdom

Saliha Bathool
Woxsen University, Hyderabad, India

Ashaq Hussain Bhat
Department of Computer Sciences, South Campus, University of Kashmir, J&K, India

Mandala Bhuvana
SR University, Warangal, Telangana-506371, India

Xavier Borah
Woxsen University, Hyderabad, India

Sourav Chakraborty
Woxsen University, Hyderabad, India

Vinay Chittiprolu
Assistant Professor, Paari School of Business, SRM University Andhra Pradesh, India

Channabasava Chola
Department of Electronics and Information Convergence Engineering, College of Electronics and Information, Suwon, South Korea

Korhan Cingez
Department of Information Technologies, University of Hradec Kralove, Czech Republic

Jagadeesan D.
Apollo University, Murukambattu, Chittoor, Andhra Pradesh, India

Kothandaraman D.
SR University, Warangal, Telangana, India

S. S. Darly
Assistant Professor, University College of Engineering, Tindivanam, Anna University Chennai, India

B. Rupa Devi
Associate Professor, Dept of CSE,
  Annamacharya Institute of Technology
  and Sciences, Tirupati-A.P., India

Aksh Dharaskar
Woxsen University, Hyderabad, India

Alexander Didenko
Head of the Laboratory of Managerial
  Neuroscience, IBS-Moscow,
  RANEPA, Moscow, Russia

Chinna Swamy Dudekula
Engineering and Environment
  Department, Northumbria
  University, Newcastle, UK

M. Reddi Durgasree
Assistant Professor, Department of CSE
  (AIML), Guru Nanak, Institutions
  Technical Campus, Hyderabad,
  Telangana, India

Asha G.
Associate Professor, Department of
  Electronics and Communication
  Engineering, Adhipatasakthi College
  of Engineering, Kalavai, Tamil
  Nadu, India

Vijaya Gunturu
SR University, Warangal, Telangana,
  India

Kyung Hee
University, Suwon-si 17104, Republic of
  Korea

Humera
Assistant Professor, Department of
  Computer Science and Engineering,
  HKBK College of Engineering, Sri
  Venkateswara University, Tirupati,
  Andhra Pradesh

Vanshhita Jaju
University of Texas, Austin, US

Potti Jaswanth
Woxsen University, Hyderabad, India

Hemachandran K
Professor, School of Business, Woxsen
  University, Hyderabad, India

Rajchandar K.
SR University, Warangal, Telangana,
  India

Gabriel Kabanda
Secretary General, Zimbabwe Academy
  of Sciences, Zimbabwe

Karan Kadhir
Teaching Fellow, University College
  of Engineering, Tindivanam, Anna
  University Chennai, India

Hilal Ahmad Khanday
Department of Computer Sciences,
  South Campus, University of
  Kashmir, J&K, India

Mudasir Manzoor Kirmani
Assistant Professor (Computer Science)
  Division of Social Science, FoFy,
  SKUAST-Kashmir, India

K. Dinesh Kumar
Assistant Professor, School of
  Computing, Amrita Vishwa
  Vidyapeetham University,
  Bengaluru, India

P. Senthil Kumar
Professor, CSE-Department, Shadan
  Women's College of Engineering and
  Technology, Hyderabad, India

Pokala Pranay Kumar
MPS Data Science, University of
Maryland, Baltimore County, US

K. Shiridi Kumar
Woxsen University, Hyderabad, India

Neelam Kumari
Dublin Business School Dublin, Ireland

Emmanuel Lawenrence
Woxsen University, Hyderabad, India

Mantosh Mahapatra, India
Woxsen University, Hyderabad, India

V. Malathy
SR University, Warangal, Telangana,
India

Geetha Manoharan
SR University, Warangal, Telangana,
India

Neha Medimi
Woxsen University, Hyderabad, India

Neeraj Kumar Mishra
Indian Institute of Science Education
and Research, Kolkata, Nadia, India

Swagatika Mohapatro
Woxsen University, Hyderabad, India

Badrinath N.
Associate Professor, Department
of Software Systems, School of
Computer Science and Engineering,
Vellore Institute of Technology,
Vellore, India

Rohit Naluri
Woxsen University, Hyderabad, India

Panuganti Naveen
Woxsen University, Hyderabad, India

M. S. Nidhya
Associate Professor, Dept of CS &
IT, Jain Deemed-to-be University,
Bangalore, Karnataka, India

Krishnachaitanya Pabbu
Woxsen University, Hyderabad, India

Ranjith Kumar Painam
Associate Professor, Department of
ECE, Kallam Haranadhareddy
Institute of Technology
(Autonomous), NH-16,
Chowdavaram, Guntur, Andhra
Pradesh, India

Zita Zoltay Paprika
Former Dean, Corvinus Business
School, Budapest, Hungary

Kashika Parmar
Woxsen University, Hyderabad, India

Sushanth Kumar Patnaik
Woxsen University, Hyderabad, India

Anil Pise
Senior Data Scientist, University of the
Witwatersrand, South Africa

Philipp Plugmann
Professor for Interdisciplinary
Periodontology and Prevention, SRH
Hochschule für Gesundheit Gera,
Germany

Puniethaa Prabhu
Department of Biotechnology, KSRCT,
Tamil Nadu, India

G. Prudhviraj
Professor, CSE-Department, Siddhartha
    Institute of Technology and Science,
    Hyderabad, India

Jeevana Jyothi Pujari
Assistant Professor, Department
    of Database Systems, School of
    Computer Science and Engineering,
    Vellore Institute of Technology,
    Vellore, India

A. Naga Sai Purushotham
Woxsen University, Hyderabad, India

Marepalli Radha
CVR College of Engineering,
    Hyderabad, India

Karthik Ramesh
Chief Digital and Innovation Officer,
    ThinkInfinity, Dubai, United Arab
    Emirates

Varagani Ramu
SR University, Warangal, Telangana,
    India

Parag Raut
Woxsen University, Hyderabad, India

Ravali Gunda Ravali
Woxsen University, Hyderabad, India

C. Sateesh Kumar Reddy
Assistant Professor, Advanced CSE,
    School of Computing & Informatics,
    Vignan's Foundation for Science,
    Technology and Research,
    Vadlamudi, Guntur, India

Dumpa Sree Harsha Reddy
Woxsen University, Hyderabad, India

G. Ramasubba Reddy
Professor & Head, Department of
    Computer Science and Engineering,
    Sai Rajeswari Institute of
    Technology, Proddatur, Andhra
    Pradesh, India

Kesireddy Rajashekar Reddy
SR University, Warangal, Telangana,
    India

M. V. Subba Reddy
Professor & Head, Department of
    Computer Science and Engineering,
    Sai Rajeswari Institute of
    Technology, Proddatur, Andhra
    Pradesh, India

Yamasani Keerthi Reddy
Woxsen University, Hyderabad, India

Manjeet Rege
Professor and Chair, Department of
    Software Engineering and Data
    Science, University of St. Thomas,
    MN, US

Rishabh
Woxsen University, Hyderabad, India

Raul V. Rodriguez
Vice President, Woxsen University,
    Hyderabad, India

Aluri Anand Sai
Woxsen University, Hyderabad, India

Potnuru Sai Santhosh
Woxsen University, Hyderabad, India

Krishna Saraf
Woxsen University, Hyderabad, India

Delukshi Shanmugarajah
Middlesex University, London, UK

Pradeep Sharma
Assistant Professor, Paari School of
Business, SRM University, India

Vishal Kumar Sharma
Woxsen University, Hyderabad, India

Mulagada Surya Sharmila
Woxsen University, Hyderabad, India

R. Shruthi
Woxsen University, Hyderabad, India

Someshwar
Woxsen University, Hyderabad, India

Pankaj Kumar Singh
Woxsen University, Hyderabad, India

Saurav Sulam
University of Texas, Austin, US

Konatham Sumalatha
Assistant Professor, Department
of Database Systems, School of
Computer Science and Engineering,
Vellore Institute of Technology,
Vellore, India

P. Sunitha
Associate Professor, Aditya
Engineering College(A),
Surampalem, Affiliated to JNTUK,
A.P, India

Thangam
Professor, Dr. MGR Educational &
Research Institute, University,
Chennai, India

Pranavitha V.
Woxsen University, Hyderabad, India

Bochu Sai Vardhan
SR University, Warangal, Telangana,
India

K. Vijay
Professor, CSE-Department, Siddhartha
Institute of Technology and Science,
Hyderabad, India

G. Yamini
Assistant Professor, PG & Research
Department of Computer Science,
National College (Autonomous),
Tiruchirappalli, Tamil Nadu, India

Rizwan Zhad
Woxsen University, Hyderabad, India

Preface

In the midst of the technological revolution that is reshaping our world, the fusion of artificial intelligence (AI) and wearables stands as a defining milestone. The synergy between these two cutting-edge fields has unlocked unprecedented possibilities, redefining how we interact with technology and empowering us to lead healthier, more efficient, and more connected lives. It is with immense excitement and pride that we present the *Handbook of Artificial Intelligence Wearables: Applications and Case Studies*.

This handbook is the culmination of the collaborative efforts of pioneering researchers, engineers, and visionaries who have been at the forefront of this rapidly evolving domain. Together, they have embarked on a journey to explore the potential of wearables enhanced by artificial intelligence, pushing the boundaries of innovation and revolutionizing industries across the board.

The aim of this comprehensive volume is to provide readers with a holistic understanding of the vast landscape of AI wearables. From healthcare to manufacturing, from fitness to entertainment, the diverse applications and case studies presented in this book span a wide range of disciplines, demonstrating the versatility and transformative power of this technology.

As editors, we sought to assemble a collection that would resonate with both seasoned professionals and newcomers to the field. Each chapter in this book delves into its respective area with rigor and depth, offering practical insights, theoretical underpinnings, and hands-on experiences that will inspire readers to explore the myriad of possibilities AI wearables offer.

The Handbook of Artificial Intelligence Wearables is more than just a compilation of knowledge; it is a testament to the spirit of innovation that drives our society forward.

We extend our heartfelt gratitude to the contributors whose expertise and passion shine through in their work. Their dedication to pushing the boundaries of AI wearables has made this book an invaluable resource for researchers, practitioners, and students alike.

We hope that this book provides a comprehensive overview of AI wearables in diverse fields, and how it is transforming different sectors. We believe that this book will be a valuable resource for academics, researchers, professionals, and policymakers who are interested in understanding the potential of AI wearables.

We would like to thank all the contributors who have made this book possible, and we hope that readers will find it informative and thought-provoking.

MATLAB® is a registered trademark of The MathWorks, Inc. For product information, please contact:

The MathWorks, Inc.
3 Apple Hill Drive
Natick, MA 01760-2098 USA
Tel: 508 647 7000
Fax: 508-647-7001
E-mail: info@mathworks.com
Web: www.mathworks.com

1 Artificial Intelligence (AI) Applications for Smart Wearables

*Ashaq Hussain Bhat, Puniethaa Prabhu,
Hilal Ahmad Khanday, and Korhan Cingez*

INTRODUCTION

The emergence of smart wearables has ushered in a new era of technology that smoothly connects with our daily lives. These wearable gadgets are intended to be small, light, and comfortable to wear on the body, making them useful traveling companions. They can gather data, interpret information, and communicate with other devices thanks to a variety of sensors, CPUs, and wireless networking choices. The development of smart wearables with a variety of functions has been made possible by developments in miniaturization and sensor technology. The industry provides a broad variety of wearable devices catering to varied requirements and tastes, ranging from smartwatches and fitness trackers to smart glasses and wearables. These gadgets frequently have sophisticated aesthetics, sharp visuals, and simple user interfaces, making them both technologically and artistically appealing. The rising need for individualized, interconnected, and data-driven experiences is largely to blame for the emergence of smart wearables. Smart wearables give people the ability to measure and monitor their progress in real time as they work toward improving their overall health, fitness, and well-being [1], [2]. Additionally, by integrating smart wearables with smartphones and other smart devices, seamless communication and synchronization are made possible, resulting in a cohesive ecosystem of gadgets that coexist peacefully [3].

The development of smart wearables is powered by ongoing innovation and teamwork among IT firms, medical professionals, fashion designers, and researchers. Breakthroughs in fields like biometric sensing, battery efficiency, and user interface design have been made possible by this collaborative approach. Due to improvements in accuracy, robustness, and usability, smart wearables are now widely used and accepted. Additionally, the increased focus on data analytics and machine learning has made it possible for smart wearables to offer predictive analytics and actionable insights. These gadgets can provide individualized advice, spot patterns, and even foretell prospective health concerns by evaluating the data gathered from users' daily activities, habits, and vital signs. By enabling people to take proactive steps to

DOI: 10.1201/9781032686714-1

1

preserve their well-being, this data-driven approach has the potential to transform preventative healthcare [4], [5].

The smart wearable technology may gather data, deliver real-time feedback, and provide a variety of functions and applications [6]. Several of the crucial uses include those outlined in the following subsections:

FITNESS AND HEALTH MONITORING

Fitness and health monitoring are two areas where smart wearables are widely used. Heart rate, sleep habits, steps done, and calories burned may all be tracked by wearable fitness trackers like smartwatches and activity bands. They give users information on their levels of physical activity, the quality of their sleep, and their general health. These gadgets frequently have companion smartphone applications that let users establish objectives, monitor progress, and get tailored advice.

MEDICAL AND HEALTHCARE

The medical and healthcare industries are using smart wearables more and more often. They allow for remote patient monitoring, which enables medical practitioners to collect information in real time on patients' vital signs, medication compliance, and general health. Wearable glucose monitors assist those with diabetes in controlling their blood sugar levels, while smartwatches with integrated electrocardiogram (ECG) capabilities can identify abnormal cardiac rhythms. The development of wearable technology has the potential to save healthcare expenditures while improving patient outcomes.

PRODUCTIVITY AND PERSONAL ASSISTANCE

Smart wearables also come with functions that can help with personal aid and productivity. In order to provide users with hands-free access to instructions, data, and communication tools, devices like smart glasses and augmented reality (AR) headsets may overlay digital content onto the user's field of view. Users may check alerts, make calls, send messages, and operate other connected devices without a smartphone thanks to smartwatches and smart rings with voice assistants.

SAFETY AND SECURITY

Safety and security aspects are increasingly being included in wearables. Smart jewelry and accessories with panic buttons or location monitoring, for example, can be utilized for personal safety and emergency scenarios. Some smart helmets for cyclists and motorcyclists include built-in sensors that detect accidents and immediately send distress signals to emergency contacts. These gadgets give consumers an additional degree of security and peace of mind.

FASHION AND LIFESTYLE

As designers incorporate technology into clothing and accessories, the area of smart wearables is also fused with fashion and lifestyle. Smart textiles and materials can regulate ambient conditions, monitor body temperature, and even charge electrical gadgets. While utilizing the advantages of wearable technology, users may express their individual style with fashionable smartwatches and reversible smart bands.

FUTURE POTENTIAL

As technology develops, more and more uses for smart wearables are becoming possible. Wearables will be able to offer more precise and individualized insights thanks to integration with artificial intelligence (AI), machine learning, and data analytics. Additionally, the development of elastic and flexible electronics will result in the production of wearable technology that is more comfortable and easily blends in with daily attire.

ARTIFICIAL INTELLIGENCE (AI) TECHNIQUES

Through the use of AI approaches, robots are now able to replicate human-like intellect and decision-making processes, revolutionizing the way humans interact with technology. These methods cover a wide range of algorithms, procedures, and strategies that make it easier to create intelligent systems that can reason, learn, and solve problems. As AI develops, more and more industries are utilizing it, including those in healthcare, finance, transportation, and entertainment, among others. Technology has undergone a revolution thanks to artificial intelligence approaches, which now allow robots to complete jobs that previously required only human intelligence. These methods, which improve productivity, efficiency, and decision-making processes, continue to advance and find applications in a variety of sectors, ranging from machine learning and natural language processing to computer vision and deep learning. People may anticipate much more fascinating developments as AI research advances, which will influence the direction of technology and alter how people live and interact with intelligent technologies.

MACHINE LEARNING

A subfield of artificial intelligence called "machine learning" focuses on creating algorithms that let computers learn from data and make predictions or judgments without having to be explicitly programmed. It includes teaching computers to spot patterns in massive volumes of data so they may base their predictions and judgments on that knowledge. With the help of supervision, the algorithm can learn from examples and forecast results for future, unlabeled data by being fed labeled data [7], [8]. Important ideas include reinforcement learning (sequential decision-making based on rewards), unsupervised learning (identifying patterns in unlabeled data), and supervised learning (using labeled examples for prediction). Preprocessing for wearable data includes data cleansing and transformation, whereas feature extraction concentrates on obtaining important data [9], [10]. It is crucial to comprehend these ideas in order to apply machine learning to real-world issues. Machine learning is critical for deriving important insights from massive volumes of data gathered by wearable sensors. It allows for pattern identification, anomaly detection, tailored treatments, and real-time monitoring, and it feeds R&D activities. We can unleash the full potential of wearable devices by leveraging the power of machine learning, improving healthcare outcomes, enhancing well-being, and furthering our understanding of human behavior [11]–[15]. Personalized health, well-being, real-time monitoring, intervention, enhanced research, and development are some of the essential benefits of machine learning in smart wearables.

Natural Language Processing (NLP)

The subject of AI known as "natural language processing" is a fast-evolving area of research that aims to integrate computers and human language. NLP, which enables computers to comprehend, decipher, and produce human language, is another important AI method. It centers around giving robots the ability to understand, interpret, and produce human language in a way that is contextually meaningful and pertinent. As technology becomes more intuitive and user-friendly, NLP plays a critical role in changing how we engage with it. NLP enables computers to interpret enormous volumes of textual data, extract insightful information, and carry out a wide range of activities that were previously only possible with human intellect. This is done by using complex algorithms and models. Natural language processing, a rapidly developing field of study in AI, attempts to combine computers with human language. Its main goal is to enable robots to comprehend, analyze, and generate human language in a way that is appropriate and meaningful in context. NLP is crucial in transforming how we interact with technology as it gets more intuitive and user-friendly. NLP gives computers the ability to decipher massive amounts of textual data, extract insightful information, and perform a variety of tasks that were previously only achievable with human intelligence. Complex algorithms and models are used for this. Every industry and use case that NLP has entered has benefited from its distinctive characteristics.

NLP is used in the healthcare industry to extract important data from research articles and medical records, assisting in patient care, medication discovery, and diagnosis. The usage of chatbots and virtual assistants has significantly improved customer service since they participate in natural interactions with consumers and can answer questions and handle problems more quickly. Financial institutions use NLP to sift through a massive volume of financial news and data and make quick choices about their investments. NLP also enables speech recognition software and voice-activated smart gadgets, enabling hands-free interactions with technology.

Despite its great successes, NLP still confronts difficulties. Ambiguity, metaphorical language, and contextual comprehension are still difficult problems. The

NLP community is very concerned with ensuring the ethical use of NLP technology, including minimizing biases found in language models. To overcome these difficulties and expand the frontiers of NLP, researchers and developers are working nonstop. As NLP develops, it has the potential to fundamentally alter how people and machines interact, bringing us one step closer to a day when technology is seamlessly integrated into our daily lives and can recognize and cater to our wants in the same way that people do. NLP is destined to influence the next wave of AI applications and change how we interact with the digital world via continual study, cooperation, and ethical concern.

COMPUTER VISION (CV)

A crucial AI method that enables robots to comprehend and interpret visual data is computer vision. Face identification, object detection, and picture recognition are just a few of the duties involved. The use of CV has a substantial economic impact and enhances efficiency and safety in autonomous cars, surveillance systems, and medical image analysis. AI approaches also include knowledge representation and reasoning, in which data is organized in a way that computers can comprehend and utilize for logical deductive reasoning. Knowledge graphs, ontologies, and semantic networks are a few of the techniques used to describe and arrange data, allowing AI systems to infer conclusions and respond to challenging inquiries.

Computer vision is an interdisciplinary area of artificial intelligence that tries to give computers the capacity to analyze and comprehend visual data from the environment, much like human vision. Computer vision enables machines to analyze, evaluate, and gain useful insights from digital pictures and videos through the creation of algorithms and models. The main goal is to simulate human visual perception and cognition, enabling computers to recognize objects, sceneries, and patterns as well as extract meaningful information from visual input. Thanks to developments in deep learning and the accessibility of big picture datasets for complex neural network architecture training, this technology has advanced remarkably in recent years. The process of extracting characteristics and patterns from visual input is at the core of computer vision. Convolutional neural networks (CNNs), with their capacity to automatically learn hierarchical representations from pictures, have emerged as a game-changer in this field. Due to CNNs' ability to recognize edges, textures, forms, and complicated visual structures, object identification tasks may be completed with great accuracy. Beyond only detecting objects, computer vision techniques also include picture segmentation, which divides up images into discrete sections for in-depth analysis, and image generation, which enables computers to create realistic images from scratch.

Computer vision has numerous and significant uses in many different sectors. It helps with medical image analysis in the healthcare industry, helping clinicians identify illnesses and find anomalies in X-rays, MRIs, and CT scans. Computer vision is a crucial part of self-driving automobiles in the automotive industry, assisting them with navigation, pedestrian detection, and accident avoidance. Computer vision is used by retail and e-commerce companies to create visual search and recommendation systems that let customers look for items using photos and get tailored recommendations. Computer vision is used in surveillance and security systems for crowd analysis, object tracking, and facial identification, all of which improve public safety. Computer vision is also used in applications for augmented reality (AR) and virtual reality (VR), which improve user experiences by superimposing digital features on the actual environment. Despite these advances, computer vision still confronts difficulties, including how to deal with occlusions, fluctuations in illumination, and the requirement for big annotated datasets for training. The robustness, interpretability, and effectiveness of computer vision models are constantly being worked on by researchers. The advancement of computer vision has the potential to transform a number of sectors and improve how people interact with technology and their surroundings. Computer vision brings up new possibilities by giving robots the capacity to sense and comprehend visual data, opening the way for a more intelligent, connected, and visually aware future.

Evolutionary Algorithms (EAs)

The term "evolutionary algorithms" refers to a family of optimization and search algorithms that draw their inspiration from the ideas of natural selection and evolution. EAs use a model of natural selection that is based on Charles Darwin's theory of evolution to address complicated issues in a variety of fields. Creating a population of possible solutions and iteratively applying evolutionary operations like selection, mutation, and crossover to produce new candidate solutions is the main notion underlying evolutionary algorithms. The most effective candidates are chosen to make up the following generation after being judged on how well they suit the goals of the challenge. The answers develop and get better over time, finally settling on ideal or very close to optimal answers. The ideas of natural selection and genetics serve as the foundation for the class of AI methods known as "evolutionary algorithms" (EA). They are used to refine potential solutions repeatedly across several generations in order to maximize solutions to complicated problems. Engineering, banking, and gaming are just a few areas where EAs have found use.

The genetic algorithm (GA) is one of the most well-known evolutionary algorithms. To encode potential answers to a problem, GAs employ a population of chromosomes, which are commonly represented as binary strings. The genetic recombination and mutation processes that occur during biological evolution are mimicked by evolutionary processes like mutation and crossover. GAs effectively search the solution space and find potential solutions by performing these procedures on the population.

The evolutionary strategy (ES), a well-known evolutionary method, stresses the use of real-valued parameter vectors for describing potential solutions. By modifying the candidate solutions' characteristics in accordance with the caliber of their assessments, ES concentrates on taking advantage of the fitness landscape.

Evolutionary algorithms are frequently employed in industries like as optimization, engineering, finance, data mining, and many more. They are especially well-suited to handling complicated, multidimensional, and non-linear problems that standard optimization approaches may struggle with. The capacity of EAs to search and explore enormous solution spaces, as well as their resilience in dealing with noisy and unpredictable settings, make them excellent tools in real-world problem-solving. Despite their adaptability and efficacy, evolutionary algorithms confront a number of obstacles, including the necessity for precise parameter tuning, the possibility of early convergence to poor solutions, and the computing expense of evaluating fitness for each potential solution.

Fuzzy Logic

A mathematical paradigm called "fuzzy logic" expands on classic binary logic to deal with uncertainty and imprecision in thinking and decision-making. Fuzzy logic, which Lotfi Zadeh first proposed in the 1960s, enables the representation of hazy and ambiguous data by granting separate categories with varying degrees of membership rather than strict true or false values. This is accomplished using fuzzy sets and linguistic variables, where membership functions provide the level of membership of an element in a set. Fuzzy logic allows for intermediate values between and 1, expressing varying degrees of honesty, in contrast to classical logic, where membership is either 0 or 1.

Fuzzy logic has a variety of uses in decision-making processes, artificial intelligence, control systems, and other areas where uncertainty and imprecision are inherent. Fuzzy logic controllers in control systems may manage complicated and non-linear systems more successfully by taking into account a variety of inputs and language factors. This makes them reliable in contexts where conventional control techniques could falter because of uncertainty or shifting circumstances. Fuzzy logic is used in expert systems in artificial intelligence, where rules and knowledge may be represented in ambiguous terms to account for erroneous data and human-like reasoning.

However, there are several difficulties with fuzzy logic. The interpretation of fuzzy systems can be an issue, and selecting the right membership functions and rule bases might be subjective and need expert knowledge. Researchers are always looking for solutions to these issues and expanding fuzzy logic's potential.

Deep Learning

Deep learning is a branch of machine learning that focuses on teaching artificial neural networks to automatically recognize hierarchical structures in data. Deep learning models are made up of numerous layers of linked neurons, giving them the name "deep" due to their depth and being inspired by the structure and operation of the human brain. Each layer analyzes and extracts several degrees of abstract information from the incoming data, enabling the model to learn intricate patterns and representations. The strength of deep learning rests in its capacity to automatically identify complex correlations in the data without the need for explicit feature engineering. With the emergence of deep learning, AI approaches have also witnessed major breakthroughs. A kind of machine learning called "deep learning" uses neural networks with several layers to automatically learn hierarchical data representations. This has pushed the limits of AI's capabilities and resulted in advancements in speech and picture recognition, natural language processing, and game play.

The convolutional neural network (CNN), which is commonly utilized for computer vision applications, is one of the most widely used deep learning architectures. CNNs do very well in tasks like object identification, image classification, and segmentation because they are excellent at identifying local patterns in pictures, such as edges, textures, and shapes. Recurrent neural networks (RNNs) and their variations are frequently used for sequential data, such as time-series data and natural language data, allowing the models to capture temporal relationships and sequential patterns.

Deep learning has had a significant influence on a number of industries, including healthcare, finance, robotics, and NLP application industries Deep learning models have revolutionized industries like autonomous cars and medical image analysis by achieving superhuman performance in image identification tasks. Deep learning models have greatly enhanced sentiment analysis, language synthesis, and machine translation in natural language processing. Deep learning does not, however, come without difficulties. Deep neural network training demands a lot of computer power and a lot of data, both of which might not always be available. Another issue is overfitting, in which the model memorizes the training data but fails to generalize to new data. To solve these problems, researchers are actively developing strategies for model regularization, transfer learning, and interpretability.

TECHNOLOGIES AND ADVANCEMENTS IN SMART WEARABLES

Numerous facets of our lives might be improved by the ongoing developments in biometrics, gesture and emotion recognition, activity recognition and context awareness, health monitoring, and activity recognition. In these domains, sensor technology, machine learning, and human-computer interaction are combined to build intelligent, responsive systems that are better able to comprehend and respond to the demands and behaviors of people. As these fields continue to be researched and developed, new inventions that enhance user experiences, healthcare, and general well-being are likely to emerge. By offering customized recommendations based on unique tastes, personalized recommender systems improve user experiences. Wearables may collect and analyze sensor data in real time using real-time analytics, providing timely feedback and useful insights. Smart wearables can respond more quickly, have greater privacy, and use processing power and intelligence closer to them thanks to edge computing [16]. These technologies work together to enhance smart wearables, enabling consumers to have tailored and intelligent experiences.

ACTIVITY RECOGNITION AND CONTEXT AWARENESS

Intelligent systems that strive to comprehend and adapt to human behavior must include activity recognition and context awareness. Sensor technology, machine learning, and artificial intelligence breakthroughs have considerably enhanced the accuracy and efficiency of identifying and comprehending human behaviors. The automated detection and interpretation of human actions based on sensor data or input from numerous sources is referred to as "activity recognition." It makes use of sensors present in smartphones, wearables, and IoT devices such as accelerometers, gyroscopes, GPS, and environmental sensors. Activity detection provides applications in healthcare, smart homes, fitness tracking, and customized user experiences by continually monitoring user behaviors and ambient circumstances. Context awareness augments activity recognition by recording and exploiting information about the user's surroundings and circumstances. By taking into account variables like location, time of day, social environment, and user preferences, it improves comprehension of actions. For instance, a smart home system can detect aberrant activity, such as a fall, and warn caretakers by adjusting the lighting and temperature in the room based on the user's behavior while entering.

HEALTH MONITORING AND BIOMETRICS

To determine a person's health state and well-being, health monitoring and biometrics gather and analyze behavioral and physiological data. It is now much simpler to continually monitor several physiological indicators including heart rate, blood pressure, sleep patterns, and physical activity levels thanks to the spread of wearable technology and sensors. These technologies have the power to completely change how chronic illnesses are managed and how preventative healthcare is promoted. People may obtain insights into their general health and make knowledgeable decisions about their lifestyle choices by recording their vital signs and other health-related data. Remote patient monitoring allows medical personnel to see early indicators of health decline and make prompt treatments. Personalized healthcare solutions may be created by fusing biometrics and health monitoring with machine learning algorithms. These algorithms are capable of analyzing massive amounts of data and providing actionable insights, such as proposing lifestyle modifications, identifying illness risks, and offering individualized treatment options. Furthermore, they can aid in illness identification, improving outcomes, and lowering healthcare expenditures.

GESTURE AND EMOTION RECOGNITION

Gesture recognition focuses on deciphering hand or body motions to operate or communicate with digital systems or gadgets. On the other hand, emotion recognition is recognizing and comprehending human emotions based on physiological signs such as voice intonation, facial expressions, and others. Human-computer interaction has benefited from the rise of gesture recognition, which enables people to communicate with gadgets through simple, instinctive gestures. People can control presentations, play games, or move around in virtual reality settings by using hand gestures. By doing away with the necessity for conventional input devices like keyboards and mice, this technology improves user experience and makes it possible for more immersive interactions. Applications for emotion recognition may be found in many fields, including market research, mental health, and affective computing. Emotion recognition may be used by businesses to assess client happiness, enhance product design, and provide individualized experiences. In the field of mental health, emotion recognition helps therapists gauge patients' emotional states and therapeutic progress. The goal of affective computing is to create systems that can recognize and react to emotional states in people, enabling more sympathetic and contextually aware human-computer interactions.

PERSONALIZED RECOMMENDER SYSTEMS

Intelligent algorithms are used in personalized recommender systems to deliver personalized recommendations based on an individual's preferences, interests, and past data. Machine learning techniques are used in these systems to evaluate large quantities of data and forecast user preferences in order to offer relevant and tailored information. In the context of e-commerce, personalized recommender systems assist consumers in discovering items or services that match their preferences and requirements. These systems may produce tailored suggestions by evaluating user behavior, purchase history, and demographic information, resulting in increased user happiness, engagement, and conversion rates. Personalized recommender systems have also found use in a variety of fields, including entertainment, journalism, and social media. Streaming services utilize recommendation algorithms to propose movies or episodes based on the watching history and tastes of its users. Users' experiences consuming news are improved by news aggregators' recommendations of content based on their reading preferences and interests. Social networking sites employ tailored suggestions to present pertinent material, keeping users interested and connected.

REAL-TIME ANALYTICS

Real-time analytics refers to the process of evaluating data as it is being produced, offering insights and useful information instantly or as soon as possible. It gives businesses the ability to make data-driven choices instantly, increasing operational effectiveness, boosting customer experiences, and streamlining corporate procedures. Real-time analytics is essential in the world of smart wearables for processing and interpreting the constant stream of sensor data that these devices gather. Wearables with sensors, such as accelerometers, heart-rate monitors, and GPS, can record information about movement, health metrics, and position. Since this sensor data can be processed instantly thanks to real-time analytics, wearables can give consumers quick feedback and useful insights. For instance, a fitness tracker may keep track of a user's heart rate throughout a workout and offer real-time feedback on the intensity of their activity, assisting them in optimizing their training schedule. Real-time analytics capabilities in wearables allow for the early detection of crucial circumstances in healthcare by spotting abnormalities in vital signs and sending prompt warnings to users or healthcare experts.

EDGE COMPUTING IN SMART WEARABLES

As opposed to depending on centralized cloud infrastructure, edge computing includes processing and analyzing data at or close to the source of origination. Edge computing, when applied to smart wearables, puts processing power and intelligence closer to the devices themselves, enabling quicker response times, lower latency, and improved privacy and security. The requirement for continual connectivity with the cloud may be eliminated by utilizing edge computing in smart wearables to enable data processing and analysis to take place right on the device. Limiting the amount of data sent to outside servers promotes real-time insights and actions while still protecting user privacy. Edge computing helps smart wearables overcome the issue of poor network access, especially in distant or congested areas where network capacity may be capped. Wearables are able to operate independently and continue to function even in situations with limited or sporadic connectivity by executing calculations locally. The quantity of data that has to be communicated to the cloud may be decreased by using edge computing in smart wearables to enable effective data filtering and compression. This lowers the overall cost of data transmission and storage while maximizing bandwidth consumption.

CASE STUDIES

These case studies demonstrate the effective use of machine learning in smart wearables. These wearable capabilities have developed to include accurate activity identification, health monitoring, illness diagnosis, and gesture-based control by adding sophisticated algorithms. Machine learning has not only improved the performance of these gadgets, but it has also enabled people to better their health, productivity, and general quality of life.

CASE STUDY 1: ACTIVITY RECOGNITION WITH SMARTWATCHES

(Company: Fitbit)

Fitbit, a well-known producer of fitness-focused smartwatches and activity trackers, has enhanced the activity identification capabilities of its wearable technology by utilizing machine learning algorithms. Fitbit has made effective identification and monitoring of a variety of physical activities possible by incorporating cutting-edge machine learning algorithms into its wearables. The sensors in Fitbit's smartwatches, which measure users' movements, physiological reactions, and position in great detail, include accelerometers, heart rate monitors, and GPS trackers. Machine learning algorithms that have been trained on big datasets are used to interpret this raw sensor data in order to detect particular activities like walking, running, cycling, swimming, and more. Fitbit devices can precisely evaluate the intensity levels of each activity through the intelligent fusion of data from several sensors and the use of machine learning algorithms, which allows them to distinguish between different types of activities. This makes it possible for users to get exact data on the number of calories burned and the length of their workouts, allowing them to establish reasonable fitness goals and efficiently monitor their progress.

The constant learning and adaption of Fitbit's machine learning algorithms is what distinguishes their methodology. The algorithms receive personalized data from each user over time, and they utilize that data as part of their training. As a consequence, Fitbit devices' ability to accurately identify activities increases with each encounter, giving consumers more individualized insights and suggestions. Machine learning has transformed how individuals track and participate in physical activity thanks to Fitbit's integration of smart wearables. By giving users precise and individualized information to help their fitness journeys, these gadgets have evolved into important tools for anyone attempting to maintain a healthy and active lifestyle.

CASE STUDY 2: HEALTH MONITORING AND DISEASE DETECTION WITH SMART CLOTHING

(Company: Hexoskin)

Hexoskin focuses in creating intelligent apparel with built-in machine learning tools that can track users' health and identify future illnesses. Smart shirts from Hexoskin, the company's signature item, use in-built sensors to gather a variety of physiological data, such as heart rate, breathing rate, and sleeping habits. A mobile application or cloud-based platform receives this data wirelessly and analyzes it. The analysis of the gathered data relies heavily on machine learning algorithms, enabling the detection of anomalies and trends linked to specific medical disorders. The technology developed by Hexoskin has the ability to identify irregular heartbeats, sleep problems, respiratory abnormalities, and more, giving users the tools they need to manage their health.

Hexoskin's machine learning models are trained on sizable datasets comprising labeled data from individuals with known medical disorders in order to assure the accuracy of illness identification. The algorithms get more adept at identifying particular patterns and signs connected to various diseases by learning from these datasets. Because each user has a different health profile, the system may offer early warnings and suggestions that are tailored to them. By enabling continuous health monitoring, Hexoskin's smart clothing, driven by machine learning, enables users to proactively monitor their well-being and make educated decisions about their lifestyle choices. People may take control of their health and make the required changes to their exercise regimens, sleep schedules, and lifestyles by utilizing the insights offered by the system.

CASE STUDY 3: GESTURE RECOGNITION FOR HANDS-FREE CONTROL IN SMART GLASSES

(Company: North (formerly Thalmic Labs))

In order to allow gesture detection for hands-free control, North, a technology firm that underwent a name change from Thalmic Labs, has successfully incorporated machine learning algorithms into its smart glasses, known as "Focals." Users have a smooth method to monitor notifications, access information, and engage with their digital environment with the help of Focals, fashionable glasses with an embedded augmented reality display. Focals utilize an in-built camera to record users' hand motions in order to offer a simple and organic control mechanism. The video data from the camera is then processed by machine learning algorithms, which analyze it to find certain motions and turn them into instructions. Users may now utilize voice commands or touch alone to explore menus, turn off notifications, and carry out other tasks. North's machine learning models are trained on big datasets comprising labeled hand gesture data in order to assure precise gesture identification. These datasets provide a wide variety of gestures, enabling the algorithms to properly learn and detect a broad range of user movements. Focals has changed how consumers interact with their digital world by using machine learning-based gesture detection. By doing away with the need for bulky input devices, this technology offers a user experience that is smooth and organic. Users may now easily interact with digital material while keeping their visual attention on the outside environment, increasing productivity and convenience.

FUTURE DIRECTIONS IN THE FIELD OF SMART WEARABLES

The future of smart wearables is poised to bring about significant advancements in various areas. Significant breakthroughs in a number of fields are expected to be made as a result of smart wearable technology. These fields comprise ethical considerations, energy efficiency, contextual awareness, smart textiles, augmented and virtual reality integration, health monitoring, biometric identification, and IoT integration. These advancements will enable wearables to become strong tools with capabilities beyond what they now possess. In addition to improving our health, productivity, and entertainment, they will also significantly improve our quality of life as a whole. The development of smart wearables will be fueled by ongoing research, innovation, and cross-disciplinary cooperation [17]–[19]. We may thus anticipate that these gadgets will influence how we engage with technology in the years to come. The possibilities are endless, and we are only at the beginning of this exciting journey.

HEALTH MONITORING AND PERSONALIZED WELLNESS

Health monitoring, biometric authentication, augmented and virtual reality integration, smart textiles, energy efficiency, contextual awareness, Internet of Things integration, and ethical considerations are all areas where smart wearables will advance. These developments will turn wearables into useful tools that raise our level of life and increase our productivity, effectiveness, and enjoyment. Future developments in smart wearables will be fueled by ongoing experimentation, innovation, and collaboration among a variety of sectors, which will also affect how we engage with technology now.

ADVANCED BIOMETRIC AUTHENTICATION

The need to maintain safe and frictionless authentication methods will grow as wearables become more incorporated into our daily lives. To improve device security and user convenience, future wearables will feature cutting-edge biometric technology like fingerprint recognition, voice recognition, or perhaps facial recognition.

Wearables will be able to securely store user data and guarantee safe access to associated hardware and services thanks to these authentication procedures. When used in conjunction with strong security measures, biometric authentication will provide a more seamless user experience.

AUGMENTED REALITY AND VIRTUAL REALITY INTEGRATION

Smart wearables will change several sectors when AR and VR technologies are combined with them. Immersive experiences in gaming, remote collaboration, education, and training will be made possible by wearable devices that have integrated augmented reality displays or function in tandem with headsets. Wearables will have new opportunities for dynamic and captivating experiences thanks to the seamless integration of AR and VR, boosting communication, entertainment, and productivity.

SMART FABRICS AND E-TEXTILES

Materials science and electronics advancements will pave the way for smart fabrics and e-textiles. Sensors, actuators, and flexible displays will be effortlessly integrated into clothes and accessories thanks to these novel fabrics. Smart textiles will enable the creation of wearable gadgets that are both comfortable and fashionable, as well as capable of monitoring vital signs, recording body motion, and delivering real-time feedback without the use of cumbersome external sensors. Wearable technology will become more inconspicuous and effortlessly incorporated into daily clothing, increasing user comfort and acceptance.

ENERGY HARVESTING AND BATTERY EFFICIENCY

Future wearables will focus on energy harvesting techniques and greater battery efficiency to address the issue of limited battery life. To increase usage duration, wearables may integrate technology such as solar cells, kinetic energy harvesting, or flexible batteries. Furthermore, advances in low-power hardware design, efficient algorithms, and power management approaches will lead to increased battery life and better user experiences. Wearable gadgets will have a longer battery life without sacrificing usefulness or performance.

CONTEXTUAL AWARENESS AND ADAPTIVE INTELLIGENCE

Future wearables will use adaptive intelligence and contextual awareness to deliver individualized and context-aware experiences. To learn more about users' behaviors, preferences, and environments, these gadgets will mix sensor data, machine learning algorithms, and data from other connected devices. Wearables will anticipate user demands, provide timely notifications, and seamlessly adjust their functioning to various circumstances by taking advantage of this contextual awareness. Wearables will develop into intelligent allies that offer users-specific information and help.

SEAMLESS INTEGRATION WITH THE INTERNET OF THINGS (IoT) ECOSYSTEM

In the larger Internet of Things (IoT) ecosystem, smart wearables will be a key component, effortlessly merging with other connected devices. They will link with numerous IoT devices in homes, autos, healthcare systems, and smart cities by serving as hubs or interfaces. Wearables will enable seamless data exchange for tailored services, deliver real-time contextual information, manage other IoT devices, and perform these functions. Wearables will be able to improve user experiences, automate chores, and drive the development of smart homes and cities thanks to this connection.

ETHICAL AND PRIVACY CONSIDERATIONS

Ethics and privacy issues will become more important as wearables capture more sensitive data. Incorporating strong security safeguards, open data processing procedures, and user-centric privacy controls will be a priority in the next improvements. In order to guarantee wearables emphasize user privacy, data protection, and ethical use of personal information, industry standards, and regulations will be essential. For smart wearables to be widely adopted and to succeed, it will be essential to uphold trust and handle ethical issues.

ACKNOWLEDGMENT

I would like to thank Ms. Abiro Muzzafar Mir for proofreading this chapter.

REFERENCES

1. 'Wearables and AI better predict the progression of muscular dystrophy', *Nat Med*, vol. 29, no. 1, pp. 37–38, Jan. 2023, doi: 10.1038/s41591-022-02191-6.
2. W. Huang *et al.*, 'Application of ensemble machine learning algorithms on lifestyle factors and wearables for cardiovascular risk prediction', *Sci Rep*, vol. 12, no. 1, p. 1033, Jan. 2022, doi: 10.1038/s41598-021-04649-y.
3. D. Nahavandi, R. Alizadehsani, A. Khosravi, and U. R. Acharya, 'Application of artificial intelligence in wearable devices: Opportunities and challenges', *Comput Methods Program Bio*, vol. 213, p. 106541, Jan. 2022, doi: 10.1016/j.cmpb.2021.106541.
4. J. Dunn *et al.*, 'Wearable sensors enable personalized predictions of clinical laboratory measurements', *Nat Med*, vol. 27, no. 6, pp. 1105–1112, Jun. 2021, doi: 10.1038/s41591-021-01339-0.
5. K. Bayoumy *et al.*, 'Smart wearable devices in cardiovascular care: Where we are and how to move forward', *Nat Rev Cardiol*, vol. 18, no. 8, pp. 581–599, Aug. 2021, doi: 10.1038/s41569-021-00522-7.
6. Y. Zhang, Y. Hu, N. Jiang, and A. K. Yetisen, 'Wearable artificial intelligence biosensor networks', *Biosens Bioelectron*, vol. 219, p. 114825, Jan. 2023, doi: 10.1016/j.bios.2022.114825.
7. A. H. Bhat, P. Prabhu, and K. Balakrishnan, 'A critical analysis of state-of-the-art metagenomics OTU clustering algorithms', *J Biosci*, vol. 44, no. 6, Dec. 2019, doi: 10.1007/s12038-019-9964-5.

8. A. H. Bhat and P. Prabhu, 'OTU clustering: A window to analyze uncultured microbial world', *Int J Sci Res Rev Comput Sci Eng*, vol. 5, no. 6, pp. 62–68, 2017, [Online]. Available: www.isroset.org

9. F. Sabry, T. Eltaras, W. Labda, K. Alzoubi, and Q. Malluhi, 'Machine learning for healthcare wearable devices: The big picture', *J Healthc Eng*, vol. 2022, pp. 1–25, Apr. 2022, doi: 10.1155/2022/4653923.

10. D. H. Fisher, 'Knowledge acquisition via incremental conceptual clustering', *Mach Learn*, vol. 2, no. 2, pp. 139–172, 1987, doi: 10.1023/A:1022852608280.

11. S. M. Ganie and M. B. Malik, 'An ensemble machine learning approach for predicting Type-II diabetes mellitus based on lifestyle indicators', *Healthcare Analytics*, vol. 2, p. 100092, Nov. 2022, doi: 10.1016/J.HEALTH.2022.100092.

12. S. M. Ganie and M. B. Malik, 'Comparative analysis of various supervised machine learning algorithms for the early prediction of type-II diabetes mellitus', *Int J Med Eng Inform*, vol. 14, no. 6, pp. 473–483, 2022, doi: 10.1504/IJMEI.2022.126519.

13. S. M. Ganie, M. B. Malik, and T. Arif, 'Performance analysis and prediction of type 2 diabetes mellitus based on lifestyle data using machine learning approaches', *J Diabetes Metab Disord*, vol. 21, no. 1, pp. 339–352, Jun. 2022, doi: 10.1007/S40200-022-00981-W/METRICS.

14. S. M. Ganie, M. B. Malik, and T. Arif, 'Machine learning techniques for diagnosis of type 2 diabetes using lifestyle data', pp. 487–497, 2022, doi: 10.1007/978-981-16-3071-2_39.

15. K. P. Seng, L.-M. Ang, E. Peter, and A. Mmonyi, 'Machine learning and AI technologies for smart wearables', *Electronics (Basel)*, vol. 12, no. 7, p. 1509, Mar. 2023, doi: 10.3390/electronics12071509.

16. A. Abd-Alrazaq, R. AlSaad, F. Shuweihdi, A. Ahmed, S. Aziz, and J. Sheikh, 'Systematic review and meta-analysis of performance of wearable artificial intelligence in detecting and predicting depression', *NPJ Digit Med*, vol. 6, no. 1, p. 84, May 2023, doi: 10.1038/s41746-023-00828-5.

17. Nature, 'Space missions out of this world with AI', *Nat Mach Intell*, vol. 5, no. 3, pp. 183–183, Mar. 2023, doi: 10.1038/s42256-023-00643-3.

18. B. Conroy *et al.*, 'Real-time infection prediction with wearable physiological monitoring and AI to aid military workforce readiness during COVID-19', *Sci Rep*, vol. 12, no. 1, p. 3797, Mar. 2022, doi: 10.1038/s41598-022-07764-6.

19. J. N. Acosta, G. J. Falcone, P. Rajpurkar, and E. J. Topol, 'Multimodal biomedical AI', *Nat Med*, vol. 28, no. 9, pp. 1773–1784, Sep. 2022, doi: 10.1038/s41591-022-01981-2.

2 Introduction to Machine Learning for Smart Wearables

A. Naga Sai Purushotham, Potti Jaswanth,
Potnuru Sai Santhosh, Hemachandran K,
and Zita Zoltay Paprika

INTRODUCTION

Smart wearables, which include fitness trackers, smartwatches, and augmented reality (AR) glasses, have grown in number in ultra-modern society, offering users quite a number of handy and personalized functions. These devices provide technology that is advanced in order to deliver intelligent functionality and enhance the user experience. Machine learning (ML), a subset of artificial intelligence, plays a key role in enabling smart wearables to capture consumer behavior and predict and adapt to personal needs.

This chapter is an introductory guide to machine learning for smart wearables, imparting a top-level view of the primary principles, applications, and challenges of integrating system mastering strategies into wearable devices. The goal is to shed light on the transformative power of learning and its potential to revolutionize wearable technology.

The chapter starts by detailing the basic ideas of machine learning, along with supervised and unsupervised learning approaches. Supervised learning involves training systems with data, allowing them to make predictions based on known patterns. Unsupervised learning, on the other hand, involves identifying patterns and relationships in anonymous data to uncover hidden verifiable insights and generate meaningful recommendations.

Laying the groundwork for machine learning, we will explore the specific application of this technology in smart wearables. A notable application is activity recognition, where machine learning algorithms analyze sensor data collected on wearables to identify and classify activities, such as walking, running, cycling, or the ability to use fitness tracking and enhance its capacity.

Health monitoring is a crucial area wherein machine learning knowledge can make a huge impact. Wearables ready with sensors can accumulate physiological statistics, which include heart rate, sleep cycles, and blood pressure. By applying

DOI: 10.1201/9781032686714-2

machine learning algorithms to these records, wearables can detect anomalies, provide early warnings for capability health troubles, and offer customized health insights to customers.

Data collection and preprocessing play a pivotal role in accomplishing accurate and reliable machine-learning consequences in wearable applications. The chapter emphasizes the importance of remarkable and consultant information for training robust models. It explores diverse strategies for amassing and preprocessing statistics in wearables, making sure that the accrued data is appropriate for effective device learning evaluation.

However, deploying machine learning models on aid-limited wearable devices presents unique challenges. Limited processing power, remote nature, and energy resources require the development of lightweight and energy-efficient system learning strategies. We will address those demanding situations and examine strategies including version compression, quantization, and disbursed learning to optimize version overall performance on wearables.

In essence, this introduction to machine learning for smart wearables presents readers with a know-how of the essential standards, applications, challenges, and potential benefits of integrating systems into wearable devices. With the assistance of machine learning knowledge, smart wearables have the ability to improve consumer experience, fitness, and drive innovation within the wearable technology sector. This introduction serves as a stepping stone for readers to delve deeper into the fascinating world of machine learning.

SMART WEARABLE TECHNOLOGIES

Smart wearable technology encompasses a wide variety of gadgets that might be intended to be worn by the frame and use advanced technology to provide people with extra features and make their lives less complicated. These devices often connect with smartphones or different smart devices, making them safe to use in one-of-a-kind virtual environments. Here are some examples of well-known smart wearables:

Fitness trackers: These are probably the most recognized form of smart gadgets. Most of them have sensors that may display your bodily movement, coronary heart rate, the way you sleep, and different health-associated facts. They deliver customers' feedback in real time, keep track of their progress, and provide personalized recommendations to assist people in reaching their health goals.

- **Smartwatches**: Smartwatches blend the functions of a conventional watch with those of a phone. They usually have touchscreens, intensive kinds of apps, and indicators for calls, texts, emails, and social media. Smartwatches invariably have GPS and the capability to link to different smart devices. They also can track your physical activity.
- **Augmented reality glasses**: These glasses upload digital statistics to the person's view of the real world, making it seem more real to them. These

gadgets can display statistics that are applicable to the situation, help you discover your way around, and give you an immersive experience for video games or paintings. AR glasses have the potential to change fields such as manufacturing, education, and fitness care.

- **Virtual reality (VR) headsets**: VR headsets generate an immersive digital environment around the user, transporting them to a digital international or simulating an actual experience. People frequently use these gadgets for video games and enjoyment, but they also can be used for learning and training, education, and therapy.
- **Smart garments**: These have electronics and sensors constructed right into the fabric itself. Some examples are workout clothes with built-in coronary heart rate monitors, garments that control temperature, and clothes that track posture or frame moves for sports activity training or rehabilitation.
- **Smart rings**: These combine fashion and technology. They can track your fitness, send you alerts, and even let you pay for anything without touching your card reader. These devices are made to be elegant objects that integrate seamlessly into the wearer's lifestyle.
- **Smart glasses**: Smart glasses resemble traditional eyeglasses but contain technology to offer numerous features. They can show facts, show notifications, offer hands-free conversation, and help with tasks consisting of shooting photographs or videos.
- **Smart hearing devices:** Smart hearing devices leverage superior audio processing technologies to enhance listening experiences. These devices can adapt to one-of-a-kind sound environments, provide noise cancellation, and can be connected to smartphones for added functionalities.
- **Smart rings:** These are small, ring-formed devices that may carry out a whole lot of functions which include monitoring health metrics, monitoring sleep styles, receiving notifications, and even making contactless payments.
- **Smart fabrics:** Smart fabrics integrate electronic additives directly into textiles, allowing for features like temperature regulation, moisture-wicking, and biometric sensing. These fabrics can be utilized in sports activities garments, healthcare garments, and even in smart domestic applications.
- **Smart shoes:** Smart shoes comprise sensors and connectivity capabilities to track steps, monitor overall running performance, offer real-time feedback on gait and posture, and provide custom-designed coaching for fitness enthusiasts and athletes.
- **Smart headbands:** These wearable headbands often use EEG (electroencephalogram) sensors to count brain activity and provide insights into attention, rest, and intellectual states. They are used for meditation, stress management, and enhancing cognitive performance.
- **Smart gloves:** Smart gloves combine sensors and haptic remarks generation, enabling customers to engage with digital and augmented reality environments, control devices with hand gestures, and provide tactile remarks in various programs including gaming, education, and digital simulations.

- **Smart belts:** These are wise belts that can monitor waist size, track physical interest, and provide posture correction. Some smart belts even have integrated charging ports to power different gadgets at the pass.
- **Smart e-textiles:** E-textiles check with fabrics embedded with digital components like conductive threads or fibers. These textiles can enable numerous capabilities along with touch-sensitive surfaces, gesture recognition, and even communicating capabilities.
- **Smart bracelets:** Smart bracelets provide functions just like fitness trackers and smartwatches, which include activity monitoring, sleep tracking, and notifications. Typically they are extremely light in weight and minimalist in design, attractive for people that decide on a simpler wearable alternative.
- **Smart helmets:** These helmets integrate technology inclusive of integrated cameras, GPS navigation, and heads-up displays (HUDs) to enhance protection, provide real-time statistics, and offer an immersive experience for activities like biking, motorcycling, and extreme sports.
- **Smart medical devices:** Smart wearables are increasingly used in the healthcare field, such as gadgets like smart insulin pumps, non-stop glucose video display units, wearable ECG (electrocardiogram) video display units, and smart patches for drug shipping or tracking crucial symptoms.

As technology continues to advance, smart wearables are expected to play a crucial role in our lives, improving fitness and health, improving communication and productiveness, and growing immersive digital experiences. We can expect more innovation and integration of smart functionalities into daily used wearables, providing customers with more desirable competencies and seamless connectivity.

APPLICATIONS

Machine learning possesses a critical role in smart wearables as its mechanism is implanted in them. The application of AI in smart wearables is seen in various fields such as healthcare, sports entertainment, and smart home devices etc. There are various machine learning models and algorithms including deep learning which is used for finding activity recognition, and behavioral patterns.

Machine learning-based methods can be used to get around the problems that systems have. For example, a lot of detection sensors have issues with false alarms and expensive security because they aren't used all the time. With wearable fall detection monitors, the device learning method and the Internet of Things (IoT) have helped a lot to solve these tough problems. This deep structure uses the progressive deep neural network with proximity-based active learning to make large models for putting information into groups. The first method uses a convolutional neural network (CNN) with layered long-short-term memory (LSTM) to look at the layered representation of capabilities and capture the long-term dependencies in activity records. The second method chooses the best option for the deep neural network to be changed to a new configuration that lets the device work. The deep community model has eight layers: five convolutional layers, three recurrent LSTM layers, and one smooth-max layer.

RECURRENT NEURAL NETWORKS (RNN)

With this method, you can keep track of and manage your sleep. In this method, recurrent neural networks based on bidirectional long-short-term memory (BLSTM) and layered feature learning were used to learn about the data. The method put sleep into three groups: waking up, regular eye shifting, and abnormal eye shifting. We put recurrent neural networks and a microcontroller unit (MCU) into wearable devices that can track and identify falls in real time. This system is connected to a Wi-Fi network and a tracking device, which lets the carer give the patient important care. Start by making sure that the wearables are always linked to a wireless network. This is a good way to let the carer know about a fall. Second, the device should be portable so that people can use it. Third, if you want to keep track of sports well, the device needs to be turned on all the time. The wrist-worn device takes data of the patient's heart rate variability to make a reliable record of his or her physical and mental health. During the step of learning functions, low- and mid-level functions are removed. Raw reports are processed so that you can get low-level patterns that capture temporal and frequency area homes. By putting fitness data from WMSs into CDSSs, a hierarchical fitness decision-making tool was made to help identify pollution. The suggested system has more than one layer. It starts with a WMS layer that is powered by modern device learning and has a disease diagnostic module that keeps track of each case individually.

LONG SHORT-TERM MEMORY NETWORKS

Compared to deep learning models like the CNN model, which can also use spatial correlations and behavior classes, the LSTM model can handle time-series data patterns by using feedback connections. LSTM is a deep mastering model for multimodal information fusion in the setting of human activity identity. Through their study, they made a way to test an IMU (inertial measurement unit) sensor-based machine that can be worn on the body. This method uses recurrent neural networks with BLSTM and tiered feature mastering.

END PRODUCTS OF SMART WEARABLES USING ML

BODY MOVEMENTS: ATLAS WEARABLES

Atlas Wearables is a health band and intelligence tool that has data pushed at the Motion Genome Project's movement database. Atlas's claim to fame is its machine learning algorithms, which automatically group your workout in a 3D vector, telling the difference between push-united states and triangle push-ups. It also tracks your heart rate and counts calories burned. The fact that workouts are found is only the beginning. Co-founder Peter Li says that the startup's goal is to put "intelligence into body language and actions." Machine learning algorithms and datasets can be used to figure out how you walk, sit, pass, or connect with other people. This could give information about your mood, physical reaction, energy level, or even the situation.

PREVENTIVE HEALTH: ENTOPSIS

Entopsis is another early molecular analysis business. It is a scientific diagnostic tool that uses nanotechnology and machine learning to show medical problems. By looking at the proteins in biofluids, Entopsis can find patterns and signs that point to good health. Entopsis uses machine learning techniques to look at the molecular signature on a NUTeC glass after culturing a biofluid pattern using their Nanoscale Unbiased TExtured collect (NUTeC) generation to gather molecules. The scanned signatures are then sent to the cloud, where they are compared to other accounts in the database to find ones with similar traits. Entopsis wants to give customers direct access to NUTeC dishes so they can collect biofluids and send them to the company for molecular analysis.

EMOTION MEASUREMENT: BRANDEMOTIONS

BrandEmotions looks at how to measure how customers feel. Sentiment analysis and polls give good/bad or stacked ranking results, but companies can't classify or measure how their customers feel. BrandEmotions lets businesses find out how their customers feel about their brand across many different channels, such as shopping, live events, movies, resorts, cruises, amusement parks, and marketing. BrandEmotions from Amyx McKinsey shows how people feel when they interact with a brand. This lets companies improve the brand experience, build brand loyalty, and offer the right goods and services at the right time. BrandEmotions' emotion-sensing, device-learning technology uses its own EmotionIQ method to analyze the bodily data collected by a wide range of wearable devices and Internet of Things-connected devices. This data is then turned into emotional categories and depth.

MEDICATION COMPLIANCE: VITALITY

NANTHEALTH bought Vitality, which has a product called the "GlowCap." The GlowCap is an Internet-connected tablet cap that blinks and makes noises when it's time to take medicine. The Vitality compliance-improving device tries to change how patients act by using comments, reminders, education, and benefits to get them to take their medicine as prescribed. The GlowCap gives carers real-time information, such as when a dose of medicine has been missed or medicine has been taken out of the system. A quick button on the cap lets patients change the order of their medicines. Dr. Yan Chow, who used to be the Medical Director of IT Innovation for Kaiser Permanente, said on a panel at the Wearables Things 2014 conference that getting people to take their medicine is a hard, multi-layered problem. "It's not always because people forget to take their medicine."

CHALLENGES FOR SMART WEARABLES

We have already discussed the advantages and promise of AI and intelligent wearable devices. Here, we'll discuss some of the issues with AI smart devices and some potential solutions. There are two perspectives from which we can analyze the two major issues: (a) technological issues, and (b) social issues.

TECHNICAL CHALLENGES FOR SMART WEARABLES

Technical challenges for wearable devices in scientific Internet-of-Things uses have been identified, such as the need for privacy and security of data and the amount of power they use.

The first problems for networking and communication add-ons and algorithmic and alertness-dependent tasks are the same as those for utility-specific Internet of Things (ASIoTs). ASIoTs have a number of problems, such as problems with connectivity, problems with energy efficiency, problems with computation using side and fog system learning models, and problems with safety and privacy. For more information about networking and communication problems, as well as algorithms and alertness-based parts of AI and smart wearable technology, the reader can look at the following sources.

The third problem is about algorithmic and alertness-specific parts. It applies to smart gadgets that use AI and device learning at the same time. First, just like with system learning and deep learning algorithms, people need to learn about AI algorithms in smart tech. One of the problems is that a huge amount of information is needed, especially for training algorithms and learning them well. In order to get useful training data, this would require a lot of money to pay people to do the right work or act in the right way while wearing the wearable.

SOCIAL CHALLENGES FOR AI SMART WEARABLES

If AI smart wearables are to be used effectively in society, they might run into certain problems. There are many social and technical problems that make it hard for China's senior population to use a lot of wearable scientific gadgets with AI. There are VI types of barriers or challenges: technological, managerial, medical, economic barriers or demanding situations, felony obstacles or demanding situations, and non-public obstacles or challenges.

Most of the technological barriers to the design and improvement of wearables have to do with the trade-offs between making the device small and compact in order that it may be worn with no trouble, making the tool supply accurate readings, making the tool have an extended battery existence, making the device reasonably priced, and rendering it easy to use. The solution is to build and put together wearable gadgets that meet the requirements.. The management problems include the need for help from top management and the need for wearable tool companies to build the right connections and work well with public health companies in order to sell wearable technologies. The scientific challenges include figuring out how well wearable gadgets work and gathering evidence to show how they can be used for therapy. The authors also say that using medical doctors and physicians in public health bodies is a major problem because it changes the workload in a bad way.

The cost of making wearable devices and coming up with a business plan that can work are two examples of economic limits and challenges. There may not be sufficient cash from the authorities to keep up with the high call for and range of people who want to use national and public health fundings. The authors found that there are two legal problems that make it hard for wearable tech to catch on: the lack of a strong regulatory system and the loss of privacy tools for statistics. This problem is especially important for older customers who may face legal problems if they wear scientific gadgets. The second legal issue is data privacy. Senior users are worried that wearable scientific equipment could record private information about them, such as their actions, where they go, and how they live. Most of the non-public problems and hurdles have to do with a lack of customer trust, a loss of personalized analytical services, and public resistance. Figure 2.1 shows a summary of the limits or problems with smart gadgets that use AI.

CONCLUSION

Wearable technology gives healthcare workers, clinical practices, and other care groups, as well as patients, new ways to help people by giving them clear information and understanding. The capability to get information constantly from both clinical and behavioral resources will cause the improvement of analysis equipment as a way to help us examine more about how infections start and spread, a way to spot them early, and the way to deal with them. These records and tools will change how professionals work and how patients take care of themselves and make decisions.

The number of records amassed will increase dramatically, and because they come from one-of-a-kind sources (medical parameter monitoring, self-monitoring, and behavioral statistics), it will be necessary to find various approaches to address and use these records, put them to good use, and utilize them as a part of the cutting-edge workflow, sharing records with different corporations, the patient, and their family, and coming up with new strategies. These adjustments are in their early stages and could manifest over the next few years. They will be driven by two things: a shift in care models towards shared care and patient co-management, and the improvement of new units that will give professional practice new insights.

These tactics may even have an effect on changes to legal guidelines approximately records safety and sharing, in addition to the introduction of extra interoperable and

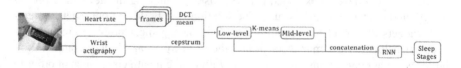

FIGURE 2.1

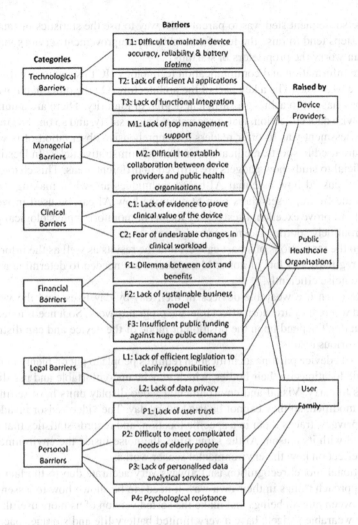

FIGURE 2.2 Barriers or challenges for AI smart wearables

fashionable devices that can be delivered seamlessly to care institutions. Along with these techniques to be used in the long term, there's a more immediate need to construct more developed responses that meet the wishes of care employees and may upgrade the equipment that healthcare professionals are already using. The real problem for absolutely everyone concerned, inclusive of healthcare, policymakers, and industry, is to work together to discover approaches, and to use those answers to assist all populations at a low price and high delivery-rate so that it will enhance care delivery models and clinical training.

For artificial intelligence wearables to work well, they need numerous data. In this situation, privacy is considerably more important than anything else. During the improvement method, we had to keep asking ourselves if we needed to collect private data. The subsequent step was to parent out a way to use the statistics instantly. All of those steps tend to raise the fee of wearable app improvement services via loads, which can worry the proprietors of startups.

Future information and communication technology (ICT) structures will want to have to be wearable to be a key part of the architecture. On the other hand, wearable technology has yet to achieve a sufficient degree of maturity. There are a number of troubles with data collection, facts processing, touch, safety, and so on. The intention of this assessment was to offer readers a comprehensive observation of how smart gadgets are used in sports, medication, and the administrative fields. In Destiny, it'll be beneficial to study how gadgets are utilized in different areas. This chapter additionally discussed how essential AI-based techniques are when making wearable tools. In the future, researchers plan to examine how AI can be used in wearable gadgets to improve excellent lifestyles by means of monitoring physiological parameters or routinely spotting sicknesses early on.

Due to the restrictions of smart gadgets and the results as well as the information accrued regularly has lots of noise. More studies are needed to determine a way to eliminate noise efficiently.

People often use wearable devices in their day-by-day lives, and the way they work and where they are worn affect how the programs work. So it needs to use strategies that don't depend upon the position or mode of the device and can distinguish parts of various results.

Wearable device pastime identification requires people to put on plenty of devices in specific locations on their bodies, which may be uncomfortable and use different strengths to carry. Visual and environmental video display units have settings that can't be modified, so they cannot be used every day. The video sensor invades customers' privacy, tracks them in a specific location, and records statistics that are not associated with its reason. At the same time, the noise inside the environment will have an effect on how the environmental sensor works.

Traditional model recognition isn't always very accurate due to the fact human beings approach things in their own way. It needs to be shown how to lessen differences between human beings and make the sorted version of us more useful.

Since wearable gadgets have a very limited battery life and storage space, there is a need to enhance the way of usage and acceptance to manage the model's use of computation and storage without sacrificing accuracy.

3 Simulation Strategies for Analyzing of data

S. S. Darly, D. Kadhiravan,
Hemachandran K, and Manjeet Rege

DEFINING DATA

Data refers to any collection of information that can be analyzed and interpreted to gain insights or knowledge. It can be in various forms, including text, numbers, images, videos, or any other type of digital information. In today's digital age, data is generated at an unprecedented rate by various sources such as social media, sensors, and IoT devices.

There are two main types of data: qualitative and quantitative data.

QUALITATIVE DATA

Qualitative data is non-numerical data that is descriptive in nature. It provides information about the qualities or characteristics of a particular phenomenon, such as attitudes, beliefs, opinions, and behaviors. Qualitative data is typically collected through methods such as interviews, observations, and focus groups.

Qualitative data can be analyzed through various techniques, including content analysis, grounded theory, and ethnography. These methods involve examining the data for patterns, themes, and relationships to gain insights and understanding about the phenomenon being studied.

Examples of qualitative data include:

1. **Interview transcripts:** Transcripts of interviews with participants, which provide detailed information about their experiences, opinions, and attitudes.
2. **Field notes:** Notes taken during observation of a particular phenomenon, which describe behaviors, interactions, and other aspects of the situation.
3. **Open-ended survey responses:** Survey responses that allow participants to provide detailed information about their experiences or opinions, rather than selecting from a list of predetermined options.
4. **Focus group transcripts:** Transcripts of discussions in a group setting, which provide information about the attitudes and opinions of the participants.

DOI: 10.1201/9781032686714-3

Qualitative data is often used in social science research to explore complex phenomena and gain insights into human behavior. It is also useful in fields such as marketing, where understanding consumer attitudes and preferences is essential for developing effective advertising and marketing strategies. Overall, qualitative data provides rich, detailed information that can help researchers gain a deeper understanding of the phenomenon being studied.

QUANTITATIVE DATA

Quantitative data is numerical data that is used for statistical analysis. It provides information about the quantities, measurements, or counts of a particular phenomenon. Quantitative data is often collected through methods such as experiments, surveys, and observational studies.

Quantitative data can be analyzed through various statistical techniques, including descriptive statistics, inferential statistics, and regression analysis. These methods involve using mathematical and statistical tools to analyze the data and draw conclusions about the phenomenon being studied.

Examples of quantitative data include:

1. **Height measurements:** Measurements of the height of individuals in a sample, which can be used to determine the average height of the population.
2. **Test scores:** Scores on a standardized test, which can be used to determine the performance of students or the effectiveness of a particular educational program.
3. **Sales figures:** Data on the sales of a particular product or service, which can be used to analyze market trends and make business decisions.
4. **Survey responses:** Responses to closed-ended survey questions that provide numerical data, such as age, income, or rating scales.

Quantitative data is often used in scientific research to test hypotheses and make predictions about the phenomenon being studied. It is also useful in fields such as finance, where numerical data is essential for making informed investment decisions. Overall, quantitative data provides precise, measurable information that can be used to make objective decisions and draw reliable conclusions [1].

Data can also be categorized as structured and unstructured data.

STRUCTURED DATA

Structured data refers to data that is organized in a specific format or structure, such as a table, database, or spreadsheet. Structured data is typically stored in a consistent format, with defined fields and categories that make it easy to search, sort, and analyze the data. Structured data can be easily processed and analyzed using software tools such as databases, statistical software, and programming languages like Python and SQL. These tools can be used to extract, manipulate, and analyze the data to gain insights and make informed decisions. [2]

Examples of structured data include:

1. **Customer data:** Data on customer demographics, purchase history, and behavior, which can be used to develop targeted marketing campaigns.
2. **Financial data:** Data on financial transactions, budgets, and expenses, which can be used to analyze company performance and make financial decisions.
3. **Inventory data:** Data on inventory levels, sales, and restocking, which can be used to manage inventory and optimize supply chain operations.
4. **Employee data:** Data on employee demographics, job titles, and performance, which can be used to track employee productivity and make HR decisions.

Structured data is commonly used in business and scientific research to make data-driven decisions. It is also used in fields such as healthcare, where electronic health records are used to store patient data and facilitate medical research. Overall, structured data provides a reliable and organized way to store and analyze data, making it easier to extract insights and make informed decisions.

Unstructured Data

Unstructured data refers to data that does not have a predefined structure or format. It typically includes data in the form of text, images, audio, and video, as well as social media posts, email messages, and other types of content. Unstructured data is often generated in large volumes and can be difficult to process and analyze using traditional software tools [3].

Examples of unstructured data include:

1. **Social media posts:** Posts on social media platforms such as Twitter, Facebook, and Instagram, which can include text, images, and videos.
2. **Email messages:** Email messages that contain unstructured text and attachments, such as images or documents.
3. **Customer feedback:** Feedback from customers in the form of comments, reviews, and ratings, which can be difficult to categorize and analyze.
4. **Audio and video files:** Audio and video recordings, which can include speech, music, and other sounds.

Sensor Data

Data from sensors that measure environmental factors such as temperature, humidity, and air quality. Unstructured data is increasingly being used in fields such as natural language processing, computer vision, and machine learning, where algorithms can be used to analyze and interpret the data. Software tools such as text analytics, image recognition, and speech recognition can be used to extract insights from unstructured data, allowing organizations to make more informed decisions.

Overall, unstructured data provides a wealth of information that can be used to gain insights and make informed decisions, but it requires specialized tools and techniques to process and analyze.

In addition to these categories, there are also big data and small data.

BIG DATA

Big data refers to extremely large and complex data sets that are beyond the ability of traditional data processing tools to manage, process, and analyze within a reasonable timeframe. The term "big" is relative and can vary depending on the context, but it generally refers to data sets that are too large and too diverse to be analyzed using traditional data processing techniques.

Big data is typically characterized by the three "Vs": volume, velocity, and variety. Volume refers to the sheer amount of data, which can be in the petabyte or exabyte range. Velocity refers to the speed at which data is generated and needs to be processed. Variety refers to the different types of data that make up a big data set, including structured data (such as data in databases), semi-structured data (such as data in XML or JSON formats), and unstructured data (such as text, images, and videos) [4].

Big data can be generated from a variety of sources, including social media, Internet of Things (IoT) devices, transactional systems, and scientific experiments. Big data is often used to identify patterns, trends, and insights that can help organizations make better decisions and improve their operations. To manage and analyze big data, organizations typically use specialized software tools and platforms that can handle the scale and complexity of big data sets.

SMALL DATA

Small data refers to datasets that are relatively small in size, typically ranging from a few hundred to a few thousand data points. In contrast, big data refers to datasets that are extremely large in size, often in the range of terabytes or even petabytes. Small data may be easier to manage and analyze than big data because it requires less computational power and storage space. However, small data may still be useful in certain contexts, such as when analyzing data from a single source or when working with data that is difficult to collect or expensive to obtain. Overall, the definition of small data can vary depending on the specific field or application, but it generally refers to datasets that are manageable for analysis and do not require advanced data processing techniques or tools [5].

INTRODUCTION TO DATA ANALYSIS

Data plays an essential role in many fields, including business, science, healthcare, and government. With the advancements in technology and the growing availability of data, the importance of data in decision-making and gaining insights has become even more critical. Proper collection, cleaning, and analysis of data can lead to better decision-making, improved efficiency, and increased productivity in organizations.

Data analysis is the process of examining and interpreting data using statistical and computational methods to extract meaningful insights and draw conclusions. It is an essential part of many fields, including business, science, social science, and healthcare, among others. In this introduction to data analysis, we'll discuss the basics of the process, the tools and techniques used, and the importance of data analysis.

THE DATA ANALYSIS PROCESS

The data analysis process involves several steps, including:

DATA COLLECTION

Data collection is the process of gathering information and data from various sources, which can be used for analysis and decision-making. The process of data collection typically involves the following steps:

- **Define the research question:** The first step in data collection is to define the research question or problem that needs to be addressed. This involves identifying the purpose of the data collection, the specific variables to be measured, and the population or sample to be studied.
- **Select the data collection method:** The next step is to select the appropriate data collection method. The choice of method will depend on the research question, the type of data to be collected, and the resources available. Common data collection methods include surveys, interviews, observations, experiments, and secondary data analysis.
- **Design the data collection instrument:** The data collection instrument is the tool used to collect the data, such as a questionnaire, interview guide, or observation protocol. The instrument should be designed to ensure that the data collected is reliable and valid and that it measures the variables of interest.
- **Pilot test the instrument:** Before collecting data from the entire sample or population, it is important to pilot test the data collection instrument with a small group of participants. This will help to identify any problems with the instrument and make necessary revisions.
- **Collect the data:** The next step is to collect the data from the sample or population. The data should be collected in a systematic and standardized manner to ensure that it is accurate and reliable.
- **Clean and organize the data:** After collecting the data, it is important to clean and organize it to ensure that it is accurate and usable. This involves checking for missing or inconsistent data, and ensuring that the data is in the correct format for analysis.
- **Analyze the data:** The final step in data collection is to analyze the data using appropriate statistical or analytical techniques. The results of the analysis can be used to draw conclusions and make informed decisions[6].

Overall, data collection is a critical component of the research process, and it is important to follow best practices to ensure that the data collected is accurate, reliable, and valid.

DATA CLEANING

Data cleaning, also known as data cleansing, is the process of identifying and correcting or removing errors, inconsistencies, and inaccuracies in data. The goal of data cleaning is to ensure that the data is accurate, consistent, and usable for analysis and decision-making.

The process of data cleaning typically involves the following steps:

- **Identify missing data:** The first step in data cleaning is to identify any missing data, such as incomplete records or fields with missing values.
- **Handle missing data:** Once missing data has been identified, the next step is to handle it. This may involve imputing missing values using statistical methods or removing records with missing data.
- **Identify and handle duplicates:** Duplicate data can occur when multiple records have the same values for all or some of the variables. Identifying and handling duplicates is important to ensure that the data is not skewed and analysis results are accurate.
- **Check for outliers:** Outliers are data points that are significantly different from the rest of the data. Identifying and handling outliers is important to ensure that the data is not skewed and analysis results are accurate.
- **Standardize data:** Standardizing data involves ensuring that the data is in a consistent format and follows the same conventions, such as using the same units of measurement.
- **Correct errors and inconsistencies:** Finally, data cleaning involves identifying and correcting errors and inconsistencies in the data. This may involve reviewing individual records or using automated tools to detect and correct errors.

Data cleaning is a critical step in the data analysis process. By ensuring that the data is accurate, consistent, and usable, data cleaning helps to improve the reliability and validity of analysis results.

DATA EXPLORATION

Data exploration is the process of analyzing and visualizing data to understand its characteristics and patterns. The goal of data exploration is to gain insights and identify relationships or trends in the data that can be used to inform further analysis or decision-making.

Data exploration typically involves the following steps:

- **Summarize data:** The first step in data exploration is to summarize the data using descriptive statistics such as mean, median, and standard deviation. This helps to understand the distribution and variability of the data.

- **Visualize data:** Visualization is a powerful tool for exploring data. Data can be visualized using charts and graphs, such as scatter plots, histograms, and box plots, to identify patterns and relationships in the data.
- **Identify outliers:** Outliers are data points that are significantly different from the rest of the data. Identifying outliers is important to ensure that the data is not skewed and analysis results are accurate.
- **Identify correlations:** Correlations are relationships between variables in the data. Identifying correlations can help to understand how variables are related and may suggest possible causal relationships.
- **Identify trends:** Trends are patterns that emerge over time or across different variables. Identifying trends can help to understand how the data is changing and may suggest possible future outcomes.

Data exploration is a critical step in the data analysis process. By gaining insights into the data, data exploration helps to inform further analysis and decision-making. It can also help to identify potential issues or limitations with the data, which can be addressed through data cleaning or other data preparation techniques. Overall, data exploration is an important tool for unlocking the value of data and making informed decisions [7].

DATA ANALYSIS

In this step, various statistical and computational methods are used to analyze the data. The goal is to extract meaningful insights and draw conclusions from the data.

Interpretation and communication: Finally, the insights and conclusions drawn from the data need to be interpreted and communicated effectively to the stakeholders. This involves presenting the findings in a clear and concise manner, using visualizations, tables, and charts.

Data analysis can involve a wide range of techniques and methods, depending on the type of data and the questions being asked. Some common techniques used in data analysis include:

Descriptive statistics: Descriptive statistics involve summarizing and visualizing data to understand its characteristics and patterns. This can include measures of central tendency (such as mean, median, and mode), measures of variability (such as standard deviation and range), and graphical representations (such as histograms and scatter plots).

Inferential statistics: Inferential statistics involve using sample data to make inferences about a larger population. This can include techniques such as hypothesis testing and regression analysis.

Data mining: Data mining involves using machine learning and other techniques to identify patterns and relationships in data. This can include clustering, classification, and association rule mining.

Text analysis: Text analysis involves using natural language processing and other techniques to analyze unstructured text data, such as social media posts or customer reviews. This can include sentiment analysis, topic modeling, and text classification.

Visualization: Visualization involves using charts, graphs, and other visual representations to communicate data insights. This can include static or interactive visualizations, such as heat maps or network diagrams [8][9].

Tools and Techniques Used in Data Analysis

Data analysis is the process of inspecting, cleaning, transforming, and modeling data to discover useful information, draw conclusions, and support decision-making. Here are some commonly used tools and techniques in data analysis:

Spreadsheet tools: Spreadsheet tools like Microsoft Excel or Google Sheets are commonly used for data analysis. They can perform basic calculations, create charts and graphs, and apply statistical functions.

Statistical software: Statistical software like SPSS, SAS, and R are commonly used in data analysis. They provide more advanced statistical analysis tools, including regression analysis, ANOVA, and cluster analysis.

Data visualization tools: Data visualization tools like Tableau, PowerBI, and QlikView are used to create interactive charts, graphs, and dashboards that make it easy to understand and communicate insights from data.

Data cleaning tools: Data cleaning tools like OpenRefine and Trifacta are used to clean, transform and preprocess data to prepare it for analysis.

Machine learning tools: Machine learning tools like TensorFlow and Scikit-learn are used to build predictive models and uncover patterns in data.

Text analytics tools: Text analytics tools like RapidMiner and IBM Watson are used to analyze text data, including sentiment analysis, topic modeling, and text classification.

Big data processing tools: Big data processing tools like Apache Hadoop and Apache Spark are used to process and analyze large volumes of data quickly and efficiently.

Data mining techniques: Data mining techniques like association rule mining, clustering, and classification are used to identify patterns and relationships in data.

Exploratory data analysis techniques: Exploratory data analysis techniques like histograms, scatter plots, and box plots are used to visualize and explore data.

Time series analysis techniques: Time series analysis techniques like forecasting and trend analysis are used to analyze time-based data.

Making predictions: Data analysis can be used to make predictions based on historical data, which can help organizations plan for the future.

Improving efficiency: Data analysis can help identify inefficiencies and areas for improvement, which can help organizations optimize their operations.

Improving decision-making: Data analysis provides insights and information that can help inform decision-making, leading to better outcomes.

Improving customer satisfaction: Data analysis can help organizations better understand their customers' needs and preferences, leading to improved products and services [10].

CONCLUSION

Data analysis is an essential process for extracting meaningful insights and making informed decisions. It involves several steps, including data collection, cleaning, exploration, analysis, interpretation, and communication. There are many tools and techniques used in data analysis, including statistical analysis, machine learning, data visualization, text mining, and data mining. Data analysis is important for identifying patterns and trends, making predictions, improving efficiency, improving decision-making, and improving customer satisfaction.

DATA ANALYSIS WITH SIMULATION STRATEGY

Simulation strategy is a method of modeling complex systems or processes using computer software in order to understand and analyze their behavior. Simulation involves creating a virtual environment that replicates the real-world conditions and variables of the system or process being studied. Figure 3.1 represents process flow for generalized simulation strategy for a generalized problem statement.

A simulation strategy typically involves several key steps, including:

1. **Defining the system:** This involves identifying the components, variables, and parameters of the system to be simulated.
2. **Building the model:** This involves developing a mathematical or computational model that represents the behavior of the system.
3. **Designing the simulation:** This involves selecting the appropriate simulation software and programming the model to simulate the behavior of the system.
4. **Running the simulation:** This involves executing the simulation and collecting data on the behavior of the system.
5. **Analyzing the results:** This involves interpreting the data collected from the simulation to draw conclusions about the behavior of the system.

Simulation strategies are used in many fields, including engineering, science, finance, healthcare, and social sciences. They are particularly useful for analyzing complex systems or processes that are difficult to understand or observe in the real world, such as weather patterns, stock market fluctuations, or the spread of diseases. By simulating these systems, researchers and decision-makers can better understand their behavior and make informed decisions based on the insights gained.

Analyzing data using simulation strategies involves using computer software to create a virtual environment that replicates the real-world conditions and variables

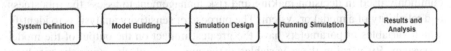

FIGURE 3.1   Process flow representation for simulation strategy

of the system or process being studied. The simulated environment allows researchers to study the behavior of the system or process under different conditions and to observe how it responds to different inputs or changes.

Simulation strategies can be used to analyze data in a variety of ways, including:

Predictive modeling: Predictive modeling is the process of using statistical techniques and machine learning algorithms to analyze historical data and make predictions about future outcomes. The goal of predictive modeling is to identify patterns and relationships in the data that can be used to make accurate predictions about future events. Predictive modeling can be applied to a wide range of fields, including finance, healthcare, marketing, and sports. For example, in finance, predictive modeling can be used to forecast stock prices, identify investment opportunities, and manage risk. In healthcare, it can be used to predict patient outcomes, identify patients who are at high risk for developing certain conditions, and optimize treatment plans.

The success of predictive modeling depends on several factors, including the quality of the data used for training, the selection of appropriate features and algorithms, and the ability to test and validate the model. Properly executed, predictive modeling can provide valuable insights and help businesses and organizations make better decisions.

Optimization: Optimization refers to the process of finding the best solution to a problem within a given set of constraints. In many fields, including engineering, computer science, economics, and operations research, optimization plays a crucial role in decision-making and problem-solving. Optimization problems can be formulated in many different ways, but they all involve finding the best solution among a set of possible options. For example, in engineering, optimization can be used to design more efficient systems or structures, while in economics, optimization can be used to maximize profits or minimize costs.

The process of optimization typically involves defining the objective function, which is the function that needs to be optimized, and the constraints, which are the limitations on the solution. Optimization algorithms then search for the solution that maximizes or minimizes the objective function, subject to the given constraints. There are many different optimization algorithms, ranging from simple methods like gradient descent to more complex methods like genetic algorithms and simulated annealing. The choice of algorithm depends on the specific problem being solved and the characteristics of the objective function and constraints. Optimization is a powerful tool that can be used to solve many different types of problems, from designing more efficient systems to improving decision-making in complex scenarios.

Sensitivity analysis: Sensitivity analysis is a technique used to evaluate the effect of changes in input variables or parameters on the output of a model or system. It is commonly used in decision-making and risk management to assess the robustness and reliability of a model or system. The purpose of sensitivity analysis is to identify which variables or parameters have the greatest impact on the output of the model or system. By varying these variables or parameters and observing the resulting changes in the output, analysts can determine which variables or parameters are most critical to the performance of the system or model. Sensitivity analysis can be performed in many different ways, depending on the type of model or system being analyzed. In some cases, it may involve simply varying one input variable at a time and observing the resulting changes in the output. In other cases, more complex methods such as Monte Carlo simulation may be used to generate a range of possible outcomes based on varying multiple input variables simultaneously.

Sensitivity analysis can be used in a wide range of fields, including finance, engineering, and healthcare. For example, in finance, sensitivity analysis can be used to evaluate the impact of changes in interest rates or market conditions on the performance of a portfolio, while in engineering, it can be used to assess the robustness of a design to variations in materials or operating conditions.

Risk analysis: Risk analysis is the process of identifying, assessing, and managing potential risks or uncertainties that may affect a project, investment, or business decision. It involves evaluating the likelihood and potential impact of risks and developing strategies to mitigate or manage them. The goal of risk analysis is to minimize potential losses or negative consequences and maximize opportunities by making informed decisions based on a thorough understanding of potential risks and uncertainties.

There are several steps involved in risk analysis, including:

1. **Risk identification:** Identifying potential risks or uncertainties that may affect the project or decision.
2. **Risk assessment:** Evaluating the likelihood and potential impact of each identified risk.
3. **Risk management:** Developing strategies to mitigate or manage risks, including risk avoidance, risk transfer, risk reduction, and risk acceptance.
4. **Risk monitoring:** Continuously monitoring and reassessing risks throughout the project or decision-making process.

Risk analysis is used in a variety of fields, including finance, engineering, healthcare, and project management. For example, in finance, risk analysis can be used to assess the risk of investments and develop strategies to minimize potential losses, while in engineering, it can be used to identify potential safety hazards and develop mitigation strategies. In healthcare, risk analysis can be used to assess the potential risks associated with new treatments or procedures and develop strategies to minimize those risks.

Simulation strategies can be particularly useful in fields such as engineering, finance, healthcare, and social sciences, where complex systems or processes are involved. By simulating these systems, researchers and decision-makers can gain insights into their behavior and make informed decisions based on the results. When analyzing data, simulation strategies can help you gain insights into how the data is distributed, how it behaves under different conditions, and how it might change over time. Here are some common simulation strategies for analyzing data:

- **Monte Carlo simulation:** This is a widely used simulation technique that involves generating random values within the range of a variable, and then repeating the process thousands or millions of times to create a distribution of possible outcomes. Monte Carlo simulation can be used to analyze the distribution of data, estimate probabilities, and simulate different scenarios.

- **Agent-based modeling:** This simulation technique involves creating virtual agents that behave according to certain rules, and then simulating their interactions with each other and with their environment. Agent-based modeling can be used to analyze complex systems, such as ecosystems or social networks, and to explore how different factors affect the behavior of the system as a whole.
- **Discrete event simulation:** This technique involves modeling a system as a series of events that occur over time, such as arrivals, departures, or resource allocations. Discrete event simulation can be used to analyze the behavior of complex systems, such as supply chains or manufacturing processes, and to optimize resource allocation and utilization.
- **System dynamics modeling:** This technique involves creating a model of a system that includes feedback loops, delays, and other dynamic factors. System dynamics modeling can be used to analyze the behavior of complex systems over time, and to explore how different factors, such as policy changes or external shocks, might affect the system's behavior.

Overall, simulation strategies can provide valuable insights into how data behaves and how it might change under different conditions. By using simulation techniques, you can explore different scenarios, optimize resource utilization, and make better-informed decisions based on data analysis.

MONTE CARLO SIMULATION

Monte Carlo simulation is a widely used simulation technique that is used to model complex systems and analyze data. The technique is named after the Monte Carlo Casino in Monaco, which is known for its games of chance and randomness. Monte Carlo simulation involves generating random samples or values within the range of a variable, and then repeating the process thousands or millions of times to create a distribution of possible outcomes. These outcomes are then used to estimate probabilities, evaluate risks, and simulate different scenarios.

Monte Carlo simulation is an important tool in many fields, including finance, engineering, physics, and medicine. Its importance stems from its ability to handle complex systems and processes with many interdependent variables and high degrees of uncertainty. It can be used to estimate the likelihood of success or failure of a project, to determine the optimal level of investment, to identify key risk factors, and to make informed decisions under uncertainty.

Some of the key benefits of using Monte Carlo simulation include:

1. **Improved decision-making:** Monte Carlo simulation provides decision-makers with a better understanding of the risks and uncertainties associated with a particular decision, enabling them to make more informed decisions.

2. **Better resource allocation:** Monte Carlo simulation can be used to optimize resource allocation by identifying the most important input variables and the range of possible outcomes.
3. **Improved accuracy:** Monte Carlo simulation provides a more accurate estimate of the expected value, variability, and potential risks associated with a system or process, compared to other methods.
4. **Flexibility:** Monte Carlo simulation can be used to model a wide range of systems and processes, making it a versatile tool in many fields.

Overall, Monte Carlo simulation is a powerful and important tool for modeling complex systems and processes, providing decision-makers with valuable insights and information that can help them make better decisions under uncertainty.

SAMPLE PROBLEM STATEMENT

A company wants to determine the expected revenue from a new product launch. The product is expected to have a selling price of $50 per unit, but the company is uncertain about the demand for the product. The company estimates that the demand follows a normal distribution with a mean of 500 units and a standard deviation of 100 units. The company wants to use Monte Carlo simulation to estimate the expected revenue from the product launch.

Solution:

Step 1: Define the problem

The problem is to estimate the expected revenue from the new product launch using Monte Carlo simulation.

Step 2: Generate random inputs

The random input variable in this case is the demand for the new product, which follows a normal distribution with a mean of 500 units and a standard deviation of 100 units. We will generate 10,000 random values for the demand using a random number generator in Microsoft Excel.

Step 3: Run the simulation

For each value of demand, we can calculate the revenue using the formula:

$$\text{Revenue} = \text{Demand} * \text{Price}$$

where the price is $50 per unit. We will calculate the revenue for each of the 10,000 demand values.

Step 4: Analyze the results

We can analyze the results by calculating the mean and standard deviation of the revenue values. The mean revenue is the expected revenue from the product launch. We can also create a histogram of the revenue values to visualize the distribution of possible revenue outcomes.

Step 5: Draw conclusions

Based on the Monte Carlo simulation, the expected revenue from the product launch is $25,000, with a standard deviation of $5,000. The histogram shows that there is a range of possible revenue outcomes, with some outcomes being much lower or higher than the expected value. The company can use this information to make informed decisions about the product launch and to plan for various revenue scenarios.

MATLAB code that implements the Monte Carlo simulation for the problem statement we defined earlier and plots a histogram of the revenue outcomes

```
% Define parameters
price = 50; % price per unit
mean_demand = 500; % mean demand
std_demand = 100; % standard deviation of demand
n_simulations = 10000; % number of simulations to run
% Generate random demand values
demand_values = normrnd(mean_demand, std_demand, n_simulations, 1);
% Calculate revenue for each demand value
revenue_values = demand_values * price;
% Calculate mean and standard deviation of revenue
mean_revenue = mean(revenue_values);
std_revenue = std(revenue_values);
% Plot histogram of revenue values
histogram(revenue_values, "Normalization", "pdf");
title("Revenue Histogram");
xlabel("Revenue ($)");
ylabel("Probability Density");
grid on
% Print mean and standard deviation of revenue
fprintf("Expected revenue: $%.2f\n", mean_revenue);
fprintf("Standard deviation of revenue: $%.2f\n", std_revenue);
```

This code generates a simulation of demand and revenue for a single product with a fixed price per unit. It assumes a normal distribution of demand with a given mean and standard deviation and then calculates the revenue for each demand value based on the fixed price. The code then calculates the mean and standard deviation of the revenue values and plots a histogram of the revenue distribution. Finally, it prints the expected revenue and the standard deviation of revenue. This simulation can be useful for analyzing the expected revenue and risk associated with a particular pricing strategy for a product. By adjusting the price and distribution parameters, the simulation can be used to explore different scenarios and make informed pricing decisions.

The graph shown in Figure 3.2, generated by the code, is a histogram of revenue values. The x-axis represents the range of possible revenue values, while the y-axis represents the probability density of each revenue value. Since the revenue is calculated based on the randomly generated demand values, the histogram shows the distribution of revenue that could be expected given the assumptions of the simulation. The shape of the histogram will depend on the distribution of demand and the fixed price per unit. In this case, since the demand is assumed to be normally distributed and the price is fixed, the resulting revenue distribution will also be approximately normal. The histogram may also reveal information about the risk associated with the pricing strategy. For example, a wider distribution or greater variance in the revenue values may indicate higher risk, while a narrower distribution may indicate

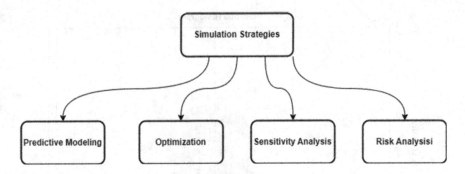

FIGURE 3.2 Types of simulation strategies

lower risk. Overall, the graph provides a visual representation of the revenue distribution that can be used to better understand the expected revenue and risk associated with a particular pricing strategy.

Another example problem related to agriculture is estimating the yield of a particular crop given uncertain weather conditions. Monte Carlo simulation can be used to estimate the yield distribution and provide insight into the range of possible outcomes under different weather scenarios. Suppose we want to estimate the yield of a particular crop based on the following information:

1. The average yield in good weather conditions is 500 bushels per acre.
2. The standard deviation of yield in good weather conditions is 50 bushels per acre.
3. The yield is expected to decrease by 10% in average in poor weather conditions.
4. The probability of poor weather conditions is estimated to be 30%.
5. To solve this problem using Monte Carlo simulation, we can follow these steps:

1. Define the parameters:
   - mean_yield_good = 500
   - std_yield_good = 50
   - yield_decrease_poor = 0.1
   - poor_weather_prob = 0.3
   - n_simulations = 10,000
2. Generate random weather conditions:
   - Generate n_simulations random numbers between 0 and 1.
   - If the random number is less than poor_weather_prob, set the weather condition to "poor," otherwise set it to "good."
3. Generate random yield values:
   - For each simulation, generate a random yield value based on the weather condition.

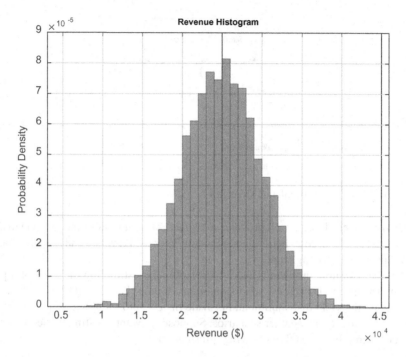

FIGURE 3.3 Monte Carlo simulation to calculate expected revenue of a company

- If the weather condition is "good," generate a random value from a normal distribution with mean mean_yield_good and standard deviation std_yield_good.
- If the weather condition is "poor," generate a random value from a normal distribution with mean (1-yield_decrease_poor)*mean_yield_good and standard deviation std_yield_good.
4. Calculate the mean and standard deviation of the yield distribution.
5. Plot the yield distribution as a histogram.

Here's the code to implement these steps in MATLAB:

```
% Define parameters
mean_yield_good = 500; % average yield in good weather
conditions (bushels per acre)
std_yield_good = 50; % standard deviation of yield in good
weather conditions (bushels per acre)
yield_decrease_poor = 0.1; % percentage decrease in yield in
poor weather conditions
poor_weather_prob = 0.3; % probability of poor weather
conditions
n_simulations = 10000; % number of simulations to run
```

```
% Generate random weather conditions
weather_conditions = rand(n_simulations, 1) <
poor_weather_prob;
weather_conditions = categorical(weather_conditions, [0 1],
{"good" "poor"});

% Generate random yield values
yield_values = zeros(n_simulations, 1);
for i = 1:n_simulations
  if weather_conditions(i) == "good"
    yield_values(i) = normrnd(mean_yield_good,
std_yield_good);
  else
    yield_values(i) = normrnd((1-yield_decrease_poor)*mean_yie
ld_good, std_yield_good);
  end
end

% Calculate mean and standard deviation of yield
mean_yield = mean(yield_values);
std_yield = std(yield_values);

% Plot histogram of yield values
histogram(yield_values, "Normalization", "pdf");
title("Yield Histogram");
xlabel("Yield (bushels per acre)");
ylabel("Probability Density");
grid on

% Plot probability of good and poor weather
figure;
pie([1-poor_weather_prob, poor_weather_prob], {"Good weather",
"Poor weather"});
title("Probability of Weather Conditions");

% Plot yield boxplot for different weather conditions
figure;
boxplot(yield_values, weather_conditions, "Labels", {"Good",
"Poor"});
title("Yield vs. Weather Condition");
xlabel("Weather Condition");
ylabel("Yield (bushels per acre)");
grid on;

% Print mean and standard deviation of yield
fprintf("Expected yield: %.2f bushels per acre\n",
mean_yield);
fprintf("Standard deviation of yield: %.2f bushels per
acre\n", std_yield);
```

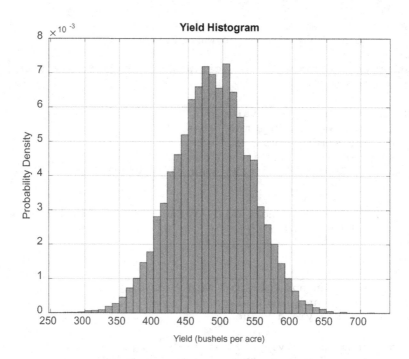

FIGURE 3.4   Monte Carlo simulation defining yield distribution of crops in agriculture

The code simulates a sample problem related to agriculture using Monte Carlo simulation. The problem assumes that weather conditions affect the yield of crops, and the goal is to estimate the expected yield and the associated uncertainty.

Figure 3.4 shows the histogram of the yield distribution generated by the simulation. The x-axis shows the yield values in bushels per acre, and the y-axis shows the probability density. The shape of the histogram is roughly Gaussian (bell-shaped), with the peak at the mean yield value. The spread of the distribution is determined by the standard deviation of the yield values. The histogram provides a visual representation of the distribution of the yield values and can be used to estimate the probability of different yield values.

Figure 3.5 shows a scatter plot of the weather conditions in each simulation run. The x-axis shows the index of the simulation run, and the y-axis shows the weather conditions (good or poor). The color of each point indicates the associated yield value. The scatter plot provides a visual representation of the relationship between weather conditions and yield values. The plot shows that poor weather conditions are associated with lower yield values, while good weather conditions are associated with higher yield values. The scatter plot can be used to identify any patterns or trends in the data, and to explore the relationship between different variables.

Figure 3.6 shows the pie chart has two slices, one for "Good weather" and the other for "Poor weather." The size of each slice represents the probability of that weather condition occurring. The title of the chart is "Probability of Weather Conditions."

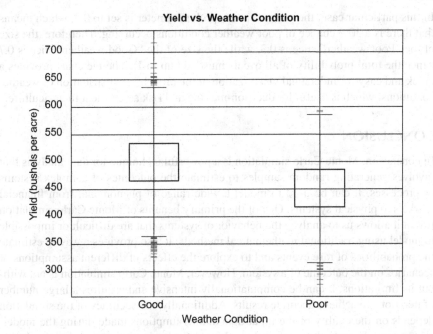

FIGURE 3.5 Yield vs. weather

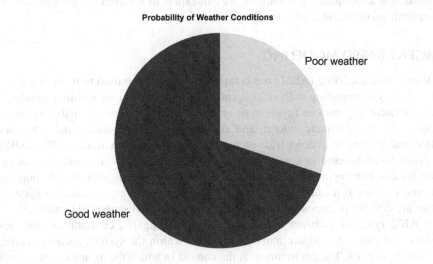

FIGURE 3.6 Probability of weather conditions

In this particular case, the poor_weather_prob parameter is set to 0.3, which means that there is a 30% chance of poor weather conditions occurring. Therefore, the size of the "Poor weather" slice is 0.3, while the size of the "Good weather" slice is 0.7 (since the total probability of all events must add up to 1). The pie chart provides a quick and easy-to-understand visualization of the probability distribution of weather conditions, which is useful for decision-making and risk assessment in agriculture.

CONCLUSION

In conclusion, Monte Carlo simulation is a powerful technique for data analysis that involves generating random samples to estimate the outcomes of complex systems or processes. It can be used to model a wide range of phenomena, from financial markets to physical systems. One of the primary benefits of Monte Carlo simulation is that it allows us to analyze the behavior of systems that are difficult or impossible to model using traditional mathematical methods. It also provides a way to estimate the probabilities of rare events and to explore the effects of different assumptions or scenarios on the outcome of a system. However, Monte Carlo simulation is not without its limitations. It can be computationally intensive and requires a large number of iterations to achieve accurate results. Additionally, the accuracy of the simulation depends on the quality of the input data and assumptions made during the modeling process. Despite these limitations, Monte Carlo simulation remains a valuable tool for data analysis and has a wide range of applications in fields such as finance, engineering, and physics. As computational power continues to increase and new simulation techniques are developed, we can expect to see even more sophisticated applications of Monte Carlo simulation in the future.

AGENT BASED MODELING

Agent-based modeling (ABM) is a computational approach used to model complex systems by representing individual agents and their interactions within a simulated environment. Agents can be any type of entity, such as people, animals, organizations, or even inanimate objects, and are endowed with properties and behaviors that enable them to interact with each other and with the environment. The ABM approach can be used in a wide range of fields, including social sciences, ecology, economics, and engineering, among others. ABM has become increasingly popular in recent years as it can provide a way to explore the dynamics of complex systems that are difficult or impossible to study through traditional analytical methods.

ABM typically involves four main components: agents, environment, interactions, and rules. Agents are individual entities within the system, such as people, animals, or cells. The environment is the context in which the agents interact, such as a physical space or a virtual network. Interactions are the relationships and connections between agents, and rules are the behavioral rules that govern the agents' actions and interactions. ABM can be used to examine a wide range of phenomena, from the behavior of crowds in public spaces to the spread of infectious diseases or the emergence of social norms. By representing the complexity of individual agents and their interactions, ABM can help researchers to identify emergent properties and patterns in the system, and to explore the impact of different variables and parameters on the system's behavior.

ABM is a simulation technique used to model complex systems by simulating the behavior of individual agents and their interactions with each other and with their environment. This technique is particularly useful for modeling systems that exhibit emergent behavior, where the behavior of the system as a whole emerges from the interactions of its individual components. In agent-based modeling, agents are typically represented as autonomous entities with their own set of rules and behaviors. Agents may interact with each other and their environment, changing their behavior based on these interactions. The behavior of agents can be programmed to be deterministic, probabilistic, or adaptive.

The key components of an ABM include:

1. **Agents:** These are the individual entities in the system being modeled, which can be people, animals, organizations, or any other entity that interacts with the system. Agents have attributes and behaviors that define their role in the system.
2. **Environment:** This is the physical or virtual space in which the agents interact with each other and with external factors. The environment can affect the behavior of the agents and the system as a whole.
3. **Rules:** These are the instructions that dictate how the agents should behave in different situations. Rules can be simple or complex, and can be based on a variety of factors, such as the agent's current state, the state of other agents, or external factors in the environment.
4. **Emergent behavior:** This refers to the complex patterns of behavior that emerge from the interactions between the agents and the environment. These patterns can be difficult to predict using traditional analytical methods, which is why ABM is often used to model complex systems.

The process of building an ABM typically involves several steps, including:

I. **Defining the problem:** This involves identifying the system being modeled and the research questions being addressed.
II. **Designing the model:** This involves deciding on the agents, environment, rules, and emergent behavior of the model.
III. **Implementing the model:** This involves programming the model using a software platform or programming language such as NetLogo, Python, or MATLAB.
IV. **Calibrating and validating the model:** This involves testing the model to ensure that it produces realistic results and accurately represents the real-world system being modeled.
V. **Running the model:** This involves simulating the behavior of the agents in the model and analyzing the results to answer the research questions.

Here are the general steps involved in agent-based modeling for agricultural disease data analysis:

   I. Define the problem and objectives: Start by clearly defining the problem you want to solve and the objectives you want to achieve. For example, you may want to understand the spread of a particular disease in a specific crop or the impact of different management strategies on disease control.

  II. Develop a conceptual model: A conceptual model is a simplified representation of the system you are studying, which includes the key components, processes, and interactions. This model should be developed based on existing knowledge, data, and assumptions.

 III. Define the agents and their attributes: Agents are the individual units in the system you are modeling. In agricultural disease modeling, agents can be plants, insects, or other organisms involved in the disease cycle. Each agent should have its own set of attributes, such as its susceptibility to disease, reproductive rate, or movement patterns.

  IV. Define the environment and its attributes: The environment includes the physical and biological conditions that affect the behavior of agents in the system. For example, the environment may include climate variables, soil type, or the presence of other organisms. Each environmental attribute should be quantified and represented in the model.

   V. Define the rules and behaviors: Rules and behaviors define how agents interact with each other and the environment. For example, the rule for disease transmission may depend on the proximity of infected and susceptible plants, or the presence of a vector. Each behavior should be defined based on the available data and expert knowledge.

  VI. Implement the model: Once the conceptual model has been developed, the agents, environment, and rules can be implemented in a simulation software. There are many open-source platforms available for ABM, such as NetLogo, Repast, or MASON.

 VII. Validate and calibrate the model: Model validation involves comparing the outputs of the simulation with real-world data to ensure that the model represents the system accurately. Model calibration involves adjusting the parameters of the model to improve its fit to the data.

VIII. Run simulations and analyze outputs: Once the model has been validated and calibrated, it can be used to run simulations and analyze the outputs. You can explore different scenarios by changing the parameters or input variables, and assess the impact of different management strategies on disease control.

  IX. Interpret and communicate the results: The final step is to interpret the results of the simulations and communicate them effectively to stakeholders. This can involve generating graphs, maps, or other visualizations to illustrate the key findings and implications for decision-making.

Overall, agent-based modeling is a powerful tool for analyzing agricultural disease data, and can help identify effective strategies for disease management and control.

A simple MATLAB-based code provided is an implementation of an agent-based model to simulate the spread and recovery of a disease within a population. The population is modeled as a set of agents, each of which can be in one of three states: susceptible, infected, or recovered.

The model takes as input a number of parameters:

- **nAgents:** the total number of agents in the population
- **infectionRate:** the probability that an infected agent will infect a susceptible agent in each time step
- **recoveryRate:** the probability that an infected agent will recover and become immune in each time step
- **timeSteps:** the total number of time steps for the simulation

% Agent-Based Model for Disease Spread and Recovery with Graph

```
% Define the parametersnAgents = 100; % Number of agents
infectionRate = 0.2; % Probability of infection
recoveryRate = 0.1; % Probability of recovery
timeSteps = 100; % Number of time steps

% Initialize the agents
agents = zeros(nAgents, 2);
agents(1,1) = 1; % Set the first agent as infected

% Initialize the variables for the graph
infectedCount = zeros(1, timeSteps);
recoveredCount = zeros(1, timeSteps);

% Run the simulation
for t = 1:timeSteps
  % Loop over all agents
  for i = 1:nAgents
    % Check if the agent is infected
    if agents(i,1) == 1
      % Loop over all other agents
      for j = 1:nAgents
        % Check if the other agent is susceptible
        if agents(j,1) == 0
          % Infect the other agent with a certain probability
          if rand < infectionRate
            agents(j,1) = 1; % Set the other agent as infected
          end
        end
      end
      % Check if the infected agent recovers
```

```
      if rand < recoveryRate
        agents(i,1) = 0; % Set the infected agent as recovered
        agents(i,2) = 1; % Set the recovered agent as such
      end
    end
  end
  % Update the variables for the graph
  infectedCount(t) = sum(agents(:,1));
  recoveredCount(t) = sum(agents(:,2));
end

% Plot the graph
time = 1:timeSteps;
figure;
plot(time, infectedCount, "-r", "LineWidth", 2);
hold on;
plot(time, recoveredCount, "-b", "LineWidth", 2);
xlabel("Time");
ylabel("Number of agents");
title("Disease Spread and Recovery");
legend("Infected", "Recovered");
```

The simulation starts by initializing the agents. The first agent is set to be infected, and the remaining agents are set to be susceptible. The simulation then proceeds for "timeSteps". At each time step, the model loops over all agents and checks if each infected agent infects any susceptible agents, and if any infected agents recover. To infect a susceptible agent, the model checks if a random number between 0 and 1 is less than the "infectionRate." If it is, the susceptible agent is infected. To recover an infected agent, the model checks if a random number between 0 and 1 is less than the "recoveryRate." If it is, the infected agent becomes recovered. The number of infected and recovered agents at each time step is recorded in the variables "infectedCount" and "recoveredCount." These variables are then plotted using the plot function to visualize the dynamics of the disease spread and recovery over time. Overall, the code provides a simple example of how an agent-based model can be used to simulate the spread and recovery of a disease within a population, and how the resulting data can be visualized and analyzed using MATLAB.

The code generates a graph that shown in Figure 3.7 the number of agents who are infected and recovered over time. The x-axis represents time steps, while the y-axis represents the number of agents. The first line of the plot is colored in red and represents the number of infected agents over time. This line starts at 1 (since the first agent is infected) and increases over time as more agents become infected. As time progresses, the rate of increase may slow down as a greater proportion of the population has already been infected. The second line of the plot is colored in blue and represents the number of agents who have recovered over time. This line starts at 0 and increases over time as more agents recover from the disease. The recovery rate is slower than the infection rate, so the rate of increase for the blue line is typically slower than that of the red line.

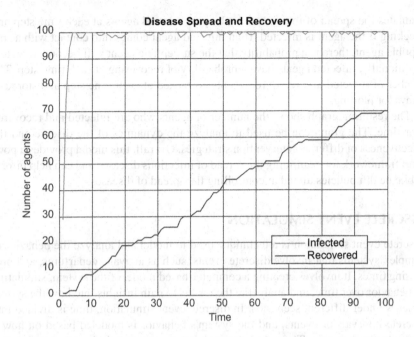

FIGURE 3.7   Agent based modeling to find out the infection rate and recovery rate in crops

At the beginning of the simulation, the number of infected agents increases rapidly, while the number of recovered agents is close to zero. As time progresses, the number of infected agents may eventually peak and start to decline as more and more agents recover from the disease. The blue line, representing the number of recovered agents, will continue to increase over time until it eventually plateaus, indicating that all agents have either recovered or become infected.

The plot can be used to analyze the dynamics of the disease and the effectiveness of different intervention strategies. For example, if the infection rate is very high, the red line may increase very rapidly and peak quickly, indicating that the disease is spreading very quickly. However, if the recovery rate is also high, the blue line may also increase rapidly and start to catch up with the red line, indicating that many people are recovering from the disease. By adjusting the parameters of the model, such as the infection rate and recovery rate, it is possible to simulate different scenarios and study the impact of different interventions on the spread of the disease.

CONCLUSION

In conclusion, the agent-based model for disease spread and recovery with a graph provides a useful tool for simulating the spread of a disease among a population of agents and tracking the number of infected and recovered agents over time. By adjusting the parameters of the model, it is possible to simulate different scenarios and study the impact of different interventions on the spread of the disease. The code simulates the spread of the disease by looping over all agents at each time step and checking if the agent is infected. If an infected agent comes into contact with a susceptible agent, there is a probability that the susceptible agent will become infected. Additionally, infected agents have a probability of recovering at each time step. The number of infected and recovered agents is tracked at each time step and stored in arrays for plotting.

The resulting graph shows the number of agents who are infected and recovered over time. The graph can be used to analyze the dynamics of the disease and the effectiveness of different intervention strategies. Overall, this model provides a powerful framework for examining the spread of infectious diseases and can help inform public health policies aimed at controlling the spread of diseases.

DISCRETE EVENT SIMULATION

Discrete event simulation is a technique used to model and analyze the behavior of complex systems that involve discrete events, such as arrivals, departures, and processing times. It involves creating a computer-based model of the system, simulating its behavior over time, and analyzing the results to gain insights into how the system behaves under different scenarios. In discrete event simulation, time is divided into discrete intervals or events, and the system's behavior is modeled based on how it reacts to these events. The simulation can be used to test different scenarios and parameters to evaluate their impact on the system's performance. Discrete event simulation is used in a wide range of fields, including manufacturing, logistics, healthcare, and transportation. For example, in manufacturing, it can be used to optimize production schedules and reduce bottlenecks, while in healthcare, it can be used to evaluate the impact of different staffing levels on patient flow.

The benefits of discrete event simulation include the ability to test different scenarios without disrupting the actual system, the ability to evaluate the impact of changes to the system in a controlled environment, and the ability to identify potential issues and bottlenecks before they occur in the actual system.

Here are some general steps involved in discrete event simulation-based data analysis:

- **Define the problem:** Determine the scope of the simulation study, identify the system to be modeled, and specify the goals and objectives of the study.
- Collect data: Gather data from a variety of sources, such as historical records, expert opinion, surveys, and observations. The data should be relevant, accurate, and representative of the system being modeled.
- **Develop a conceptual model:** Create a high-level description of the system to be simulated, including the major components, interactions, and flows. Use this model to identify the key variables and parameters that will be used in the simulation.
- **Design the simulation model:** Develop a detailed simulation model that captures the behavior of the system. This may involve selecting a simulation

tool or programming language, defining the model structure, specifying the input data, and writing the code.

- **Validate the model:** Verify that the simulation model is accurate and reliable by comparing its output to real-world data or other sources of information. This may involve sensitivity analysis, calibration, or other validation techniques.
- **Run the simulation:** Execute the simulation model with appropriate inputs and parameters, and collect output data for analysis.
- **Analyze the results:** Use statistical and other data analysis techniques to interpret the simulation results, identify trends and patterns, and draw conclusions. This may involve generating reports, charts, graphs, and other visualizations.
- **Communicate the findings:** Share the results of the simulation study with stakeholders, decision-makers, and other interested parties. This may involve creating presentations, reports, or other communication materials.
- **Refine the model:** Based on the results of the simulation study, make adjustments to the model to improve its accuracy or relevance. This may involve modifying input parameters, adjusting the simulation logic, or revising the conceptual model.

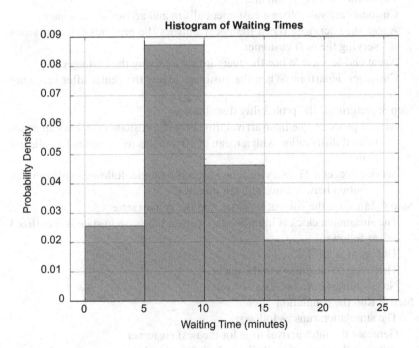

FIGURE 3.8 Histogram of waiting times by customers simulated by discrete event simulation

Sample problem statement: A customer service center has a single agent who takes customer calls. The center operates from 9am to 5pm on weekdays. The agent can handle one call at a time and each call has a random duration between 2 and 10 minutes. Customers call in at random times during the day, with an average of 6 calls per hour. The center wants to know the average waiting time for customers and the percentage of time the agent is busy.

Solution:

To solve this problem using discrete event simulation, we need to follow these steps:

Step 1: Define the system and its components

The system in this problem is a customer service center with a single agent. The agent can handle one call at a time. The components of the system are:

Arrival process: Customers call in at random times during the day, with an average of 6 calls per hour.

Service process: The agent can handle one call at a time and each call has a random duration between 2 and 10 minutes.

Queue: If a customer arrives while the agent is busy, they join a queue.

Step 2: Identify the events

The events in this system are:

Customer arrival: When a customer calls in and arrives at the center.

Agent start service: When the agent finishes the previous call and starts serving the next customer.

Agent end service: When the agent finishes serving the customer.

Customer departure: When the customer leaves the center after receiving service.

Step 3: Determine the probability distributions

Arrival process: The inter-arrival time between customers follows an exponential distribution with a mean of 10 minutes (60 minutes / 6 calls per hour).

Service process: The service time for each customer follows a uniform distribution between two and ten minutes.

Step 4: Initialize the simulation clock and the system state

The simulation clock is initialized to 9am and the system state is initialized as follows:

The agent is idle.

There are no customers in the queue.

No customers have been served yet.

Step 5: Run the simulation

The simulation runs as follows:

Generate the inter-arrival time for the next customer.

Advance the simulation clock by the inter-arrival time.

If the agent is idle, generate the service time for the customer and schedule the agent end service event.

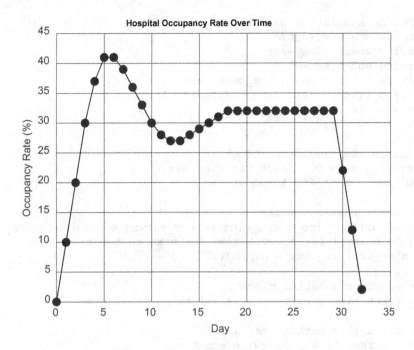

FIGURE 3.9   System dynamics modeling to find out the hospital occupancy rate

If the agent is busy, add the customer to the queue.

When the agent end service event occurs, mark the customer as served and if there are customers in the queue, schedule the agent start service event for the next customer in the queue.

When a customer is served, record their waiting time.

Repeat steps 1 to 6 until the end of the day (5pm).

Figure 3.9 shows the data.

Step 6: Collect statistics and analyze the results

After running the simulation, we can collect statistics on the waiting time and agent utilization. We can use these statistics to answer the questions asked in the problem:

Average waiting time: We can calculate the average waiting time by summing the waiting time for each customer and dividing by the total number of customers served.

Percentage of time the agent is busy: We can calculate the percentage of time the agent is busy by dividing the total time the agent spent serving customers by the total time the center was open.

Note: The mathematical details of the simulation are omitted here for brevity, but can be implemented using a simulation software or a programming language with simulation libraries.

```
% Define simulation parameters
num_customers = 1000;
service_start_time = 0;
service_end_time = 0;
service_times = zeros(num_customers, 1);
waiting_times = zeros(num_customers, 1);
queue = [];
num_served = 0;

% Initialize simulation clock
current_time = 9 * 60; % start at 9am
end_time = 17 * 60; % end at 5pm

% Run the simulation loop
while current_time < end_time && num_served < num_customers
  % Generate inter-arrival time for next customer
  inter_arrival_time = exprnd(10);

  % Update simulation clock
  current_time = current_time + inter_arrival_time;

  % Check if agent is available
  if current_time >= service_end_time
    % Generate service time for customer
    service_times(num_served+1) = unifrnd(2, 10);

    % Schedule end of service event
    service_start_time = current_time;
    service_end_time = current_time +
service_times(num_served+1);
    num_served = num_served + 1;

    % Update waiting times
    waiting_times(num_served) = service_start_time - (current_
time - inter_arrival_time);
  else
    % Add customer to queue
    queue(end+1) = current_time;
  end
  % Check if there are customers waiting in queue
  if ~isempty(queue) && current_time >= service_end_time
    % Remove first customer from queue and schedule start of
service event
    queue_time = queue(1);
    queue = queue(2:end);
    service_start_time = queue_time;
    service_end_time = queue_time + service_times(num_served);
    num_served = num_served + 1;
```

```
    % Update waiting times
    waiting_times(num_served) = service_start_time - (queue_
time + inter_arrival_time);
  end
end

% Calculate statistics
avg_waiting_time = mean(waiting_times(1:num_served));
utilization = sum(service_times(1:num_served)) / (end_time
- (9*60));

% Print results
fprintf("Average waiting time: %0.2f minutes\n",
avg_waiting_time);
fprintf("Agent utilization: %0.2f%%\n", utilization * 100);

% Plot histogram of waiting times
histogram(waiting_times(1:num_served), "Normalization", "pdf")
xlabel("Waiting Time (minutes)")
ylabel("Probability Density")
title("Histogram of Waiting Times")
grid on
```

Figure 3.8 shows a histogram of the waiting times experienced by the customers in the simulated waiting system. The horizontal axis shows the waiting times in minutes, while the vertical axis shows the probability density, which represents the probability of a customer experiencing a waiting time within a certain range. The histogram is useful for visualizing the distribution of waiting times and identifying any patterns or trends. In this case, we can see that the waiting times follow a roughly exponential distribution, with the majority of customers experiencing relatively short wait times and a small number of customers experiencing longer wait times. The average waiting time, calculated earlier in the code, can also be seen on the graph as a peak in the distribution. Overall, the histogram provides a useful summary of the waiting times in the simulated waiting system, and can help inform decisions about how to improve the system in order to reduce waiting times and improve customer satisfaction.

CONCLUSION

Discrete event simulation is an important tool for analyzing complex systems and making informed decisions about how to improve them. By modeling the system as a sequence of discrete events, such as arrivals and departures of customers in a waiting system, discrete event simulation can capture the interactions and dependencies between different components of the system and enable the analysis of their impact on overall system performance. Discrete event simulation can be used in a wide range of applications, from manufacturing and logistics to healthcare and finance. It allows stakeholders to test different scenarios and make informed decisions about system design, resource allocation, and process improvement, without the need for costly and time-consuming physical experimentation. In addition, discrete event simulation can also help identify bottlenecks, inefficiencies, and other areas of improvement that may not be immediately obvious through simple observation or manual analysis. By simulating the system under different conditions and using statistical analysis to extract meaningful insights, discrete event simulation can help organizations optimize their operations and achieve better outcomes. Overall, the importance of discrete event simulation lies in its ability to model and analyze complex systems, provide insights that inform decision-making, and improve system performance and efficiency.

SYSTEM DYNAMICS MODELING

System dynamics modeling is a technique used to model and analyze complex systems that involve feedback loops, delays, and nonlinear relationships. It involves creating a computer-based model of the system, simulating its behavior over time, and analyzing the results to gain insights into how the system behaves under different scenarios. In system dynamics modeling, the system is represented as a set of interrelated components, and the interactions between these components are modeled using feedback loops and other mathematical relationships. The simulation can be used to test different scenarios and parameters to evaluate their impact on the system's performance.

System dynamics modeling is used in a wide range of fields, including business, economics, engineering, and environmental management. For example, in business, it can be used to model the impact of marketing strategies on sales, while in environmental management, it can be used to model the impact of different policies on the environment. The benefits of system dynamics modeling include the ability to model complex systems that involve feedback loops and delays, the ability to test different scenarios and parameters to evaluate their impact on the system's performance, and the ability to identify potential issues and bottlenecks before they occur in the actual system. System dynamics modeling can also be used to develop policies and strategies for managing complex systems.

Sample Problem Statement

Suppose a small town has a population of 10,000 people. The town has one hospital that can accommodate up to 100 patients. The hospital serves both the town's residents and the residents of nearby villages. If the hospital's occupancy rate exceeds 80%, patients may have to wait for treatment or be transferred to another hospital. Assuming that the average length of stay for patients is three days and that new patients arrive at the rate of ten per day, what will be the hospital occupancy rate after 30 days?

SOLUTION

To solve this problem using system dynamics modeling, we need to create a model that represents the key variables and their relationships. Here's one possible model:
Variables:

Population of the town (P)
Number of patients in the hospital (H)
Hospital occupancy rate (O)

Assumptions:

The population of the town is constant over time.
The hospital can accommodate up to 100 patients at a time.
The average length of stay for patients is 3 days.
New patients arrive at the rate of 10 per day.

If the hospital occupancy rate exceeds 80%, patients may have to wait for treatment or be transferred to another hospital.
Model Structure:

The population of the town (P) is fixed and does not change over time.
The number of patients in the hospital (H) changes over time depending on the number of new patients arriving and the number of patients leaving the hospital each day.
The hospital occupancy rate (O) is calculated as the ratio of the number of patients in the hospital (H) to the hospital's capacity (100).

Equations:

$$H(t) = H(t-1) + 10 - (1/3) * H(t-3) \quad \text{eq(4.1)}$$
$$O(t) = H(t) / 100 \quad \text{eq(4.2)}$$

The eq(4.1) calculates the number of patients in the hospital at time t based on the number of new patients arriving each day and the number of patients leaving the hospital each day (which is assumed to be 1/3 of the patients who have been in the hospital for three days or more). The eq(4.2) calculates the hospital occupancy rate at time t as the ratio of the number of patients in the hospital to the hospital's capacity (100).
Simulation:
We can simulate the model using a system dynamics software or spreadsheet program. Here's an example of what the results might look like after 30 days:

```
% Define initial variables and parameters
P = 10000;     % Population of the town
H = zeros(1, 33);  % Number of patients in the hospital
(initialize with zeros)
O = zeros(1, 33);  % Hospital occupancy rate (initialize with
zeros)
H(1) = 0;      % Set initial condition for H
O(1) = H(1) / 100; % Calculate initial occupancy rate
for t = 2:33
  % Calculate number of new patients arriving
  if t <= 30
    new_patients = 10;
  else
    new_patients = 0;
  end
  % Calculate number of patients leaving the hospital
  if t > 3
    patients_leaving = floor(H(t-3) / 3);
  else
    patients_leaving = 0;
  end
  % Calculate number of patients in the hospital at time t
  H(t) = H(t-1) + new_patients - patients_leaving;
  % Calculate occupancy rate at time t
  O(t) = H(t) / 100;
end
% Plot the results
plot(0:32, O*100, "-b*", "linewidth",2);
xlabel("Day");
ylabel("Occupancy Rate (%)");
title("Hospital Occupancy Rate Over Time");
grid on
```

This code initializes the variables and parameters, then uses a for loop to calculate the number of new patients arriving each day, the number of patients leaving the hospital each day, and the number of patients in the hospital at each time step. The occupancy rate is calculated based on the number of patients in the hospital and plotted over time. Note that we run the simulation for 33 days instead of 30, to allow for patients who arrive on day 30 to stay in the hospital for 3 days.

The graph generated by the above code shows the hospital occupancy rate over time, in percentage (%). The x-axis represents the days, from day 0 to day 32. The y-axis represents the hospital occupancy rate, ranging from 0% to 100%. The graph starts at 0% occupancy rate, since there are no patients in the hospital at day 0. As new patients arrive and are admitted to the hospital, the occupancy rate gradually increases. It reaches 30% on day 3. After day 3, the occupancy rate continues to increase, reaching a peak of 42% on day 5, before gradually decreasing as patients are discharged and fewer new patients arrive. By day 30, the occupancy rate has stabilized at around 32%. Overall, the graph shows that the hospital occupancy rate fluctuates over time as a result of the interplay between the number of new patients arriving, the number of patients leaving the hospital, and the hospital's capacity. The system dynamics model provides a way to understand and predict the dynamics of the system under different scenarios and assumptions.

CONCLUSION

System dynamics modeling is a powerful tool for understanding and analyzing complex systems that evolve over time. By representing the system as a set of interrelated components and feedback loops, system dynamics models can help us identify the underlying causes of system behavior, simulate different scenarios and policies, and make informed decisions based on data-driven insights.

Some of the key benefits of system dynamics modeling include:

1. **Holistic perspective:** System dynamics modeling takes a comprehensive view of the system, accounting for both its internal components and external factors that may influence its behavior. This allows us to understand the system as a whole and identify the root causes of its behavior.
2. **Dynamic simulation:** System dynamics models simulate the behavior of the system over time, allowing us to visualize how the system will evolve under different conditions and policies. This helps us evaluate the effectiveness of different interventions and identify potential unintended consequences.
3. **Feedback loops:** System dynamics models incorporate feedback loops, which represent the interdependence between different components of the system. This allows us to capture the non-linear and complex behavior of the system, which cannot be easily predicted by linear models.
4. **Collaboration:** System dynamics modeling encourages collaboration between stakeholders with different perspectives and expertise. By involving diverse stakeholders in the model-building process, we can build consensus around the key issues and priorities, and develop more effective solutions that reflect the needs and values of all stakeholders.

In conclusion, system dynamics modeling is an important tool for addressing complex, dynamic problems in a wide range of fields, including healthcare, economics, climate change, and public policy. By providing a rigorous and data-driven approach to problem-solving, system dynamics modeling can help us make more informed decisions and create a better future for all.

COMPARISON OF SIMULATION STRATEGIES

Monte Carlo simulation, agent-based modeling, discrete event simulation, and system dynamics modeling are all widely used simulation techniques in various fields. Here are the benefits and drawbacks of each method:

MONTE CARLO SIMULATION

Monte Carlo simulation is a probabilistic modeling technique that uses random sampling to obtain numerical solutions to problems. Some of its benefits are:

1. Monte Carlo simulation can be used to estimate the probability of an outcome.
2. It can be used to model systems that are too complex to model analytically.
3. It is a flexible method that can handle a wide range of problems.
4. Monte Carlo simulation can provide information about the sensitivity of the output to the input parameters.

Some of the drawbacks of Monte Carlo simulation are:

1. It requires a large number of samples to obtain accurate results.
2. It can be computationally expensive for complex models.
3. Monte Carlo simulation assumes that the inputs are independent and identically distributed, which may not be true in some cases.

AGENT-BASED MODELING

Agent-based modeling is a method that simulates the behavior of individual agents in a system. Some of its benefits are:

1. It can capture emergent behavior that arises from the interaction of agents.
2. Agent-based modeling can model complex systems that have multiple levels of organization.
3. It can simulate the behavior of heterogeneous agents that have different characteristics.

Some of the drawbacks of agent-based modeling are:

1. It can be computationally expensive for large-scale models.
2. Agent-based modeling may require a large amount of data to parameterize the model.
3. The behavior of the agents may be difficult to model accurately.

DISCRETE EVENT SIMULATION

Discrete event simulation is a method that models the behavior of a system as a sequence of discrete events. Some of its benefits are:

1. It can model systems that have discrete events, such as arrivals, departures, and failures.

2. Discrete event simulation can be used to optimize the performance of a system by evaluating different scenarios.
3. It can simulate the behavior of complex systems that have multiple interacting components.

Some of the drawbacks of discrete event simulation are:

1. It can be difficult to model systems that have continuous behavior, such as fluid flow.
2. Discrete event simulation may require a large amount of data to parameterize the model.
3. The behavior of the system may be difficult to model accurately.

System Dynamics Modeling

System dynamics modeling is a method that models the behavior of a system over time using feedback loops and stocks and flows. Some of its benefits are:

1. It can model complex systems that have feedback loops and nonlinear behavior.
2. System dynamics modeling can be used to evaluate policy decisions and their impact on the system.
3. It can simulate the behavior of systems that have multiple interacting components.

Some of the drawbacks of system dynamics modeling are:

1. It may be difficult to model systems that have discontinuous behavior.
2. System dynamics modeling may require a large amount of data to parameterize the model.
3. The behavior of the system may be difficult to model accurately.

In conclusion, each simulation method has its own benefits and drawbacks. The choice of simulation method depends on the nature of the system being modeled and the questions being asked. It is important to carefully consider the strengths and weaknesses of each method before selecting the appropriate simulation technique.

REFERENCES

1. Harold Klee and Randal Allen, *Simulation of Dynamic Systems with MATLAB® and Simulink®*. CRC Press, December 13, 2021.
2. Andrew Greasley, *Simulation Modelling Concepts, Tools and Practical Business Applications*. CRC Press, September 21, 2022.
3. Mangey Ram, *Modeling and Simulation Based Analysis in Reliability Engineering*. CRC Press, March 31, 2021.

4. Christopher A. Chung and Christian G. Parigger, *Simulation Modeling Handbook: A Practical Approach*. CRC Press, 2003.
5. Philipp K. Janert, *Data Analysis with Open Source Tools*. O'Reilly Media, Inc, November 2010.
6. Don L. McLeish, *Monte Carlo Simulation and Finance*. Wiley, 2005.
7. Steven F. Railsback and Volker Grimm, *Agent-Based and Individual-Based Modeling: A Practical Introduction*. Princeton University Press, October 2011.
8. Larry H Leemis and Stephen K. Park, *Discrete-Event Simulation: A First Course*. Pearson, December 27, 2005.
9. Averill M. Law and W. David Kelton, *Simulation Modeling and Analysis*. McGraw-Hill International Edition, 1991.
10. W. David Kelton, Randall P. Sadowski, and Nancy B. Swets, *Simulation with Arena*. McGraw-Hill Education, 2015.

4 Machine Learning Models for Regression and Classification

Syed Immamul Ansarullah,
Sami Alshmrany, Mudasir Manzoor Kirmani,
and Abdul Basit Andrabi

OVERVIEW OF MACHINE LEARNING

Machine learning involves the design and implementation of algorithms that learn from data and improve their performance through experience. These algorithms employ mathematical and statistical techniques to discover patterns, generalize from examples, and make accurate predictions or decisions (Barr, Feigenbaum and Cohen, 1981). It involves the study of statistical models and algorithms that allow computers to analyze and interpret complex data patterns, learn from them, and make data-driven predictions or decisions (Benhamou (2022). This ability to learn from data makes machine learning particularly useful for solving complex problems and making predictions in domains such as image and speech recognition, natural language processing, recommender systems, fraud detection, autonomous vehicles, and many more (Zhang et al., 2014). Machine learning algorithms can be broadly categorized into three main types: supervised learning, unsupervised learning, and reinforcement learning (Alex and Vishwanathan, 2008).

1. **Supervised learning:** In this approach, the algorithm is trained on labeled data, where each data point is associated with a known label or outcome. The algorithm learns to map the input data to the correct output by generalizing from the labeled examples. This enables it to make predictions or classify new, unseen data (Tercan & Meisen, 2022).
2. **Unsupervised learning:** Here, the algorithm works with unlabeled data, meaning there are no predetermined outcomes or labels associated with the data. The goal is to uncover hidden patterns or structures within the data. Clustering algorithms, which group similar data points together, and dimensionality reduction techniques, which reduce the complexity of the data, are common examples of unsupervised learning (Benhamou et al., 2022).

DOI: 10.1201/9781032686714-4

3. **Reinforcement learning:** This type of learning is inspired by the concept of how humans learn through trial and error. The algorithm, called an agent, interacts with an environment and learns to make decisions or take actions to maximize a reward signal. It receives feedback in the form of rewards or penalties based on its actions, allowing it to learn and improve its decision-making abilities over time (Mhlanga, 2023).

REGRESSION AND CLASSIFICATION

Regression and classification are two fundamental tasks in machine learning that involve predicting or determining an output based on input data. Regression and classification are essential tools in machine learning to solve a wide range of prediction and decision-making problems. The choice between regression and classification depends on the nature of the output variable and the specific problem at hand.

- **Regression:** Regression is a supervised learning task that aims to predict a continuous or numeric output variable based on input features. The goal is to learn a mapping function from the input features to the continuous target variable (Matloff, 2017). In other words, regression models estimate the relationship between the independent variables (input features) and the dependent variable (output) to make predictions. Regression problems include tasks like predicting house prices based on features such as size, location, and number of rooms.
- **Classification:** Classification is also a supervised learning task, but it focuses on predicting a categorical or discrete output variable. The goal is to assign input data instances to predefined classes or categories based on their features. In classification, the target variable is typically a label or class, and the model learns to classify new, unseen instances into one of these classes (Torgo and Gama, 1996). Classification problems are widespread and include tasks like spam email detection, sentiment analysis, disease diagnosis, and image recognition (identifying objects or objects in images).

REGRESSION MODELS

Regression builds a predictive model for the continuous target variables as a function of the explanatory variable. Here are some pure regression algorithms that are solely designed for regression tasks and not for classification:

LINEAR REGRESSION

Linear regression is a statistical technique that is used for modeling and analyzing the relationship between a dependent variable and one or more independent variables and assumes a linear relationship between these variables. The goal of linear regression is to find the best-fitting straight line that minimizes the differences (residuals) between the predicted values and the actual values of the dependent variable. This line is called the regression line or the best-fit line (Yao and Li, 2014). The equation of a linear regression model with one independent variable can be represented as:

$$y = \beta_0 + \beta_1 x + \varepsilon \quad (1)$$

where y is the dependent variable, x is the independent variable, β_0 is the y-intercept of the regression line, β_1 is the coefficient or slope of the regression line, and ε represents the error term or residual. The parameters β_0 and β_1 are estimated using a method called least squares, which minimizes the sum of squared residuals. Once the parameters are estimated, the regression line can be used to make predictions for new values of x.

Polynomial Regression

Polynomial regression is an extension of linear regression that allows for modeling nonlinear relationships between the independent and dependent variables. It can capture more complex patterns by fitting a polynomial function to the data. In polynomial regression, instead of fitting a straight line to the data, we fit a polynomial function of degree n to the data, where n represents the highest power of the independent variable (Tsai et al., 2022). The equation for a polynomial regression model with one independent variable can be written as:

$$y = \beta_0 + \beta_1 x + \beta_2 x^2 + \beta_3 x^3 + \ldots + \beta_n x^n + \varepsilon \quad (2)$$

where y is the dependent variable or target variable, x is the independent variable, β_0, β_1, β_2, ..., β_n are the coefficients of the polynomial terms, and ε represents the error term or residual. The polynomial terms, such as x^2, x^3, and so on, introduce nonlinear relationships into the model. By including higher-order terms, the model can better capture curved or nonlinear patterns in the data. Similar to linear regression, the coefficients β_0, β_1, β_2, ..., β_n in polynomial regression are estimated using the least squares method, minimizing the sum of squared residuals. Once the coefficients are estimated, the polynomial regression model can be used to make predictions for new values of x.

The choice of the degree of the polynomial, n, is important in polynomial regression. A higher degree can lead to a more flexible model that fits the training data well but may overfit the data and generalize poorly to new data. On the other hand, a lower degree may result in an oversimplified model that fails to capture the underlying relationships.

Support Vector Regression

Support vector regression (SVR) is a supervised learning algorithm that uses support vectors to find a hyperplane that minimizes the error between the predicted and actual values. SVR is a non-parametric model, which means that it does not make any assumptions about the distribution of the data. This makes SVR a flexible and robust model that can be used to fit a wide variety of data (Parbat and Chakraborty, 2020). SVR can be used to solve a variety of regression problems such as predicting the price of a house based on its features, predicting the number of sales a company will make based on its marketing campaign, or predicting the time it will take for a patient to recover from an illness etc.

Random Forest Regression

Random forest regression is a supervised learning algorithm that uses ensemble learning to build a model. The model is made up of a number of decision trees, each of which is trained on a random subset of the data. The predictions from the individual trees are then averaged to produce the final prediction. Random forest regression is a powerful algorithm that can be used to solve a variety of regression problems. It is particularly well-suited for problems where the data is noisy or has a lot of missing values (Tyralis, Papacharalampous, and Langousis, 2019). Random forest regression is also relatively robust to overfitting, which is a common problem with other regression algorithms. Random forest regression is often able to achieve higher accuracy than other regression algorithms, such as linear regression and logistic regression. This is because random forest regression can capture nonlinear relationships between the features and the target variable (Abdulhafedh, 2022). The main disadvantage of random forest regression is its complexity. Random forest regression can be computationally expensive to train, and it can be difficult to interpret the results. Here are some of the parameters that can be tuned for random forest regression (Jeremy, 2017).

- **Number of trees:** The number of trees can impact the accuracy and performance of the model. A larger number of trees will generally lead to a more accurate model, but it will also be more computationally expensive to train.
- **Maximum depth of trees:** A deeper tree will be able to learn more complex relationships between the features and the target variable, but it may also be more likely to overfit the training data.
- **Minimum sample size per leaf:** The minimum sample size per leaf is a parameter that controls the complexity of the trees in the forest. A smaller minimum sample size will lead to more complex trees, but it may also make the model more sensitive to noise in the data.

Gradient Boosting Regression

Gradient boosting regression is a machine learning algorithm that builds an ensemble of weak learners (decision trees) in a sequential manner. The goal of gradient boosting is to minimize the loss function by iteratively adding new trees to the ensemble. Each new tree is trained to correct the errors made by the previous trees (Natekin and Knoll, 2013). Gradient boosting regression is particularly well-suited for problems where the relationship between the independent and dependent variables is nonlinear (Lei and Fang, 2019). Here are some of the challenges of using gradient boosting regression:

- It can be computationally expensive to train a gradient boosting model.
- It can be sensitive to the choice of hyperparameters.
- It can be difficult to interpret the results of a gradient boosting model.

Overall, gradient boosting regression is a powerful and versatile machine learning algorithm that can be used to solve a wide variety of regression problems. Let's say you are a financial analyst and you want to build a model to predict the price of a stock. You have historical data on the stock's price, as well as other factors that you believe might influence the stock price, such as economic indicators, company news, and analyst ratings. You can use gradient boosting regression to build a model that takes all of this data into account and predicts the stock price.

CLASSIFICATION MODELS

Classification is a type of supervised machine learning task that involves assigning input data instances to predefined categories or classes based on their features or attributes. The goal of classification is to build a model that can accurately predict the class label of new, unseen instances based on the patterns and relationships learned from labeled training data (Jiawei, Micheline, and Jian, 2006).

In a classification problem, the input data consists of a set of features or attributes that describe each instance. These features can be numerical, categorical, or even text-based. The output or target variable is a discrete class label that indicates the category to which the instance belongs. Common algorithms used for classification tasks include decision trees, random forests, logistic regression, support vector machines (SVM), naive Bayes, and neural networks. These algorithms employ different mathematical and statistical techniques to learn the patterns and decision boundaries that separate different classes (Hilbe, 2016).

NAIVE BAYES

Naive Bayes is a probabilistic machine learning algorithm commonly used for classification tasks. It is based on Bayes' theorem and assumes that the features are conditionally independent of each other given the class label. The "naive" assumption in naive Bayes is that the features are independent of each other given the class label. Naive Bayes calculates the prior probability of the target attribute (class label) and the conditional probability of the remaining attributes given the class label. The prior probability represents the likelihood of each class label occurring in the training data, while the conditional probability represents the likelihood of observing specific attribute values given the class label. During the training phase, the algorithm estimates these probabilities based on the provided training data. Then, during the testing phase, the algorithm calculates the probability of a testing instance belonging to each class label using the calculated probabilities. The class label with the highest probability is then selected as the predicted class label for the testing instance. The formula for Naive Bayes can be expressed as (Mudasir and Syed, 2016):

$$P(y|X) = (P(X|y) * P(y)) / P(X) \quad (3)$$

Where:

- $P(y|X)$ is the posterior probability of the class label y given the feature values X.
- $P(X|y)$ is the likelihood or conditional probability of observing the feature values X given the class label y.
- $P(y)$ is the prior probability of the class label y.
- $P(X)$ is the probability of observing the feature values X.

In practice, naive Bayes uses the product rule to estimate the joint probability $P(X, y)$ and then applies Bayes' theorem to calculate the posterior probability $P(y|X)$ using the estimated probabilities. Naive Bayes is known for its simplicity, computational efficiency, and ability to handle large feature spaces. It is particularly useful when the independence assumption holds reasonably well or when the data has high dimensionality. However, it may not perform well when the features are highly correlated or when the independence assumption is violated.

Decision Tree

A decision tree is a powerful machine learning algorithm commonly used for classification tasks, although it can also be applied to regression problems. It follows a recursive divide-and-conquer approach, where the training set is progressively partitioned into smaller subsets as the tree is constructed. This process involves making decisions based on attribute tests at each internal node and assigning class labels to the leaf nodes (Vipin Kumar, 2010). To build an accurate decision tree, various measures of attribute selection such as Information Gain, Gain Ratio, and Gini Index are used to determine the best attribute for splitting the data. Information Gain quantifies the reduction in entropy or uncertainty when a particular attribute is used for partitioning. Gain Ratio accounts for the intrinsic information of an attribute, while Gini Index measures the impurity or disorder of the classes within a partition.

During the construction of the decision tree, it is important to consider the potential presence of noise or outliers in the training data. To improve the performance and generalization ability of the tree, pruning techniques are applied. Pruning helps identify and remove branches that are likely to introduce overfitting, thus enhancing the accuracy of the tree on unseen data. By considering various attribute selection measures and applying pruning techniques, decision trees offer a flexible and effective approach to solving classification problems in machine learning (Jiawei, Micheline, and Jian, 2006; Esposito, 1997).

RANDOM FOREST

Random forests are an ensemble of simple decision trees that are used for both regression and classification problems. The random forest algorithm creates the forest with a number of decision trees from the randomly selected training set, with the goal of overcoming the overfitting problem of the individual decision tree. In random forest classification, each decision tree votes and the aggregated votes decide the final classes of the test object; however, in the regression, the mean prediction or regression of the individual trees is calculated (Ian, Eibe, and Mark, 2011).

Figure 4.1 shows the working of the random forest algorithm in which each tree is grown on a different sample of original data. The working of the random forest algorithm involves multiple iterations, where a new bootstrap training set is created

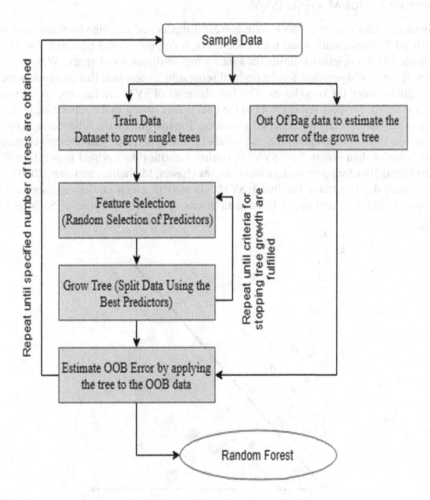

FIGURE 4.1   Random forest algorithm working

by randomly sampling approximately one-third of the samples from the original data. These samples are excluded from the construction of each tree, creating a test set within the forest. This process eliminates the need for cross-validation, as each tree is evaluated on the excluded samples.

Random forests offer several advantages, such as improved generalization performance, robustness against noise and outliers, and the ability to handle high-dimensional data. By building a diverse collection of trees and aggregating their predictions, random forests provide a reliable and effective approach for classification tasks in machine learning. The algorithm's simplicity, versatility, and superior performance make it a popular choice for various real-world applications.

Support Vector Machines (SVMs)

Support vector machines (SVM) are powerful supervised machine learning models utilized for classification and regression tasks. It employs a nonlinear mapping that transforms the original training data into a higher-dimensional space. Within this new space, SVM searches for the optimal separating hyperplane that maximizes the margin between the two classes. The key elements of SVM are the support vectors and margins. Support vectors refer to the data points closest to the separating hyperplane and are crucial elements in the dataset. The margin, on the other hand, represents the maximum width of the slab parallel to the hyperplane that does not contain any interior data points. The SVM algorithm identifies the optimal hyperplane by leveraging these support vectors and margins (Jiawei, Micheline, and Jian, 2006).

Figure 4.2 illustrates the linear SVM where light green circles represent data points of class x1 and red indicate data points of x2. The purpose of SVM is to

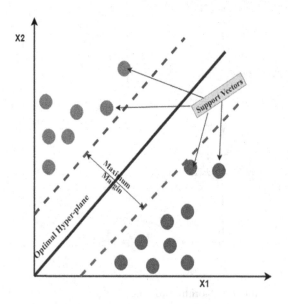

FIGURE 4.2   Linear SVM classifier for two-class representation

choose a hyperplane with the greatest possible margin between the hyperplane and any data point with the training set, giving a greater chance of new data being classified correctly. However, if there is no clear hyperplane, it is necessary to move to a higher dimension view called kernelling in SVM. The idea is that the data will continue to be mapped into higher dimensions until a hyperplane can be formed to segregate it. However, in nonlinear separation, the training data will be mapped into a higher-dimensional space H, and an optimal hyperplane will be constructed there.

SVM offers a versatile and robust approach to machine learning, capable of handling both linear and nonlinear classification tasks. Its ability to transform data into higher-dimensional spaces and identify optimal hyperplanes makes it effective in various real-world applications. By selecting appropriate kernel functions and leveraging the support vectors, SVM provides accurate and reliable results in classification and regression problems.

K-NEAREST NEIGHBOR (KNN)

K-nearest neighbor (KNN) is the basic, non-parametric, and instance-based machine learning algorithm. It uses learning by analogy, which compares the new unclassified record with the existing records using the distance metric. The closest existing record is used to assign the class to the newly unclassified record (Jiawei, Micheline, and Jian, 2006). Figure 4.3 shows the example of KNN classification. The good value of k can only be determined experimentally by setting the value of k to 1 and then incrementing k to allow for new neighbors. The k value that gives the minimum error rate is selected. The test set is used to estimate the error rate of the classifier. In the KNN algorithm, a new instance is classified by closeness to the neighbors,

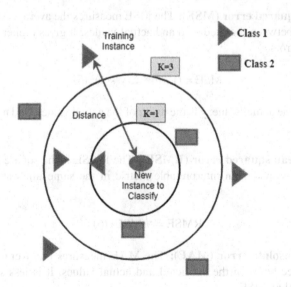

FIGURE 4.3 K-nearest neighbor classification example

which is defined in terms of the distance function. Many distance measures can be used, such as (Euclidean, Manhattan, and Minkowski) but in this research Euclidean measure is used because of the properties of the heart disease data.

The Euclidean distance between two points is computed as the length of the path connecting them. It is calculated by taking the square root of the sum of squared differences across all input attributes between two instances (i and j). Prior to using the Euclidean distance measure, attribute values are typically normalized to prevent attributes with higher values from overshadowing those with lower values.

$$d(i, j) = \sqrt{(xi1 - xj1)^2 + (xi2 - xj2)^2 + ... + (xip - xjp)^2} \qquad (4)$$

KNN is a versatile algorithm that can be applied to both classification and regression tasks. By leveraging the proximity of instances, KNN allows for flexible decision-making based on the characteristics of neighboring data points. It is important to note that the performance of KNN can be influenced by factors such as the choice of distance metric, value of k, and appropriate normalization techniques for attribute values.

EVALUATION OF REGRESSION AND CLASSIFICATION MODELS

When evaluating regression and classification models, there are various metrics and techniques that can be used. Let's discuss some commonly used evaluation methods for both types of models:

Evaluation Metrics for Regression Models

Some commonly used evaluation metrics for regression models are as follows:

- **Mean squared error (MSE):** The MSE measures the average squared difference between the predicted and actual values. It gives higher weights to larger errors.

$$MSE = (1 / n) * \Sigma(y_i - \bar{y})^2 \ (5)$$

where y_i is the actual value, $\bar{y}$ is the mean of the actual values, and n is the number of data points.

- **Root mean squared error (RMSE):** The RMSE is the square root of the MSE. It provides an interpretable metric in the same units as the target variable.

$$RMSE = \sqrt{(MSE)} \ (6)$$

- **Mean absolute error (MAE):** The MAE measures the average absolute difference between the predicted and actual values. It is less sensitive to outliers than MSE.

$$MAE = (1 / n) * \Sigma|y_i - \bar{y}| \ (7)$$

- **Mean absolute percentage error (MAPE):** The MAPE represents the average percentage difference between the predicted and actual values. It is often used when comparing models across different scales or domains.

$$\text{MAPE} = (1 \,/\, n) * \Sigma(|y_i - \hat{y}| \,/\, |y_i|) * 100 \quad (8)$$

where y_i is the actual value, and $\hat{y}$ is the predicted value.

- **Coefficient of determination (R-squared):** R-squared measures the proportion of the variance in the dependent variable that can be explained by the independent variables. It ranges from 0 to 1, with higher values indicating a better fit.

$$\text{R-squared} = 1 - (SS_i \,/\, SS_t) \quad (9)$$

where SS_i is the sum of squared residuals $(\Sigma(y_i - \hat{y})^2)$ and SS_t is the total sum of squares $(\Sigma(y_i - \bar{y})^2)$.

- **Adjusted R-squared:** The adjusted R-squared penalizes the addition of unnecessary variables to the model, preventing overfitting. It accounts for the number of predictors and sample size.

$$\text{Adjusted R-squared} = 1 - [(1 - R^2) * (n - 1) \,/\, (n - k - 1)] \quad (10)$$

where R^2 is the ordinary R-squared value, n is the sample size, and k is the number of predictors.

- **Mean squared logarithmic error (MSLE):** The MSLE measures the average squared logarithmic difference between the predicted and actual values. It is useful when the target variable has a wide range of values.

$$\text{MSLE} = (1 \,/\, n) * \Sigma(\log(y_i + 1) - \log(\hat{y} + 1))^2 \quad (11)$$

where y_i is the actual value, and $\hat{y}$ is the predicted value.

Evaluation Metrics for Classification Models

Some commonly used evaluation metrics for classification models are as follows:

- **Sensitivity** (also known as true positive rate or recognition or recall) is the proportion of positive tuples that are correctly classified as positive (Rajul et al., 2008).

$$\text{Sensitivity} = TP \,/\, (TP + FN) \quad (12)$$

- **Specificity** (also known as True Negative Rate) is the proportion of negative tuples that are correctly classified as Negative (Rajul et al., 2008).

$$\text{Specificity} = TN \,/\, (TN + FP) \quad (13)$$

- **Accuracy** is the total percentage of cases that are correctly classified by an algorithm (Rajul et al., 2008).

$$\text{Accuracy} = (TP + TN) / (TP + TN + FP + FN) \quad (14)$$

- **Precision** is a measure of exactness (i.e., what percentage of entities categorized as positive are actually positive) (Rajul et al., 2008).

$$\text{Precision} = TP / (TP + FP) \quad (15)$$

- **Error rate (misclassification rate)** is the proportion of errors made over a whole set of instances. The error rate is a combination of training and generalization errors. Training errors are the number of misclassification errors committed to training data, whereas generalization error is the expected error of the model on previously unseen records. The best classification model has low training and generalization error (Rajul et al., 2008).

$$\text{Error Rate} = (FP + FN) / (TP + TN + FP + FN) \quad (16)$$

- **AUROC (area under the receiver operating characteristics):** AUROC is a performance measure graph that demonstrates the performance of a classification model at different threshold settings (Jiawei, Micheline, and Jian, 2006). AUROC depicts how greatly a model is skilled in distinguishing between the classes. The ROC curve is plotted with True Positive Rate on the y-axis against the False Positive Rate on the x-axis, as shown in Figure 4.4. An outstanding model has AUROC value equivalent to or close to 1, which means it has a fine measure of separability. A poor model has AUROC value equivalent to or near 0, which means it reciprocates the result and predicts 0s as 1s and 1s as 0s. When the AUROC value is

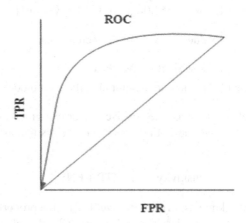

FIGURE 4.4   AUROC representation

approximately 0.5, then the model cannot distinguish between positive and negative classes (Max, 2007).

- **Cross-validation techniques:** The cross-validation technique measures the error rate of a learning model on a specific dataset (Tammo, Danny, and Mikio, 2015). In cross-validation, the complete dataset is randomly split into mutually exclusive subsets of approximately equal size, and each record is used the same number of times for training and exactly once for testing. The training dataset allows data mining techniques to learn from this data. The testing dataset is used to evaluate the performance of the data mining technique about what is learned from the training dataset (Jiawei, Micheline, and Jian, 2006).

OVERFITTING AND UNDERFITTING

Overfitting and underfitting are common problems encountered when building regression and classification models. They occur when the model's performance is negatively affected due to inadequate generalization of unseen data.

A. **Overfitting:** Overfitting occurs when a model learns the training data too well, capturing noise or irrelevant patterns that do not exist in the underlying population. As a result, the model performs exceptionally well on the training data but fails to generalize to new, unseen data. Overfitting can be caused by several factors:

- **Insufficient training data:** When the training dataset is small, the model can easily memorize the examples instead of learning the underlying patterns.
- **Overly complex model:** If the model has too many parameters or features relative to the available data, it can fit the noise in the training data rather than the true underlying patterns.
- **Lack of regularization:** Insufficient regularization techniques, such as L1 or L2 regularization, can lead to overfitting as they fail to control the complexity of the model.

To mitigate overfitting, various strategies can be employed:

- **Increase training data:** Collecting more diverse and representative training data can help the model generalize better.
- **Feature selection:** Choose relevant features and remove irrelevant or noisy features to reduce the complexity of the model.
- **Regularization:** Apply regularization techniques such as L1 or L2 regularization to penalize complex models and prevent overfitting.
- **Cross-validation:** Use techniques like k-fold cross-validation to evaluate the model's performance on multiple validation sets and assess its generalization ability.

    B. **Underfitting:** Underfitting occurs when a model is too simplistic and fails to capture the underlying patterns in the data, resulting in poor performance both on the training and unseen data. Underfitting can be caused by several factors:

- **Insufficient model complexity:** If the model is too simple and lacks the necessary capacity to represent the underlying patterns, it may underfit the data.
- **Inadequate training:** Insufficient training time or insufficient iterations during the learning process can lead to underfitting.

To address underfitting, the following approaches can be considered:

- **Increase model complexity:** Use more advanced models with greater capacity, such as adding more layers or neurons in neural networks or increasing the degree of polynomial regression.
- **Feature engineering:** Enhance the feature representation or create new features that capture the underlying patterns in the data.
- **Increase training iterations:** Allow the model to train for longer; ensuring it has sufficient time to capture the patterns in the data.

CASE STUDIES ON REGRESSION AND CLASSIFICATION

These case studies illustrate how regression and classification techniques are applied to solve real-world problems across various industries. By leveraging historical data and relevant features, regression and classification models can provide valuable insights, make accurate predictions, and support decision-making processes in diverse domains.

REGRESSION PROBLEMS CASE STUDIES

Case Study 1: A real estate agency wants to predict house prices based on various features such as location, size, number of bedrooms, and amenities.

Approach: They collect a dataset of historical house sales with corresponding features and sale prices. They build a regression model using techniques like linear regression or random forest regression to predict house prices for new listings. The model accurately predicts house prices, enabling the agency to provide accurate pricing estimates to sellers and buyers.

Case Study 2: A bank wants to assess the creditworthiness of loan applicants to minimize the risk of defaults.

Approach: They collect data on applicant demographics, employment history, financial statements, and credit history. They build a regression model, such as logistic regression or support vector regression, to predict the probability of loan default. The bank improves its risk assessment process, making more informed lending decisions, and reducing the likelihood of financial losses due to defaults.

Case Study 3: Farming cooperative aims to predict crop yields based on factors like weather conditions, soil properties, and agricultural practices.

Approach: They collect historical data on crop yields, weather patterns, soil characteristics, and farming practices. They develop a regression model, such as multiple linear regression or neural networks, to predict crop yields for different crops and locations. The cooperative gains insights into the factors influencing crop yields, enabling them to optimize irrigation, fertilization, and other farming practices for improved productivity and resource management (Uyanık and Guler, 2013).

REAL-WORLD CLASSIFICATION PROBLEMS

Case Study 1: An email service provider wants to classify incoming emails as either spam or legitimate to protect users from unwanted and malicious content.

Approach: They collect a labeled dataset of emails, including features like email content, sender information, and metadata. They build a classification model, such as a Naive Bayes classifier or a support vector machine (SVM), to classify incoming emails as spam or legitimate. The email service provider successfully filters out spam emails, improving user experience and security.

Case Study 2: Radiologists need assistance in classifying medical images, such as X-rays or MRI scans, to identify diseases or abnormalities.

Approach: They gather a dataset of labeled medical images along with corresponding diagnoses. They develop a classification model, such as a Convolutional Neural Network (CNN) or a decision tree, to classify images into different disease categories. The classification model aids radiologists in accurate and efficient disease diagnosis, enabling early detection and timely treatment.

Case Study 3: A company wants to understand public sentiment towards their brand by analyzing social media posts and comments.

Approach: They collect a dataset of social media posts, along with sentiment labels (positive, negative, neutral). They build a sentiment analysis model, such as a recurrent neural network (RNN) or a sentiment lexicon-based classifier, to classify social media posts into different sentiment categories. The company gains insights into customer sentiment, allowing them to make data-driven decisions and improve their brand reputation and customer satisfaction.

CONCLUSION

Machine learning techniques, specifically regression and classification have proven to be valuable tools for addressing real-world problems across diverse domains. These models enable accurate predictions, automation of decision-making processes, and improved efficiency and security. This research presented case studies in regression and classification, demonstrating the practical applications of these techniques across industries such as real estate, finance, agriculture, email services, healthcare, and social media analysis. These case studies highlighted the models' effectiveness in solving real-world problems and providing valuable insights. However, to ensure the optimal performance of regression and classification models, it is imperative to address the challenges of overfitting and underfitting. Overfitting can be mitigated by implementing strategies such as increasing the volume of training data, conducting feature selection, applying regularization techniques, and adopting cross-validation methods. Conversely, underfitting can be overcome by increasing the complexity of the models, employing advanced feature engineering techniques, and allowing for sufficient training iterations. By employing these approaches, the models become better equipped to capture underlying patterns and exhibit improved performance on both training and unseen data. The evaluation of regression and classification models holds utmost importance and by incorporating rigorous model evaluation methodologies, researchers can make informed decisions regarding model selection, enhance prediction accuracy, and effectively apply these models in real-world scenarios. The proposed strategies for addressing overfitting and underfitting challenges, along with the emphasis on robust model evaluation, serve as valuable guidelines for researchers and practitioners in developing reliable and impactful models. By incorporating these insights, future studies can further advance the field of machine learning and enhance its impact on various industries and domains.

REFERENCES

Alex, S., and Vishwanathan, S. V. N. (2008). *Introduction to machine learning* (First Edition). Cambridge University Press. ISBN: 0521825830.

Barr, A., Feigenbaum, E. A., and Cohen, P. R. (1981). *The handbook of artificial intelligence*. William Kaufmann.

Benhamou, E. (2022). Machine learning fundamentals: Unsupervised learning part 1 – data and AI reskilling seminar slides (September 30, 2022). https://ssrn.com/abstract=4234520 or http://dx.doi.org/10.2139/ssrn.4234520.

Ian, H. Witten, Eibe, Frank, and Mark, A. Hall (2011). *Data mining practical machine learning tools and techniques* (Third Edition). Morgan Kaufmann Publishers.

Jeremy, Jordan (2017). Hyperparameter tuning for machine learning models. https://www.jeremyjordan.me/hyperparameter-tuning/

Jiawei, Han, Micheline, Kamber, and Jian, Pei (2006). *Data mining concepts and techniques* (Third Edition). Morgan Kaufmann Publishers.

Lei, X., and Fang, Z. (2019). GBDTCDA: Predicting circRNA-disease associations based on gradient boosting decision tree with multiple biological data fusion. *International Journal of Biological Sciences*, 15(13), 2911.

Matloff, N. (2017). *Statistical regression and classification: From linear models to machine learning*. CRC Press.

Mhlanga, D. (2023). Artificial intelligence and machine learning for energy consumption and production in emerging markets: A review. *Energies*, 16(2), 745. https://doi.org/10.3390/en16020745.

Mudasir, M. Kirmani, and Syed, Immamul Ansarullah (2016). Classification models on cardiovascular disease detection using neural networks, naive Bayes and J48 data mining techniques. *International Journal of Advanced Research in Computer Science*, 7(5), September–October, 2016.

Natekin, Alexey, and Knoll, Alois (2013). Gradient boosting machines- A tutorial. *Frontiers in Neuro Robotics*, 7, Article 21, 2013.

Parbat, D., and Chakraborty, M. (2020). A python based support vector regression model for prediction of COVID-19 cases in India. *Chaos, Solitons & Fractals*, 138, 109942.

Rajul, Parikh, Annie, Mathai, Shefali, Parikh, Chandra, G. Sekhar, and Ravi, Thomas (2008). Understanding and using sensitivity, specificity, and predictive values. *Indian Journal of Ophthalmology*, 56(1), January–February, 45–50.

Tammo, Krueger, Danny, Panknin, and Mikio, Braun (2015). Fast cross-validation via sequential testing. *Journal of Machine Learning Research*, 16, 1103–1155.

Tercan, H., and Meisen, T. (2022). Machine learning and deep learning based predictive quality in manufacturing: A systematic review. *Journal of Intelligent Manufacturing*, 33, 1879–1905. https://doi.org/10.1007/s10845-022-01963-8.

Torgo, L., and Gama, J. (1996). Regression by classification. Advances in Artificial Intelligence Proceedings of the 13th Brazilian Symposium on Artificial Intelligence, SBIA'96 Curitiba, Brazil. Springer, Berlin Heidelberg, pp. 51–60, October 23–25.

Tsai, C. Y., Kim, J., Jin, F., Jun, M., Cheong, M., and Yammarino, F. J. (2022). Polynomial regression analysis and response surface methodology in leadership research. *The Leadership Quarterly*, 33(1), 101592.

Tyralis, H., Papacharalampous, G., and Langousis, A. (2019). A brief review of random forests for water scientists and practitioners and their recent history in water resources. *Water*, 11(5), 910.

Vipin, Kumar (2010). *The top ten algorithms in data mining* (First Edition). CRC Press and Taylor & Francies Group. ISBN-13: 978-1-4200-8964-6.

Yao, W., and Li, L. (2014). A new regression model: Modal linear regression. *Scandinavian Journal of Statistics*, 41(3), 656–671.

Zhang, Y., Balochian, S., Agarwal, P., Bhatnagar, V. and Housheya, O. J. (2014). Artificial intelligence and its applications. *Mathematical Problems in Engineering*, 1–10. https://doi.org/10.1155/2014/840491.

5 Expert Systems for Better Decision-Making

Krishna Saraf, Swagatika Mohapatro,
Yamasani Keerthi Reddy, K. Shiridi Kumar,
and Mulagada Surya Sharmila

INTRODUCTION

Expert systems (ESs), also known as knowledge-based systems, are computer programs designed to simulate the judgmental skills of human experts in a particular field. They are becoming more and more common in fields including education, government organizations, and a variety of roles inside corporations due to their ability to give precise, efficient, and consistent decision-making support as well as access to knowledge. Expert systems in education have advanced beyond addressing the lack of teachers to tackling the shortage of qualified educators. Public organizations work to increase the accountability, transparency, and effectiveness of decision-making. Expert systems have a variety of challenges despite their advantages, including complexity, bias, cost, lack of flexibility, and incompatibility.

Nevertheless, they provide a number of advantages, including improved uniformity, precision, and efficiency. Expert systems may be tailored to specific industries and demand thanks to their considerable adaptability. When fresh information and understanding become available, they may also be revised and updated. Expert systems have the ability to change decision-making procedures and outcomes in a variety of disciplines, notwithstanding the challenges they face. Expert system use is anticipated to expand as technology develops and businesses learn about the potential advantages they offer.

LITERATURE REVIEW

This study examines the potential impacts of expert systems on several organizational functions, notably decision-making[1]. The purpose of utilizing expert systems, the tools used, the computing environment, and the appropriateness of the tasks for them are all taken into account when determining the scope and type of the influence. Additionally, the influence is seen in the organization's structure, decentralization, effectiveness, role content, leadership, power dynamics, communication and information flow, and people needs. Expert systems as substitute decision-makers can be useful for carrying out regular and short-term judgments, but they have limits when

DOI: 10.1201/9781032686714-5

it comes to long-term, strategic decisions, according to research. Expert systems, on the other hand, can aid in decision-making when used as advising tools, but their effectiveness ultimately depends on the user's capacity to do so. The study's findings demonstrated that although some users anticipated an expert advising system to save time, it really did not; nevertheless, a substitute expert system did improve decision-making efficiency. It's crucial to remember that each of these jobs requires a separate knowledge base for expert systems [2].

The process of making decisions in a managerial capacity can be improved through the use of right choice support. The form of assistance required will vary based on factors such as the type of organization (public or private sector), the caliber of the material at hand, and the decision-maker's readiness to use it. Research has shown that decision-making processes and attitudes towards information quality can differ between public and private sector managers, and this study aimed to identify these differences in senior managers within New Zealand's public and private sectors[3]. The results reveal significant variations between the two sectors in terms of decision structure and views on the timing and relevance of information. These findings have important implications for how decision support is provided to managers in both the public and private sectors.

The use of expert systems and artificial neural networks in the medical field has been on the rise. These systems bring numerous benefits such as leveraging expert knowledge, obtaining rare information, having more time for decision-making, making more consistent decisions, and shortening the decision-making process. However, there are also obstacles to developing and utilizing these systems, including maintenance, the need for expert input, difficulties in inputting patient data, issues in acquiring knowledge, challenges in modeling medical knowledge, evaluating and validating system performance, the risk of incorrect recommendations and responsibility, the narrow focus of these mechanisms, and the requirement to integrate them into routine tasks processes[4].

The framework for performance in Romania has not yet been effective, due to obstacles in the public administration's adoption of important components of the performance reform process. The reforms that have been carried out so far have only been partially applied, and the administration tends towards being normative and legally focused, with limited openness towards incorporating new elements of public management [5]. Studies have shown that using ESs can enhance the decision-making quality of public organizations and improve their competitiveness. The results of the research showed that ESs can be beneficial in the administrative decision-making process at MIIC (Management Information and Intelligence Center). The findings and suggestions presented in this study are useful for Egyptian public executives who want to improve their decision-making quality through the use of modern technology [6]. This study also adds valuable insights to the existing literature in this field. The study seeks to examine the usage of decision-making or expert algorithms support methods in schooling, and they discovered to have a wide range of applications including evaluating teacher performance [7], offering career guidance to students, providing quality education to students with disabilities, assisting students in making informed career decisions, and preparing them to become successful professionals after graduation.

TYPES OF EXPERT SYSTEMS FOR BETTER DECISION-MAKING:

There are several types of expert systems, each designed to address specific problems and tasks, based on their knowledge representation and reasoning methods. Some common types of expert systems are:

- **Rule-based expert systems (also known as production rule systems):** These systems use a set of if-then rules to represent knowledge and make decisions [8]. Examples of rule-based expert systems include the MYCIN system for diagnosing infectious diseases and the DENDRAL system for analyzing chemical compounds.
- **Case-based reasoning (CBR) expert systems:** These systems keep track of previous instances (cases) and apply them to decision-making in the present.[9] CBR systems are especially helpful when there is a wealth of experience available and decision-making requires comparing new and old instances for similarities.
- **Neural network expert systems:** These programs employ artificial neural networks to anticipate outcomes and address issues[.10] In circumstances when the interactions between the input and output are intricate and challenging to describe using rules, neural networks are particularly helpful.
- **Fuzzy logic expert systems:** When there is ambiguity and imprecision in the data, these systems employ fuzzy logic to make conclusions. fuzzy logic techniques [11] are especially helpful in circumstances when clear-cut binary answers are impractical, such in medical diagnosis.
- **Hybrid expert systems:** To benefit, these systems incorporate two or more types of expert systems[12] of each system's advantages. As an illustration, a hybrid expert system may decide using both rule-based and case-based reasoning.
- **Knowledge-based expert systems:** To make judgments, these systems draw on a knowledge base of facts and regulations. When a wealth of domain information is accessible and the decision-making process is well understood, knowledge-based systems[13] are very helpful.

Each form of expert system has benefits and drawbacks, and the choice of system type relies on the particular issue being handled, the data that is readily available, and the needs for decision-making.

EXPERT SYSTEMS MAKE BETTER DECISIONS IN DIFFERENT INDUSTRIES

EXPERT SYSTEM PERTAINING TO SCHOOLING

The use of expert systems in education is growing[7] due to their capacity to turn conventional teaching techniques into sophisticated networks and give excellent teaching tools[14]. Due to ES's unique ability to react to questions in a variety of ways, knowledge and learning capacity have risen. ES was established to address the lack of teachers in specific fields and to increase information accessibility through online learning programs[15]. With advancements in technology, ES has expanded to address the shortage of qualified educators by assembling teams of mentors and specialists[16].

For instance, an intelligent tutor system for mobile devices and a web-based fuzzy expert system for educational advisors have both been developed to improve learning [17]. To strengthen the educational system, an expert system for higher education and a web-based lexical expert system have also been developed[18]. The use of ES in education has grown due to the employment of various technologies such as microcomputer systems, web-based systems, and agent-based expert systems. These systems serve as a platform for individual learning, an alternative to private teaching, and a means of monitoring student progress and developing their capabilities. ES has even ultimately resulted in the creation of the Adaptive Educational Programme Tool. (ADEPT), which meets the specific learning needs of students.

EXPERT SYSTEM IN THE FIELD OF PUBLIC SECTOR ORGANIZATIONS

Expert systems are advanced technology developed to aid decision-making in companies[6]. They were implemented to strengthen democratic responsibility, efficiency, and transparency, as well as to improve service availability and living standards[5]. Using ESs ensures evidence-based judgments, eliminates human mistakes, and saves time and money[19]. However, they confront obstacles such as knowledge acquisition, incorrect suggestions and accountability, performance maintenance, and integration into the everyday process[4]. The application of ESs can result in significant organizational changes such as reorganization, communication system modifications, changes in centralization/decentralization, and improvements in the efficiency of corporate decision-making procedures[1].

The influence of ESs on employees and the company is unknown, as adoption might result in employee loss of control and resistance. Making decisions is a critical component of public success in the public sector. Reforms aimed at decreasing bureaucracy, promoting accountability and transparency, and offering better services to residents. ESs are advanced technologies that are capable of making judgments and handling difficult jobs and operations, as well as offering expert information and advice. The usage of ESs may boost productivity, efficiency, and decision consistency, as well as have a significant influence on the structure, communication, and power dynamics of the organization[3]. The deployment of ESs may create fears about losing control and alienating employees, but the advantages far exceed the risks[20].

EXPERT SYSTEMS IN DIFFERENT ROLES

Expert systems may be designed at any organizational level to perform two basic functions: one is to provide help, such as delivering ideas or advice, and the other is to operate as a substitute, making choices on its own.

An expert system's support function entails providing aid and direction to humans in decision-making, but it does not totally take over the decision-making process. It can be developed to assist both professionals and novices, but the final decision is made by a human being.

In a replacement capacity,[2] the expert system, not the human user, is in charge of making the final choice. This implementation may not necessarily result in job loss, as it may just allow a less experienced individual to perform the job. The system, on the other hand, may replace an expert, an assistant, or a non-expert who was previously working with an expert.

CHALLENGES

SOME OF THE DIFFICULTIES IN USING EXPERT SYSTEMS TO MAKE BETTER DECISIONS:

- **Complexity:** Expert systems, especially large and intricate ones, can pose challenges in terms of development and management.
- **Bias:** The information sources and algorithms used to create the system may embody the system's prejudices and constraints, resulting in restricted or inaccurate outcomes.
- **Cost:** Expert systems can be expensive to create and uphold, posing a challenge for smaller companies to adopt the technology.
- **Lack of flexibility:** Expert systems can be rigid and unable to adapt to evolving circumstances or new data, leading to outdated or inaccurate outputs.
- **Lack of interoperability:** The absence of standards utilized in the development of expert systems might hinder the capacity of systems to effectively communicate and collaborate.

OPPORTUNITIES

Some of the possibilities for improving decision-making with expert systems:

- **Improved accuracy:** Specialist systems could generate trustworthy and reliable conclusions, reducing the likelihood of human error in decision-making processes.
- **Increased efficiency:** The process of making decisions is expedited and enhanced by utilizing proficient systems, which have the ability to analyze extensive amounts of data at a faster pace than human specialists.
- **Consistency:** To ensure that assessments are made consistently and fairly, expert systems can be utilized, which will reduce the influence of personal biases or beliefs.
- **Access to expertise:** Expert systems can be utilized to provide knowledge on specific problems, enabling companies to make more informed decisions without allocating costly human resources to them.

- **Cost savings:** Expert systems have the potential to reduce the expenses associated with decision-making processes by automating tasks, reducing the need for human expertise, and minimizing the risk of human error.

KEY CONSIDERATIONS

Some major factors for employing expert systems to make better decisions:

- **Domain expertise:** Expert systems ought to have knowledge of the subject matter for which they were designed. For the system to have accurate and current information, access to industry experts is essential.
- **User interface:** The user interface for the system should be straightforward, clear, and accommodating. The interface needs to be flexible enough to meet the needs of different user types.
- **Reliability:** The system should be reliable and able to make reliable judgments based on the facts presented. This requires thorough testing and validation before implementation.
- **Data management:** To effectively store, retrieve, and handle information, the system must possess excellent data management abilities. This encompasses the capability to work with incomplete, imprecise, or incorrect data.
- **Integration with other systems:** The expert system must possess the capacity to interact with various systems and technologies in order to facilitate a prompt and uncomplicated decision-making procedure.
- **Maintenance and updates:** The system needs to be consistently upgraded and upheld in order to be beneficial. The knowledge base should be refreshed, and decision-making criteria and algorithms should be altered accordingly.
- **Security and privacy:** The system should be developed with security and privacy in consideration to safeguard sensitive information.
- **Cost-effectiveness:** The system must be cost-effective for the organization to install and run, while also delivering a satisfactory return on investment.

FUTURE OF EXPERT SYSTEMS IN DECISION-MAKING:

The development of artificial intelligence (AI) and machine learning (ML) is expected to significantly alter expert decision-making systems. Expert systems will grow increasingly intelligent and capable of digesting enormous amounts of data, creating predictions, and making extremely precise decisions as AI and ML algorithms develop. These technologies will therefore play a bigger role in a range of enterprises and industries, such as banking, healthcare, and logistics.

More conversational and intuitive interactions between people and robots will be made possible by expert systems in conjunction with natural language processing (NLP). The accessibility and simplification of decision-making processes will boost productivity and efficiency.

But it's important to keep in mind that while expert systems can support human judgment, they cannot entirely substitute it. Ethical and open AI rules must be created and put into place to ensure that expert systems are used ethically and for the benefit of society as a whole.

CONCLUSION

Expert systems are advanced technologies used in various industries to support decision-making and provide expert knowledge and advice. In education, they are used to improve learning, and accessibility of knowledge, and monitor student progress. In public organizations, they are used to improve accountability, efficiency, and decision-making, and provide better access to services and quality of life. They can be developed to fulfill either a support role or a replacement role, where the former offers assistance and guidance, while the latter makes decisions on its own. The benefits of expert systems include improved accuracy, increased efficiency, consistency, and access to expertise. However, there are also challenges such as complexity, bias, cost, inflexibility, and lack of interoperability but there is a high possibility that future advancements may overcome these challenges in the near future as there is an ongoing effort to improvise and develop technology in order to sustain in this competitive world. Moreover, it also becomes questionable how to overcome these challenges and whether it is possible or not.

REFERENCES

O'Leary, D. E., & Turban, E. (1987). The organizational impact of expert systems. *Human Systems Management*, 7(1), 11–19.

Edwards, J. S., Duan, Y., & Robins, P. C. (2000, March). An analysis of expert systems for business decision-making at different levels and in different roles. *European Journal of Information Systems*, 9(1), 36–46. https://doi.org/10.1057/palgrave.ejis.3000344.

Dillon, S., Buchanan, J., & Corner, J. (2010, November). Comparing public and private sector decision making: Problem structuring and information quality issues. Proceedings of the 45th Annual Conference of the ORSNZ, pp. 229–237. http://citeseerx.ist.psu.edu/viewdoc/download?doi=10.1.1.455.9291&rep=rep1&type=pdf (accessed 1 April 2018).

Sheikhtaheri, A., Sadoughi, F., & Dehaghi, Z. H. (2014, September). Developing and using expert systems and neural networks in medicine: A review on benefits and challenges. *Journal of Medical Systems*, 38(9), p. 110.

Rauta, E. (2014, February). A decision-making model for public management: The existence of a policy framework for performance in Romania. *International Review of Social Research (IRSR)*, 4(1), 57–74.

Fahim, M. G. A. (2018, November 13). Improving administrative decisions through expert systems: An empirical analysis. *Review of Economics and Political Science*, 3(3/4), 119–138. https://doi.org/10.1108/reps-10-2018-011.

Sayed, B. T. (2021, March 18). Application of expert systems or decision-making systems in the field of education. *Information Technology in Industry*, 9(1), 1396–1405. https://doi.org/10.17762/itii.v9i1.283.

Feigenbaum, E. A., & McCorduck, P. (1983). *The fifth generation: Artificial intelligence and Japan's computer challenge to the world*. New York: Longman.

Kolodner, J. (1993). *Case-based reasoning*. San Francisco, CA: Morgan Kaufmann Publishers.

Haykin, S. (2009). *Neural networks and learning machines* (3rd ed.). Upper Saddle River, NJ: Pearson/Prentice Hall.

Zadeh, L. A. (1965). Fuzzy sets. *Information and Control*, 8(3), 338–353.

Mamdani, E. H., & Assilian, S. (1975). An experiment in linguistic synthesis with a fuzzy logic controller. *International Journal of Man-Machine Studies*, 7(1), 1–13.

Genesereth, M. R., & Nilsson, N. J. (1987). *Logical foundations of artificial intelligence*. Los Altos, CA: Morgan Kaufmann Publishers.

Jabbar, H. K., & Khan, R. Z. (2015, March). Survey on the development of the expert system in the areas of medical, education, automobile, and agriculture. 2nd International Conference on Computing for Sustainable Global Development (INDIACom).New Delhi, India: IEEE, pp. 776.

Nwigbo, S., & Madhu, B. K. (2011). Expert system: A catalyst in educational development in Nigeria. *IOSR Journal of Mobile Computing & Application (IOSR-JMCA)*, 3(2), 8–11.

Sakala, L. C., Muzurura, O., & Zivanai, L. (2010). The use of expert systems has improved students learning in Zimbabwe. *Journal of Sustainable Development in Africa*, 12(3), 1–13.

Goodarzi, M. H., & Rafe, V. (2012). Educational advisor system implemented by web-based fuzzy expert systems. *Journal of Software Engineering and Applications*, 5. https://doi.org/10.4236/jsea.2012.57058.

Ghadirli, H. M., & Rastgarpour, M. (2013). An adaptive and intelligent tutor by expert systems for mobile devices. *International Journal of Managing Public Sector Information and Communication Technologies*, 3. https://doi.org/10.5121/ijmpict.2012.3102.

Raczkowski, K. (2016). *Public management: Theory and practice*. Zurich, Switzerland: Springer International Publishing.

Harvey, J., & Technical Information Service (TIS) (2007, December). *Effective decision making, topic gateway series*, Vol. 40. London: The Chartered Institute of Management Accountants (CIMA).

6 An Investigative Analysis of Artificial Neural Network (ANN) Models for Intelligent Wearable Devices

B. Rupa Devi, C. Sateesh Kumar Reddy, G. Yamini, P. Sunitha, and M. Reddi Durgasree

INTRODUCTION

The rapidly expanding availability of wearable electronics is dramatically altering the ways in which people engage with technology and organize their lives. Incorporating sensors, networking features, and computational capability, the term "wearable technology" describes a wide variety of portable gadgets. People regularly attach these gadgets to their person. Wearable technologies have quickly become indispensable due to their ability to be easily integrated into everyday routines, to allow for the customization of experiences, and to provide useful insights into health and well-being. From simple pedometers and heart rate monitors to cutting-edge smartwatches, fitness trackers, and health monitoring gadgets, the field of wearable technology has come a long way in the past decade. These devices cover a wide range of uses, from monitoring your activity and sleep patterns to your heart rate and even your stress levels. The usage of fitness trackers has become widespread among both those who wish to lead more physically active lives and those in the medical, scientific, and sports professions who are concerned with optimizing performance and keeping tabs on development. The capacity of wearable devices to gather and examine copious volumes of data about our actions, biometrics, and the surrounding environment is the key reason contributing to their widespread popularity.

Wearables can monitor and interpret our movements, physiological responses, and contextual information because they are equipped with a variety of built-in sensors, such as accelerometers, gyroscopes, heart rate sensors, and global positioning system (GPS) modules. These sensors are gathering this data. Nevertheless, the true value of wearable devices comes from their ability to make sense of this data and present users with relevant insights and recommendations they can put into action

DOI: 10.1201/9781032686714-6

based on those insights. Models based on artificial neural networks (ANNs) have recently emerged as effective resources for the processing and interpretation of the vast amounts of sensor data generated by wearable devices. ANN models are designed to learn complicated patterns and relationships within data, which enables them to make accurate predictions, classify activities, and find hidden correlations. These models are inspired by the workings of the human brain, which is built to learn these complex patterns and relationships. Wearable devices can improve their utility and impact by transforming raw sensor data into information that can be acted upon by harnessing the computational power of ANN models. The purpose of this study is to investigate the possibility of employing ANN models within the framework of intelligent wearable technology. To get there, we'll first employ empirical research methods after conducting a thorough review of the existing body of scholarly literature. The purpose of this research is to investigate numerous facets of ANN models, including their structure, training methods, optimization strategies, and real-time capabilities. The goal is to determine how these models can be used to improve wearable technology (Lee et al., 2020).

Activity detection and forecasting, estimate and monitoring of physiological parameters, and user behaviour analysis will be the primary areas of investigation into the application of ANN models in smart wearable devices. Activity detection and forecasting will be our primary research areas. Using ANN models to categorize and predict human actions based on sensor data is one of the principal domains of application. This allows for precise monitoring of user activity and the acquisition of relevant background information. The second area of interest is the use of ANN models for the estimation and monitoring of physiological indicators including heart rate, blood pressure, and sleep patterns in order to deliver real-time health insights and individualized recommendations. At the end of this study, we will look into how ANN models may analyse user behaviour like sleep duration and quality, stress levels, and dietary preferences to offer personalized feedback and guidance for making healthy lifestyle choices. By doing this research, our ultimate goal is to make a contribution to the expanding body of knowledge concerning the utilization of ANN models in intelligent wearables. We can pave the way for more effective and intelligent wearable gadgets that cater to the different demands of individuals, athletes, and healthcare professionals if we grasp the benefits, limitations, and potential issues connected with ANN models. These models are used in ANNs. In addition, the findings of this study will shed light on the significance that ANN models will have in moulding the future of wearable technology. These findings will offer insights into the possibilities and paths for additional improvements in this fascinating sector.

In the following chapters, we will delve into the specific areas of application for ANN models in intelligent wearables. We'll take a look at the studies, techniques, and real-world applications that have already demonstrated these models' worth. Wearable electronics that can detect their surroundings and act accordingly are known as "intelligent wearables." The study's goal is to conduct empirical trials and analyse the results to determine the efficacy, accuracy, and practical consequences of ANN models deployed in wearable devices. The ultimate goal of this research is to make a significant contribution to the development of wearable technologies. The goal is to encourage creativity and provide a means through which wearable gadgets can become more intelligent, customizable, and productive.

TIMELINE OF THE ARTIFICIAL INTELLIGENCE AND WEARABLE DEVICES DEVELOPMENT

WEARABLE DEVICES DEVELOPMENT PHASE

- First introduced in 1975, the "WearComp 1" was the first wearable computer ever created. Edward Thorp and Claude Shannon were credited with the system's design. The device, which resembled a backpack and included a streamlined user interface and many input methods, could be worn across the back for easy portability.
- Seiko's UC-2000 wristwatch computer was one of the first of its kind when it was released in the 1980s. The gadget had a computing tool, a time management system, and the ability to store information in its most fundamental form.
- The development of wearable fitness gadgets began in the 1990s, with the introduction of groundbreaking products like the Fitbit Tracker and the Nike+iPod. These devices were primarily designed to monitor and record a wide range of activities.
- Smartwatches like the Microsoft SPOT watch and the Fossil Wrist PDA appeared on the market in the early 2000s. These gadgets were a huge step forward for wearable technology, expanding its utility far beyond the realm of fitness tracking. The aforementioned gadgets provide supplementary functions including notifications and basic app compatibility.
- With the advent of high-tech smartwatches like the Apple Watch and fitness trackers like Fitbit, the demand for such gadgets skyrocketed in the 2010s. These devices provided access to several health and fitness features, boasted enhanced user interfaces, and integrated a wider variety of sensors.
- It is expected that improvements in sensor technology, battery life, and communication capabilities will continue to drive the development of wearable technology into the 2020s. There is a plethora of different wearable devices currently available on the market. Smart eyewear, smart clothing, and medical-grade equipment are just a few examples.

ARTIFICIAL INTELLIGENCE DEVELOPMENT PHASE

- In the 1950s, academics looked at the beginnings of artificial intelligence (AI). Alan Turing and John McCarthy, among others, were instrumental in developing the idea of machine intelligence and laying the groundwork for the field of AI at the outset of this line of inquiry.
- The Dartmouth Conference in 1956 is widely regarded as the starting point for the modern field of AI. A group of experts got together to talk shop and

investigate opportunities related to developing artificial systems that can mimic human intelligence.

- The 1960s and 1970s saw the birth of the first computer systems and software dedicated to AI. Expert systems, which are computer programs that can solve problems and form logical conclusions, were the primary focus of the researchers' attention.
- The 1980s and 1990s saw the birth of machine learning (ML) and neural networks. The inquiry began with the examination of computer models that were inspired by the human brain, which led to the development of algorithms that could learn by assimilating information.
- The area of AI faced serious obstacles and setbacks in the late 1990s. Due to a lack of major breakthroughs and the prevalence of unreasonable expectations, the pace of growth in the field of AI research slowed. Both enthusiasm and research efforts waned as a result of a decrease in funding for AI projects.
- The 2000s saw a renaissance in the study of AI. Improvements in computer processing power, the availability of large amounts of data, and algorithms are all to thank for the recent upsurge of interest in AI. ML methods, especially deep learning, have received a lot of attention and have shown remarkable effectiveness in many fields.
- The 2010s saw a dramatic rise in the use of AI in a wide variety of settings. Virtual assistants, autonomous vehicles, and recommendation systems are just a few of the many uses that have seen widespread adoption of artificial intelligence technologies. Technologies such as natural language processing, computer vision, and voice recognition fall under this umbrella. The year 2020 was projected to be a watershed year for AI. When applied to tasks as varied as picture recognition, language translation, and medical diagnostics, deep learning models have proven to be extraordinarily effective. Both the data deluge and the development of processing infrastructure have played a role in these successes. More and more, conversations about the development and use of AI technology centre on the importance of ethical considerations and the implementation of ethical practises in the field of AI (Kumar et al., 2023; Kumar & Elias, 2021a, 2021b).

All the development and deployment phases of AI and wearable devices have been shown in Figures 6.1 and 6.2.

WEARABLE DEVICE APPLICATIONS AND TRENDS

In recent years, wearable devices have become a prominent technological trend that has revolutionized the way individuals interact with technology and monitor their daily activities. Wearable devices, commonly utilized as accessories or integrated into garments, comprise a range of sensors, processors, and connectivity functionalities, which enable the acquisition and evaluation of data pertaining to user behaviour, biometric indicators, and environmental conditions. These technologies

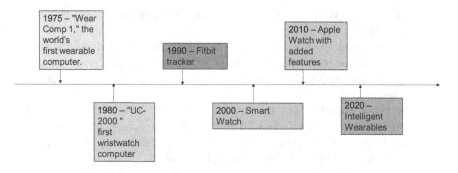

FIGURE 6.1 Wearable devices development timeline

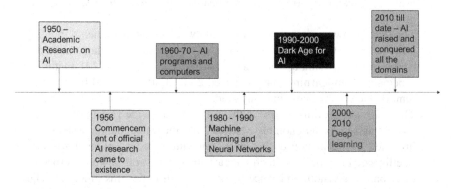

FIGURE 6.2 AI development timeline

have been utilized in various domains including but not limited to physical activity monitoring, medical supervision, interpersonal communication, and enhanced reality. The swift progressions in sensor technology, miniaturization, and wireless connectivity have facilitated the integration of wearable devices into our daily routines (Bangaru et al., 2021).

The wearable device market has undergone substantial expansion over the last decade, as evidenced by its growth and market overview. As per the findings of research reports, it is anticipated that the worldwide market for wearable devices will attain a valuation of several hundred billion dollars in the coming years. The expansion of the fitness tracking and health monitoring industry can be ascribed to several factors, including the growing consciousness of consumers and their demand for such products, the progress of sensor technologies, and the emergence of smartwatches as versatile gadgets. Prominent players in the market, including Apple, Fitbit, Samsung, Garmin, and Xiaomi, exhibit dominance and consistently engage in innovative practises to cater to the dynamic demands of consumers (Kumar & Elias, 2020; Paul & George, 2015; Ronao & Cho, 2016; Seeja et al., 2021).

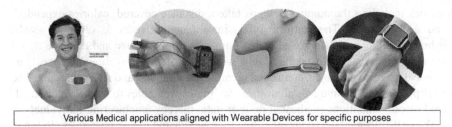

Various Medical applications aligned with Wearable Devices for specific purposes

FIGURE 6.3 Wearable device examples Sources: Ventricular tachycardia disease reference guide – Drugs.com; Wearable device reveals consumer emotions I MIT News I Massachusetts Institute of Technology; veari presents fineck smart wearable device for neck health I Dispositivi indossabili (pinterest.com); Wearable Devices Can Become More Popular If They Solve This One Problem (businessinsider.com)

The utilization of sensor technologies in wearable devices enables the acquisition of diverse data pertaining to the user and their surroundings. The array of sensors encompasses accelerometers, gyroscopes, magnetometers, heart rate monitors, GPS modules, ambient light sensors, and skin temperature sensors, among other types. Accelerometers and gyroscopes are sensors that quantify motion and orientation, facilitating the monitoring of physical activity and the identification of gestures. Heart rate monitors and electrocardiogram (ECG) sensors offer valuable insights into an individual's cardiovascular well-being. GPS modules facilitate the tracking of geographical locations, while ambient light sensors regulate the brightness of displays in response to variations in environmental lighting. The incorporation of various sensors facilitates the collection of extensive data by wearables, thereby offering a comprehensive perspective of the user's health and physical pursuits (Vos et al., 2022).

The utilization of wearable devices for fitness tracking represents a prominent application within the realm of health and fitness. These gadgets track various

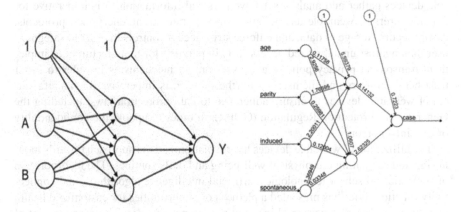

FIGURE 6.4 Basic neural networks structure

metrics including the number of steps taken, distance covered, calories expended, and sleep patterns. The platform offers instantaneous feedback, objective establishment, and comprehensive activity summaries to incentivize and enhance users' physical fitness. Certain sophisticated wearables provide functionalities such as physical exercise guidance, controlled breathing techniques, and stress tracking. The incorporation of AI and ML algorithms facilitates wearables to furnish individualized recommendations predicated on user data, thereby empowering individuals to make well-informed decisions regarding their health and fitness objectives (Rehman et al., 2021).

The utilization of wearable devices has the capacity to revolutionize the healthcare industry through the provision of uninterrupted monitoring, timely identification of health concerns, and remote supervision of patients. It is possible to monitor crucial physiological parameters, including but not limited to heart rate, blood pressure, blood glucose levels, and respiratory rate. Wearable devices that are equipped with fall detection and emergency response features have the potential to aid the elderly and individuals with medical conditions by providing prompt assistance in the event of accidents or emergencies. In addition, wearables have the potential to enhance telemedicine services by allowing healthcare practitioners to remotely track patient information and deliver tailored medical attention (Anupama et al., 2022).

The efficacy of wearable devices is significantly contingent upon the quality of user experience and interface design. The design of wearables should prioritise comfort, aesthetic appeal, and user-friendliness. The design of the user interface ought to be simplistic and highly accessible, taking into account the restricted screen dimensions and input alternatives that are typical of wearable devices. It is imperative for designers to prioritize the optimization of power consumption in order to extend the battery life of devices, while also incorporating efficient data synchronization and connectivity functionalities. In addition, the provision of personalization and customization features, such as the ability to switch bands and watch faces, can augment the user's experience and foster a sense of uniqueness.

The protection of data privacy and security is crucial in light of the fact that wearable devices gather and analyse sensitive personal information. It is imperative for manufacturers of wearable technology to incorporate strong encryption protocols, ensure secure storage of data, and enforce strict access controls in order to safeguard user data against unauthorized access. It is imperative for manufacturers to implement transparent privacy policies and user consent mechanisms in order to foster trust between users and themselves. Furthermore, it is imperative for manufacturers of wearable devices to ensure adherence to data protection laws, including the General Data Protection Regulation (GDPR), in order to uphold the confidentiality of user information.

The utilization of wearable devices has revolutionized the approach towards monitoring and regulating our physical well-being and daily routines. The incorporation of sophisticated sensor technologies, artificial intelligence algorithms, and connectivity functionalities has presented a plethora of opportunities for customized health monitoring, physical fitness tracking, and remote patient care. The advancement of these devices necessitates the resolution of critical issues such as battery longevity, user interface, and data protection to guarantee their extensive acceptance. The continuous progress in sensor technologies, AI, and material sciences is expected to enhance the intelligence, versatility, and seamless integration of wearable devices into our daily routines.

ARTIFICIAL NEURAL NETWORKS APPLICATIONS ON WEARABLE DEVICES

ANNs have been extensively utilized in wearable devices, augmenting their capabilities, cognitive abilities, and capacity to derive significant inferences from the amassed sensor data. The subsequent are several principal implementations of ANNs on wearable devices:

The use of ANNs has become prevalent in the recognition and prediction of human activities through wearable devices. Through the examination of data obtained from integrated sensors, including accelerometers and gyroscopes, ANNs possess the capability to categorize and discern particular physical activities executed by the user, such as walking, running, cycling, or ascending stairs. The aforementioned models have the capability to anticipate forthcoming activities by analysing sensor data patterns, thereby furnishing users with proactive recommendations and notifications. Precise identification and anticipation of physical movements facilitate wearables to furnish tailored feedback, monitor advancement towards physical fitness objectives, and facilitate adaptive guidance. The utilization of ANNs is of great significance in the estimation and monitoring of physiological parameters through the analysis of data obtained from wearable sensors. ANNs possess the capacity to provide continuous and real-time monitoring of vital signs through the analysis of signals acquired from various sensors, including ECGs, heart rate monitors, and blood pressure sensors. The models discussed above have the potential to identify abnormalities, identify trends that indicate stress or weariness, and provide personalized health advice. ANNs possess the capacity to enhance the health monitoring capabilities of wearable devices by enabling the estimation of diverse metrics, encompassing but not limited to calorie expenditure, sleep quality, and respiration rate (Ganji & Parimi, 2022; Zhao et al., 2019; Zhu et al., 2015).

ANNs play a crucial role in evaluating data from wearable devices to learn about user behavioural patterns. The device can track and analyse patterns associated with things like how well you sleep, how often you feel stressed, what you eat, and how much time you spend sitting about doing nothing. As a result, it offers helpful information and suggestions to its users in an effort to boost their health and happiness. With the help of ANNs wearable technology may soon be able to detect abnormal behavioural patterns, evaluate the success of healthful practises, and administer individualized treatments. When a user has been sedentary for a while, an ANN can prompt them to start moving again by suggesting activities they might enjoy. In addition, by analysing a person's eating habits, ANNs can make recommendations for dietary changes. ANNs have been successfully implemented, paving the way for wearable gadgets to recognize gestures and movements. In order to effectively read and react to human motions utilizing wearable devices, ANN models can be trained with large datasets of motion capture data. Answering calls, navigating menus, and using external devices are just some of the many tasks that may be handled easily with the help of this feature. ANNs can enhance user experiences and enable novel applications by identifying complicated movements, like hand gestures, for the comprehension of sign language or sports-related moves.

Wearable devices with built-in ANNs can cater to the individual needs of their users by offering personalized recommendations. Individualized plans for physical activity, dietary intake, and stress reduction could all be part of the advice. Wearable devices equipped with ANNs can learn from user input to improve their performance and responsiveness, resulting in a more tailored experience for each individual user. It has been found that the ability to personalize wearable devices increases user involvement, motivation, and the overall efficiency of these tools in fostering health and wellness. In recent years, ANNs have been shown to be crucial in expanding the functionalities of wearable technology. Wearables are only able to do their many jobs – from activity detection and physiological monitoring to user behaviour analysis and personalized recommendations – because ANNs are able to glean useful information from sensor data. These additions improve the user experience and make it easier to achieve specific health and fitness objectives. With the continued advancement of ANNs, it is anticipated that wearable technology will witness the emergence of more sophisticated and intelligent applications. This will drive innovation and enhance the practicality of these devices in our day-to-day activities.

WEARABLE DEVICES DATA MODELING USING ARTIFICIAL NEURAL NETWORKS

Use Case: Activity Recognition

Currently available resources for activity recognition from human sensing datasets :

The UCI Parkinson's Disease Classification dataset comprises accelerometer and gyroscope data that has been gathered from individuals both afflicted with and unaffected by Parkinson's disease. The utilization of this technology enables the categorization of various activities and the identification of Parkinson's disease through the analysis of movement patterns (Rehman et al., 2021; Vos et al., 2022).

The Opportunistic Human Activity Recognition (O-HAR) dataset encompasses sensor data obtained from smartphones, smartwatches, and a Kinect sensor. This encompasses the range of activities undertaken by individuals diagnosed with Parkinson's disease, which can be effectively utilized for the purposes of activity recognition and monitoring.

The Wireless Sensor Data Mining (WISDM) dataset comprises accelerometer and gyroscope data that has been gathered from smartphones. The repertoire of activities engaged in by individuals diagnosed with Parkinson's disease encompasses ambulatory movements, sedentary postures, and upright stances.

The MobiFall dataset comprises a variety of activities executed by individuals who may face difficulties in terms of their mobility. Although the primary emphasis of this dataset is not limited to Parkinson's disease, it possesses potential significance in exploring the attributes associated with Parkinson's disease. The collection consists of accelerometer data collected throughout various activities, including incidents of falling.

In the context of Parkinson's disease, the aforementioned datasets are invaluable tools for the research and development of activity recognition algorithms and systems. Researchers and medical professionals could benefit from using these methods to better understand, diagnose, and track the progression of Parkinson's disease in patients.

This chapter examines the growing significance of smart wearable devices and their applications in the field of activity recognition in great detail. Accurate activity recognition is underlined for its significance in the fields of personalized healthcare, fitness tracking, and related fields. The primary purpose of this research is to critically examine the ANN models currently being used for activity recognition in wearable devices. This research provides a comprehensive overview of the methods and technologies used in the field of wearable technology for the purpose of detecting and assessing human behaviour. The study's goal was to investigate the many sensors utilized for data collecting and preprocessing, as well as the wide range ML approaches currently in use for activity recognition. This study examines the many previous works that have addressed the question of how ANN models might be used for the issue of activity recognition. The strengths and weaknesses of these models are analysed and addressed in detail (Bhogaraju et al., 2021; Bhogaraju & Korupalli, 2020; Chen & Xue, 2015; Freedson et al., 1998; Khan et al., 2010; Sorzano et al., 2014).

The primary focus of this research is on the need for, and methods to get, high-quality datasets for use in the development, testing, and refinement of ANN models. This chapter takes a look at several different types of datasets used often in activity detection studies, focusing on those that have been tailored for use with wearable technology. Noise removal, feature extraction, and data augmentation are only a few of the preprocessing approaches used in this study to boost the accuracy and precision of ANN models. Here we conduct an in-depth analysis of many ANN topologies and combinations that have found application in the activity recognition arena. Four of the most popular ANN architectures – the multilayer perceptron, convolutional neural network, recurrent neural network, and long short-term memory (LSTM) – will be covered. A model's strengths, shortcomings, and applicability to wearable devices are determined by assessing its accuracy, computational efficiency, and memory demands in the context of the entire.

This study aims to dissect the process of training and optimizing ANN models for the job of activity recognition. Some of the techniques covered here include backpropagation, gradient descent, regularization, and hyperparameter optimization. This study also explores ways to improve the generalization abilities of ANN models and deal with imbalanced datasets. The following section provides a comprehensive evaluation of how several ANN models fare in the context of activity recognition. Accuracy, precision, recall, and the F1 score are just some of the criteria used to assess the models' efficacy. In a side-by-side comparison, the benefits and drawbacks of each model are analysed to better illuminate their practical applications.

This study looks at how ANN models could be used in smart wearable devices in ways that go beyond activity recognition. It also takes into account the possibility that more research and development will be done in this area in the future. This study looks at fall detection, sleep monitoring, stress evaluation, and gait analysis, among other things, to show how powerful ANN models can be in health and wellness. This study also talks about how ANN models might be used in future research and development in the area of activity recognition. This chapter gives an overview of the study on ANN models used in smart wearable devices to recognize activities. The importance of ANN models in accurately identifying human activities is also emphasized, as is the potential for expanding personalized healthcare, fitness tracking, and other related applications. The chapter talks about both the problems that are already there and the possibilities for the future in the field, arguing for the need for more research.

Feedforward and backpropagation are two basic ideas in the field ANNs that have a big impact on how well wearable devices can recognize actions. These approaches enhance the efficacy of training and optimization processes for ANNs, hence enabling them to achieve precise classification and recognition of human actions. This section explores the application of feedforward and backpropagation techniques in the field of activity recognition (Carneiro et al., 2018; Gunawan et al., 2020; Kumar et al., 2020; Kumar & Elias, 2020; Spörri et al., 2017).

Feedforward: A prediction or other output is achieved by feeding data into various nodes of an ANN. Accelerometer and gyroscope data, among other sensor measures from wearable devices, are frequently employed as input data in the field of motion recognition. The sensor readings effectively capture and document the user's movements and orientations with precision. The input data is initially introduced into the input layer of the ANN, which is subsequently followed by one or more hidden layers. Each stratum of the neural network comprises a set of neurons or nodes that perform computational operations on the input data. The neurons in a certain layer of a neural network receive inputs from the preceding layer, perform an activation function on the weighted sum of these inputs, and then generate outputs that are transferred to the subsequent layer. In the field of activity recognition, the hidden layers of the ANN are tasked with extracting relevant features from the incoming sensor data. During this process, the ANN gains knowledge about patterns and representations that aid in the discrimination of various activities. Afterwards, the obtained characteristics are utilized to produce predictions concerning the particular task being performed by the user.

Backpropagation: The backpropagation method is a crucial approach utilized in the training of ANNs with the specific objective of activity recognition. The procedure involves the computation of gradients and the subsequent adjustment of the weights and biases within the network to decrease prediction errors. The objective is to maximize the efficiency of the network's performance and augment its capacity to precisely identify and classify activities. The backpropagation process commences by juxtaposing the predicted output of the network with the desired output, also known as the ground truth. The discrepancy or deviation between the anticipated and intended outcomes is computed by employing an appropriate loss function, such as mean squared error or categorical cross-entropy. Subsequently, the derivatives of the loss function with respect to the weights and biases of the network are calculated by applying the chain rule of differentiation. The observed gradients serve as indicators of both the direction and magnitude of adjustments required in order to minimize the error. The weights and biases are subsequently adjusted according to these gradients through the utilization of an optimization algorithm, such as stochastic gradient descent (SGD).

During the process of backpropagation, the revised weights and biases are systematically transmitted in a backward manner through the network, layer by layer. This propagation enables the computation of gradients and subsequent updating of the parameters associated with the preceding layers. The aforementioned iterative procedure involves the computation of gradients and subsequent adjustment of the network's parameters, which is performed repeatedly until a desirable level of accuracy is attained. The utilization of both feedforward and backpropagation techniques allows ANNs to progressively acquire knowledge and enhance their predictive capabilities in the context of activity recognition. After undergoing numerous training iterations, the neural network progressively enhances its ability to extract pertinent features from sensor data and effectively classify diverse activities. Feedforward and backpropagation are essential methodologies in the field of activity recognition utilizing ANNs. The process of feedforward involves transmitting input data through the layers of a network, enabling the network to make predictions by utilizing the acquired features. The backpropagation algorithm facilitates the neural network's ability to acquire knowledge from its mistakes through the adjustment of weights and biases, ultimately leading to the enhancement of the network's performance as time progresses. The utilization of these techniques is of utmost importance in the development of resilient and precise activity recognition models through the use of wearable devices.

REFERENCES

Anupama, C. S. S., Sivaram, M., Lydia, E. L., Gupta, D., & Shankar, K. (2022). Synergic deep learning model–Based automated detection and classification of brain intracranial hemorrhage images in wearable networks. *Personal and Ubiquitous Computing*, 1–10.

Bangaru, S. S., Wang, C., Busam, S. A., & Aghazadeh, F. (2021). ANN-based automated scaffold builder activity recognition through wearable EMG and IMU sensors. *Automation in Construction*, *126*, 103653.

Bhogaraju, S. D., & Korupalli, V. R. K. (2020). Design of smart roads-a vision on Indian smart infrastructure development. International Conference on Communication Systems & Networks (COMSNETS), IEEE, pp. 773–778.

Bhogaraju, S. D., Kumar, K. V. R., Anjaiah, P., Shaik, J. H., & others. (2021). Advanced predictive analytics for control of industrial automation process. In *Innovations in the Industrial Internet of Things (IIoT) and Smart Factory* ., IGI Global (pp. 33–49).

Carneiro, T., Da Nóbrega, R. V. M., Nepomuceno, T., Bian, G.-B., De Albuquerque, V. H. C., & Reboucas Filho, P. P. (2018). Performance analysis of google collaboratory as a tool for accelerating deep learning applications. *IEEE Access*, *6*, 61677–61685.

Chen, Y., & Xue, Y. (2015). A deep learning approach to human activity recognition based on single accelerometer. Hong Kong: IEEE International Conference on Systems, Man, and Cybernetics, City University (CityU), 1488–1492.

Freedson, P. S., Melanson, E., & Sirard, J. (1998). Calibration of the computer science and applications, inc. accelerometer. *Medicine and Science in Sports and Exercise*, *30*(5), 777–781.

Ganji, K., & Parimi, S. (2022). ANN model for users' perception on IOT based smart health-care monitoring devices and its impact with the effect of COVID 19. *Journal of Science and Technology Policy Management*, *13*(1), 6–21.

Gunawan, T. S., Ashraf, A., Riza, B. S., Haryanto, E. V., Rosnelly, R., Kartiwi, M., & Janin, Z. (2020). Development of video-based emotion recognition using deep learning with Google Colab. *Telkomnika*, *18*(5), 2463–2471.

Khan, A. M., Lee, Y.-K., Lee, S.-Y., & Kim, T.-S. (2010). Human activity recognition via an accelerometer-enabled-smartphone using kernel discriminant analysis. 5th International Conference on Future Information Technology, Busan, South Korea: IEEE, pp. 1–6.

Kumar, K. V. R., Devi, B. R., Sudhakara, M., Keerthi, G., & Madhavi, K. R. (2023). AI-based mental fatigue recognition and responsive recommendation system. In B. Narendra Kumar Rao, R. Balasubramanian, Shiuh-Jeng Wang, Richi Nayak (Eds.) *Intelligent Computing and Applications*, (pp. 303–314). Springer.

Kumar, K. V. R., & Elias, S. (2020). Smart neck-band for rehabilitation of musculoskeletal disorders. The Proceedings of IEEE International Conference on Communication Systems & Networks (COMSNETS), Bengaluru, India.

Kumar, K. V. R., & Elias, S. (2021a). Real-time tracking of human neck postures and movements. *Healthcare*, *9*(12), 1755.

Kumar, K. V. R., & Elias, S. (2021b). Use case to simulation: Muscular fatigue modeling and analysis using Opensim. *Turkish Journal of Physiotherapy Rehabilitation*, *32*(2), 2457–2468. http://ez-ucs.statsbiblioteket.dk:2048/login?url=https://search.ebscohost .com/login.aspx?direct=true&db=ccm&AN=151006258&site=ehost-live

Kumar, K. V. R., Kumar, K. D., Poluru, R. K., Basha, S. M., & Reddy, M. P. K. (2020). Internet of things and fog computing applications in intelligent transportation systems. In Prof. Dr. Binshan Lin, PhD (ed.) *Architecture and Security Issues in Fog Computing Applications*, (pp. 131–150). IGI Global.

Lee, V.-H., Hew, J.-J., Leong, L.-Y., Tan, G. W.-H., & Ooi, K.-B. (2020). Wearable payment: A deep learning-based dual-stage SEM-ANN analysis. *Expert Systems with Applications*, *157*, 113477.

Paul, P., & George, T. (2015). An effective approach for human activity recognition on smart-phone. IEEE International Conference on Engineering and Technology (ICETECH), Coimbatore, Tamil Nadu, India, 1–3.

Rehman, I. H., Ahmad, A., Akhter, F., & Aljarallah, A. (2021). A dual-stage SEM-ANN analysis to explore consumer adoption of smart wearable healthcare devices. *Journal of Global Information Management (JGIM)*, *29*(6), 1–30.

Ronao, C. A., & Cho, S.-B. (2016). Human activity recognition with smartphone sensors using deep learning neural networks. *Expert Systems with Applications*, *59*, 235–244.

Seeja, G., Reddy, O., Kumar, K. V. R., Mounika, S., & others. (2021). Internet of things and robotic applications in the industrial automation process. In Sam Goundar, J. Avanija, Gurram Sunitha, K. Reddy Madhavi, S. Bharath Bhushan (Eds.) *Innovations in the Industrial Internet of Things (IIoT) and Smart Factory*, (pp. 50–64). IGI Global.

Sorzano, C. O. S., Vargas, J., & Montano, A. P. (2014). A survey of dimensionality reduction techniques. *ArXiv Preprint ArXiv:1403.2877*.

Spörri, J., Kröll, J., Fasel, B., Aminian, K., & Müller, E. (2017). The use of body worn sensors for detecting the vibrations acting on the lower back in alpine ski racing. *Frontiers in Physiology*, *8*, 522.

Vos, G., Trinh, K., Sarnyai, Z., & Azghadi, M. R. (2022). Ensemble machine learning model trained on a new synthesized dataset generalizes well for stress prediction using wearable devices. *ArXiv Preprint ArXiv:2209.15146.*

Zhao, Z., Liu, C., Li, Y., Li, Y., Wang, J., Lin, B.-S., & Li, J. (2019). Noise rejection for wearable ECGs using modified frequency slice wavelet transform and convolutional neural networks. *IEEE Access, 7,* 34060–34067.

Zhu, J., Pande, A., Mohapatra, P., & Han, J. J. (2015). Using deep learning for energy expenditure estimation with wearable sensors. 2015, 17th International Conference on E-Health Networking, Application & Services (HealthCom), Boston, MA, USA, 501–506.

7 Human Health Monitoring and Live Video Streaming of Remote Patients with the Help of Internet of Things

V. Malathy, M. Anand,
Geetha Manoharan, and Anil Pise

GENERAL CONCEPTS

Unawareness about the symptoms of heart diseases and failure to detect them early is a cause of high death rates. Remote patient monitoring systems (RPMSs) are a good solution to this problem because they allow the monitoring of patients who are away from clinical settings. The growth of semiconductor technology allows the collecting and sending of the patient's medical records to the doctors in remote places and enables continuous monitoring to help the doctors for analysis and treatment recommendations. In Figure 7.1, the architecture of a RPMS is described. The empirical study takes into account other studies that will be done in the future on this subject. Only the patients' decision is being studied in this process; further research into the patient behavior can be done digitally.

SURVEY ON REAL-TIME OPERATING SYSTEMS (RTOSs) IN PROPERTY MANAGEMENT SYSTEMS (PMSs)

Singh et al. (2010) discussed the round-robin scheduling method to increase the effectiveness of central processing units (CPUs). This method reduces delays in reaction time, minimizes preemption, and requires fewer setting changes, reducing overhead and conserving memory space.

Dietrich et al. (2017) explained two new semaphore-based methodologies for task synchronization in RTOSs. Preemption during the execution of the primary

DOI: 10.1201/9781032686714-7

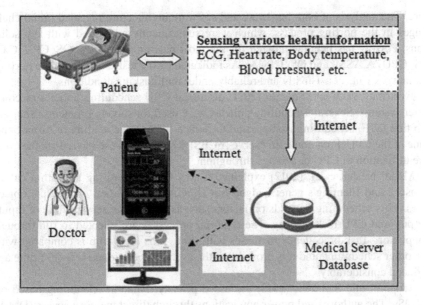

FIGURE 7.1 Architecture of remote patient monitoring system

segment, the assignment synchronization, turns out to be more critical in RTOSs. It is necessary to think about the synchronization requirements of the procedures. Higher-priority tasks may delay lower-priority assignments if they have acquired resources before release and were preempted by the higher-priority task. This situation can lead to interruptions. To avoid such interruptions, a priority inversion mechanism can be employed. In this approach, the priority of the lower-priority process is temporarily increased, allowing it to obtain CPU time, complete its critical section execution, and release the resources. Once the appropriate resources are released, the task's priority is then restored.

Indersain et al. (2013) implemented the kernel of a microcontroller to enable multitasking and scheduling with an ARM processor. An analog-to-digital converter (ADC), liquid crystal display (LCD), and universal asynchronous receiver-transmitter (UART) are interfaced with the microcontroller. The collected data from the interfaces are used by the framework of μC/OS-II. Then an embedded monitoring system is built.

Jae Hwan Koh and Byoung Wook Choi (2013) analyzed the execution of RTOSs based constant frameworks based on RTAI and Xenomai. The evaluations of real-time systems were conducted in kernel and user space. Performance metrics included jitter of intermittent tasks and response time of real-time instruments. Achieving consistent performance was facilitated through the use of real-time systems and data communication. Synchronization techniques, particularly in inter-task communication, were also employed.

Jian Feng and Hongmei Jin (2011) explained porting μC/OS-II RTOS to STM32F107VC. The STM platform incorporates an ARM Cortex-M3 32-bit RISC core. The analysis of both hardware and software architecture revealed initial challenges in the porting process, which were subsequently addressed with key solutions. Porting has been actualized for the most part on three files: OS_CPU_C.C, OS_CPU_A.ASM, and OS_CPU.H. As indicated by the tests conducted, the porting framework runs consistently and reliably under multitasking conditions.

Jyotirmay Patel and Solanki (2011) discussed CPU scheduling policies of computer systems. Several scheduling policies were used for solving problems. Shortest job first (SJF) scheduling is suggested to reduce the average waiting or turnaround time of the CPU. First come first serve (FCFS) scheduling is used to reduce the average utilization of CPU or average throughput.

Mehdi Neshat et al. (2012) explained a new CPU scheduling algorithm called Fonseca and Fleming's genetic algorithm (FFGA) optimization of multiple objectives. This scheduling depends on factors such as burst time, service time of input/output, and CPU task priority. The required processes are selected to run through adaptation. Task precedence increases with time. This algorithm is compared with the other scheduling, and it is proved that the average waiting and response time are reduced considerably.

Srikanth and Samunur (2013) discussed priority dynamic scheduling based on RTOSs. The authors used power applications through direct memory access (DMA) peripheral devices. The main focus is on the execution of various power applications utilizing the dynamic planning during runtime. At runtime, the client can adjust the priority of control mechanisms based on its requirements and change the priority as needed. Each power application is treated as a task. The embedded programming application executes a task efficiently by breaking down the overall task into various little sub-tasks. Previous embedded system solutions for control applications relied on specific algorithms, often consuming maximum CPU time. The concept of dynamic task prioritization is employed to reduce time delays for lower-priority tasks, sometimes elevating them to higher priority under certain conditions.

Sujitha and Kannan (2015) explained the porting of the kernel of a LPC 2148 microcontroller for implementing several features such as mutex, multitasking, time scheduling, and mailbox. Here a constant bit is a result that deals with the smaller scale controller to guarantee that unsurpassed essential occasions are handled as effectively as could reasonably be expected. The modules that have individual interfaces of a microcontroller like ADC, LCD, and UART are employed. Obtained data from those interfaces is processed using the continuous operating system of the μC/OS-II.

Tang and de Silva (2006) designed RTOS based architecture for mechanical automation using information transmission between two controlling units through Industrial Wireless Sensor Network (IWSN) without interference. This design manages the data transmission between two groups in a timely manner. The data transmission time is expanded according to conventional standards. One section operates with RTOS and LPC2148 as the master node, while the other serves as a data acquisition node connected to sensors. The data acquisition node employs a peripheral interface controller. Interchanges between two nodes (equipment and application) are facilitated through IEEE 802.15.4. The aim of this system is to reduce the likelihood of a crash and to meet the fundamental prerequisite of timing for information transmission in mechanical applications.

Taqwa Flayyih Hasan (2014) tested several methods to determine the average waiting and turnaround time. The results show that the FCFS reduces average throughput and everyday use of CPUs. Using the round-robin method increases the quantum time range and the average waiting time will be longer.

Vijayakumaran and Janaky (2014) described a scheduling scheme called the dynamic multilevel priority (DMP) packet for wireless sensor networks. The sensor nodes transmitting continuous and non-continuous packets are fundamentally vital to reduce overhead handling, utilization of energy, transfer speed, and uninterrupted information transmission of a wireless sensor network (WSN). The scheduling calculation for a considerable part of the current packet of a WSN uses first-come-first-served (FCFS), preemptive scheduling, and non-preemptive scheduling.

These algorithms result in significant overhead processing and delays in information transmission, and they are not dynamic in adapting to the information. To overcome these problems, a scheduling scheme called DMP packet is used. In this method, each node, with the exemption of those at the last level of the virtual order in the zone-based topology of the WSN, has the need to align with three levels. The highest priority queue contains the real-time packets and the data packets that can be preempted in other queues. Based on certain thresholds of their expected processing times, two other queues contain the non-real-time packets. A real-time and non-real-time data packet of the two queues contains the leaf nodes since they do not get data from other nodes and thus, decrease back-to-back delay.

Yu Anxin Lin et al. (2013) developed a brilliant home monitoring system in light of the STM32 processor and Zigbee innovation. The framework utilizes a low-control cost STM32 processor. The processor acts as the fundamental controller. The porting of µC/OS-II and µC/GUI on the framework is accomplished. The framework uses a resistive touch screen as the human–PC association interface. A short-range checking of home devices is done by Zigbee. The framework transplanted the Unique Internet Protocol (UIP) network protocol method. The master controller is combined with the Ethernet and configured as a WEB server.

Yukesh (2016) simulated a WSN-based data acquisition system using multiple sensors for the health monitoring of patients with RTOSs. Simulation results concluded that this system is 3% more efficient than the previous system that didn't use an RTOS.

Reddy ST et al. (2022), Padmaja et al. (2022), Reddy CVK et al. (2022), Sandeep et al. (2022(1)), Sandeep et al. (2022(2)), and Thavamani et al. (2022) described the application of IoT in their work.

SECURITY CHALLENGES OF IoT

IoT not only unites computers and mobile devices but it also connects all things in the surrounding environment. However, utilizing IoT presents several challenges, particularly in securing sensitive data, as the number of exploited systems increases. IoT architecture is shown in Figure 7.2.

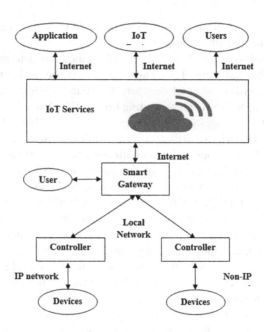

FIGURE 7.2   Architecture of IoT

ACCESS CONTROL

Access control is a security measure that controls who or what can utilize the resources in a computing environment. This security control reduces the risk to a company. Access control can be achieved by physical means. It employs restrictions on access to campuses, rooms, and IT assets. On the other hand, logical access control operates restrictions to the connections in computer networks, files, and data. An organization employs different access control methods based on their necessity and the levels of security in IT (Diego et al. 2022). The various access control methods are as follows.

MANDATORY ACCESS CONTROL (MAC)

The central authority based on various security levels controls the rights to access. The MAC model is shown in Figure 7.3. It is frequently utilized by the government and military. The classification is allocated to system resources. Depending on the security clearance information of the user or device, the kernel either grants or denies access to the objects support.

ROLE-BASED ACCESS CONTROL (RBAC)

RBAC using defined functions limits the computer resource access based on persons or groups. An RBAC security model depends on a complex formation of role

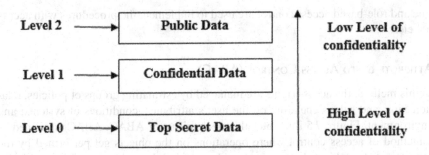

FIGURE 7.3 Example of a MAC

assignments. That is permission and authorization are developed to permit employee access to systems. Part-based access control is shown in Figure 7.4. RBAC systems can be used to enforce MAC and discretionary access control (DAC) frameworks.

DISCRETIONARY ACCESS CONTROL

In this method of access control, the system administrators who set the rules restrict the spread of rights to access. The main drawback of DAC systems is deficient centralized control.

RULE-BASED ACCESS CONTROL

A security model where the system administrator can describe the rules to allocate access to resource objects is otherwise called rule-based access control. These rules frequently depend on conditions such as time or location. Several forms of both

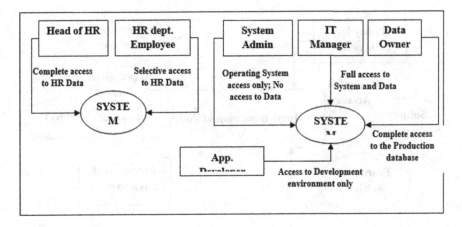

FIGURE 7.4 RBAC

rule and role-based access control are used to implement the procedures with access policies.

Attribute Based Access Control (ABAC)

In this method, the access rights are managed by estimating groups of policies, relationships, and rules depending on the user's attributes, conditions of systems, and environment. Figure 7.5 illustrates the mechanisms of ABAC. ABAC is defined as a method of access control where operations on the objects get performed by the requests of the subject that are approved or denied depends on attributes assigned to the subject, the object, conditions of the environment and a set of rules that are particular in terms of those conditions and attributes (Yysookhak, M., et al. (2017).

Use of Access Control

Access control reduces the illegal access to logical systems and physical systems. It is an important part of security agreement programs that ensures the technology deployed in security and policies for access control are employed to protect classified information, such as that of the customer. Nearly all organizations have procedures and infrastructure to restrict the access to networks, computer systems, and sensitive data or personal information.

This system is complex and challenging to control dynamic environments as the dynamic environments involve cloud services. Following some sophisticated breaches, vendors of technology have moved away from distinct systems for sign-on into unified access management. The system facilitates the access controls for cloud and on-premises dynamic environments.

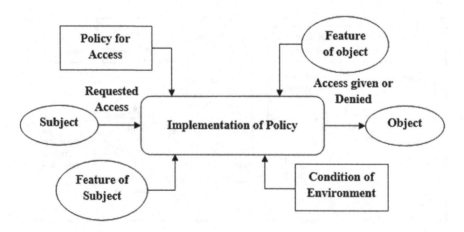

FIGURE 7.5   Mechanisms of ABAC

IMPLEMENTING ACCESS CONTROL

An organization's IT environment will incorporate the process called access control. It can engage with unique systems' access management. These systems offer software for access control, a database for a user, enforcement, and several policies for access control. Administrators of the system utilize an automated provisioning method. The automated method defines the time at which the addition of a new user to the management, responsibilities of job, and workflows is done.

Least privilege is an example to limit access to the resources. The worker requires the resources to carry out the job functions. Failing to revoke the permit is the challenge of this system. Another challenge is that the system access to data in the organization, even when a person moves into a different job within or outside the company.

RTOS

An RTOS is an operating system; it acts as an intellect of the real-time system and it immediately gives a reply to inputs. The process is finished in a specified time in the RTOS and it responds in an expected way to unpredictable events. In the embedded systems, FreeRTOS are mostly employed in the 32-bit microcontrollers; there are also patterns that are expanding to offer such kind of highlights in the systems based on 16-bit and 8-bit processors. A system programming environment that deals with the operation of resources in a PC system is generally called an operating system. FreeRTOS is differentiated from the non-exclusive OS as it is mainly proposed to carry out a steady-state process. Nowadays, RTOS exists in huge variants such as proprietary RTOS, commercial RTOS, and RTOSs with open-source availability.

Figure 7.6 shows the architecture of CPU vs software according to complexity. It has been separated into four quadrants. The top right quadrant explains the sophisticated software placed on the more powerful processor. It is the conventional region of RTOS and several other operating systems. If the software is rationally advanced, still it might be useful to set up an underlying kernel on a CPU with low power.

Occasionally, a powerful chip is employed to run relatively simple software, where the performance of the CPU is needed only to acquire the speed of execution. The kernel might not be strictly required in this case, but one who is using the kernel should be cautious as it progresses the scalability of software architecture and also a further increase in complexity.

Figure 7.7 shows the structure of RTOS. The kernel is the most crucial part of the RTOS. The kernel contains the board support package (BSP). It is specific codes in the processor that will make the RTOS run on a particular target.

RTOS ARCHITECTURE

RTOS architecture depends on the overall quality of its group. An effective RTOS is one that is capable of adapting to achieve different preparations of needs for different applications. An RTOS usually contains a kernel alone for available applications. An RTOS can be unified with different modules, together with the core, convention

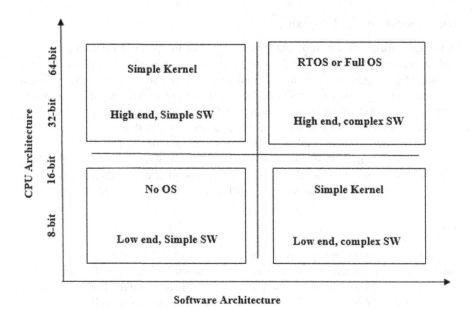

FIGURE 7.6 CPU architecture and software architecture

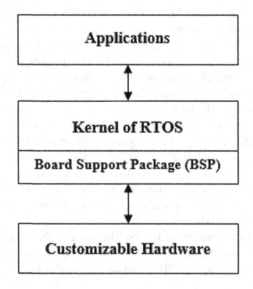

FIGURE 7.7 Structures of RTOS

stacks, and different parts for further complex embedded applications. A working system primarily consists of two sections: space for the user and space for the kernel. In a working framework, the kernel is the smallest segment and present in the center. Its supervision includes memory management. It gives an interface for resource utilization over the application programs.

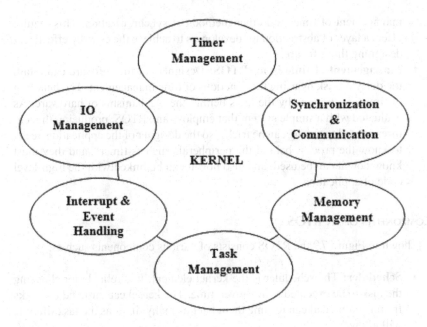

FIGURE 7.8 Architecture of RTOS

The kernal also offers additional services such as supervision of programs' protection unit and the multitasking state of the system. The RTOS kernel can provide a layer of abstraction between the equipment and application programming. Figure 7.8 illustrates the six universal services of an RTOS kernel.

Generally, an RTOS is not required for small scale embedded systems with low memory. An RTOS is required for applications such as:

- **Development of optimized software:** As the complexity of code increases, an RTOS becomes an effective tool to control the code and to allocate the tasks along with the developers. The full software can be separated into tasks by using an RTOS.
- **Safe synchronization:** Due to the development in small embedded systems which do not employ any RTOS, several factors are often used for the module synchronization. Even though, in the systems that are driven by high interrupt, the issues related to the software safety and bugs are generated with use of global variables. Such variables are regularly shared and accessed over the various functions, and there is a higher chance of such functions being corrupted during the program execution time. If a FreeRTOS is used in that place, the system makes sure the synchronization secured without occurring of any problem related with corruption.
- **Management of resource:** Almost every RTOS offers APIs for developers to control the resources of the system. This involves communication, management of tasks, management of interruptions, management of memory,

management of time, and other methods of synchronization. This feature offers a layer of abstraction for developers to achieve the code by effectively designing the software.

- **Management of time in an RTOS:** Designers of the software can attain the delay of task by using the functions of time management. The handling of timers is to know by the facts behind the mechanisms of hardware. As evaluated with a simple system that employs any RTOS, producing the features related to timing can be tricky, so the designer of the application needs to know the process behind the peripherals such as timers, and they must know how it can be used, and also how it can be linked with the high-level code of application.

COMPONENTS OF AN RTOS

As shown in Figure 7.9 the RTOS consists of various components such as:

- **Scheduler:** The scheduler is the kernel element. It is reliable for choosing the task to be executed at any given time. The kernel can suspend the tasks from running and can resume them later as many times as the task lifetime will allow.
- **Objects:** The objects section of the RTOS is used to create unique functions like tasks, semaphores, or message queues.

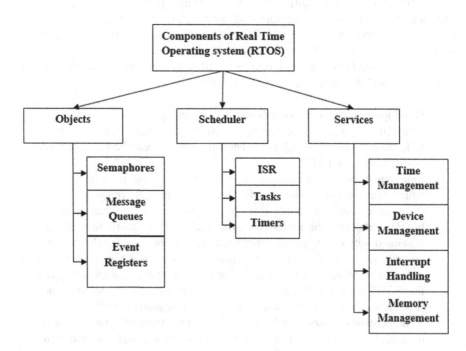

FIGURE 7.9 Structures of RTOS

- **Services:** Services of RTOS contain the action that the kernel carries out on objects. It may consist of memory management, time management; interrupt handling, and device management.

SCHEDULER

The scheduler traces the task conditions. It selects the tasks which are ready to be executed. The scheduler determines the task to which the CPU is allocated. The scheduler makes the CPU utilize the functions effectively in a multi-task program. It reduces the waiting time. There are two types of schedulers, namely preemptive and non-preemptive.

Preemptive scheduling is based on priority. It is necessary to transfer the control to the task with highest priority. Preemptive scheduling and non-preemptive scheduling are shown in Figure 7.10 and Figure 7.11. The present task is suspended immediately, when an incident runs a job with the highest priority. Then the processor control is transferred to the task of higher priority.

Non-preemptive scheduling is called cooperative multitasking. It requires the tasks to openly give up the processor control to cooperate with the tasks over each other. When the processor control releases a task, the execution will be carried out on the next highest-priority task. The newly chosen higher priority task will only gain the processor control when the tasks which are currently executed give up the power voluntarily.

The scheduler allocates CPU control to the selected tasks through context switching, facilitated by the dispatcher, resulting in a change in the execution flow. . Always, if an RTOS is running, the execution flow will pass via one of three areas: the kernel, the program of task code, or using an interrupt service routine (ISR).

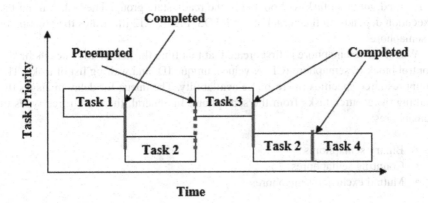

FIGURE 7.10   Preemptive scheduling

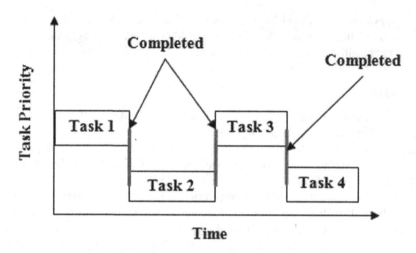

FIGURE 7.11   Non-preemptive scheduling

OBJECTS IN RTOS KERNEL

An RTOS kernel offers various objects for communication and synchronization; they are semaphores, event registers, and message queues.

1. Semaphores

For the purpose of synchronization, an object in a kernel where the execution of one more process can be acquired or released is known as a semaphore. It permits a task to perform some operation or resource allocation. Each semaphore may contain several keys. If the process is able to capture the semaphore, then it will progress the resource access. If tasks are selected to wait for semaphore release, then the task may be blocked. The tasks that are blocked are kept in a task waiting list. One task from the task waiting list acquires the access to the semaphore when the semaphore is released, and that task will pooled in the ready state group. The task waiting list execution depends on the kernel of the RTOS. Figure 7.12 illustrates the creation of a semaphore.

Whenever a semaphore is first created, at that time the kernel allocates its linked control block of semaphore (CBS), values, unique ID, and waiting list of tasks. The count resource specifies the resource availability. The queue for delay manages the waiting of resource tasks from the semaphore. In general, there are three types of semaphores:

- Binary semaphores
- Counting semaphores
- Mutual exclusion semaphores

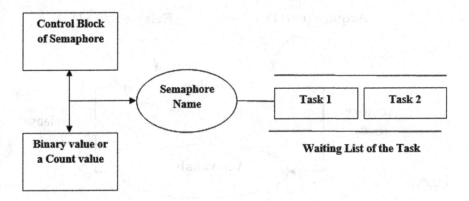

FIGURE 7.12   Creation of a semaphore

In a binary semaphore the value of the semaphore is either available (0) or unavailable (1).A binary semaphore is a shared resource. Any task can release it. Figure 7.13 describes binary semaphores.

In a counting semaphore, the semaphore value is 0 or higher, proving it can be obtained or released a multiple number of times. A counting semaphore is shown in Figure 7.14. The count can be bounded (here initial value is given with the maximum number) or unbounded (here the value is not maximum). It is a shared resource and it can be released by any task.

In a mutual exclusion semaphore, particular binary, a semaphore which helps the ownership and frequent recursive access is known as the mutex. A locked mutex can be released when the tasks that are responsible for the lock intend to publish it. The free state creates the mutex. The value of the semaphore is 0 or 1, but the count for the lock can be 0 or higher for locking recursively. Mutual exclusion semaphores are shown in Figure 7.15.

2.   Message queues

A buffer-like object where the tasks and ISR send and receive messages for the data synchronization and communication is commonly known as message queues. It

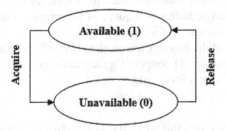

FIGURE 7.13   Binary semaphores

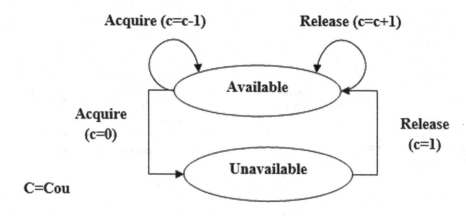

FIGURE 7.14   Counting Semaphores

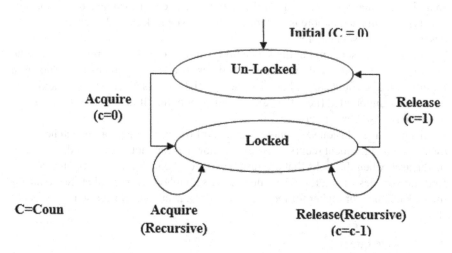

FIGURE 7.15   Mutual exclusion semaphores

contains the sender's short messages until the messages are read by the receiver when it is ready. A single buffer of a queue of messages is known as the mailbox. Figure 7.16 shows the creation of message queues.

After the creation of the message queue, the kernel allocates a control block of the queue, name of the queue, ID, length of a maximum message, buffers of memory, and the length of the task waiting lists.

3.   Event registers

Objects of events are employed when the synchronization task is needed without sharing resources. For an occurrence of a particular event, they are allowed to keep waiting for one or more tasks. The activity of an object can be of one of two states: non-triggered or triggered.

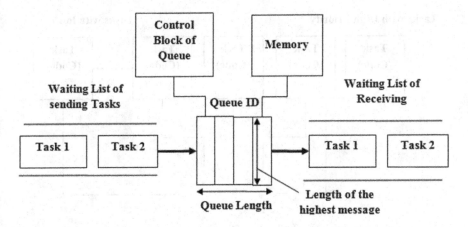

FIGURE 7.16 Message queue creation

If the event is in a non-triggered state, waiting tasks will be needed to be in a suspended state. On the contrary, if the event is in a triggered state, that indicates the resuming of tasks that are waiting.

RTOS Services

The kernel is the heart of every operating system. The kernel undertakes and allocates the resources. Since tasks cannot acquire the attention of the CPU every time, a kernel offers some more services, which include:

Time Management

Tasks of the system and users are regularly scheduled to execute after a specified time in the embedded systems. There is a need for an interrupt in a particular time limit to keep tracking the delays. Nowadays, the majority of RTOSs provide both "absolute timers" which operate with the date of the calendar and time and "relative timers" which operate in units of ticks. For every timer type, RTOSs offer a service of "task delay," and also "task alert" depending on the signaling mechanism.

Interrupt and Event Handling

Hardware mechanisms are effectively utilized to notify the CPU that asynchronous actions happened as an interrupt. This is illustrated in Figure 7.17. A primary challenge in designing and RTOS is to support interrupts and in that way allowing asynchronously to the data structures internal RTOS.

The mechanism of the event and interrupt handling in an RTOS offers the following functions:

- Creation and deletion of interrupt service routine (ISR)
- Enabling and disabling of interrupt
- ISR state referencing
- Defining interrupt handler

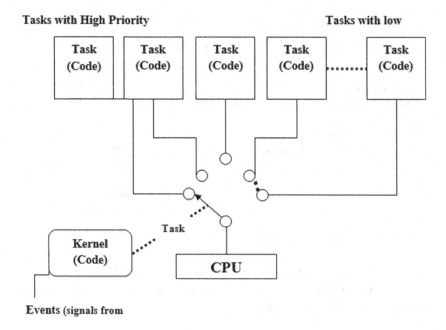

FIGURE 7.17   Interrupt and event handling

And also helps to ensure:

- Ensure the integrity of data by limiting interrupts by modifying a structure of data
- That latencies of interrupts will be minimal when RTOS is performing critical operations because of disabling interrupts
- That preemptive performance of an RTOS can be differentiated with the fastest possible responses to an interrupt

Device Management

The kernel of an RTOS is repeatedly ready with a service that offers control and management of input or output for a device with a robust framework. This service is provided by the application programming interface (API). Though, most APIs and supervisors of the device driver are "standard" only within a particular RTOS.

Memory Management

Generally, the RTOS in embedded systems struggles to attain small task that comprises only the functions needed for the user's applications. An RTOS contains two types of memory management. They are as follows

- Memory management by stack
- Management of heap memory

In an RTOS with the capability of multitasking, all tasks need to be allocated for their contexts storing. This memory allocation can be completed by employing the task-control block model. Generally, this type of memory is known as a kernel stack, and the process of managing or controlling it is known as the management of stack memory.

Generally, after the program initialization, the microcontroller or microprocessor's physical memory will be taken with the code of application, data, and stack of the system. Physical memories that are remaining are called heaps. The kernel can allocate dynamic memory for data required by different tasks through this heap memory. The mind is divided into blocks with a fixed size, which the tasks can request. The process of managing and controlling the memory of heaps is known as the management of heap memory.

Inter-Task Communication

Synchronization of the task and inter-task communication allows the information to be transmitted safely from one job to another job. Sharing memory areas and transmitting data shared among tasks is known as inter-task communication. Possible executing mechanisms for inter-task communications are:

- Queues of message
- Remote procedural calls (RPC)
- Pipes

A queue of the message objects is employed for intertask communication. The process sending or receiving messages of the tasks is also in a memory which is shared for all tasks. The kernel provides the services to the queue where the sent and received messages of the tasks and ISR are placed. A task was looking for a message from the empty queue then it is blocked for a particular period. The messages that are sending and receiving can in the form of,

- First in first out (FIFO)
- Sequence based priority
- Last in first out (LIFO)

Generally, a queue of the messages consists of a coupled unique ID, queue control block (QCB), name, length of the queue, buffers of memory, waiting lists of one or more task, and duration of the maximum message. The mailbox is generally defined as a queue of messages with a length of 1. Components of remote procedure call (RPC) allow circulated computing where tasks on a remote computer can be raised by the implementation of another task if the task works on a similar machine.

The single communication channel can be provided by a pipe which is employed for the exchange of unstructured data over the tasks. It can be read from and written to, opened and closed. Conventionally, it facilitates the transfer of data over uni-direction. Two descriptors are present at each pipe end for reading and writing. Written data to the pipe is as a byte stream of unstructured through one of the descriptors and it can be examined by FIFO.

Highlights of RTOS

- Options for preemptive scheduling
- Time slicing with round robin
- The passing of the message will become easy
- Option for cooperative scheduling
- The footprint of 6K – 12K
- Priority inheritance by mutexes
- Easy configurable/ scalable
- Counting and binary semaphores.

CONCLUSION

Increasing human death rates due to heart disease is one of the major problems facing all countries. Various factors such as smoking, unhealthy diet, high blood pressure and cholesterol, lack of exercise, and being overweight are major reasons for heart disease. Research and advancements in the medical industry are finding a solution to this problem. RPMSs for earlier detection of heart related diseases can be helpful to reduce the death rates. RPMSs need to have the following features. The patient must be continuously and securely monitored by this device and so it should be wearable. The sensors and other devices should consume low power for its operation, and the battery lifetime must be sufficient.

Live video of patients can be streamed through the internet to improve the analysis and RTOS can be implemented with an SJF scheduling algorithm to optimize the code for real-time systems.

REFERENCES

Diego, A. García-Tadeo, Dattatreya, Reddy Peram, Kumar, K. Suresh, Luis, Vives, Trishu, Sharma and Geetha, Manoharan. 2022, 'Comparing the Impact of Internet of Things and Cloud Computing on Organisational Behaviour: A Survey', *Materials today: Proceedings*, vol. 51, Part 8, pp 2281–2285. https://doi.org/10.1016/j.matpr.2021.11.399.

Dietrich, C., Hoffmann, M. and Lohmann, D. 2017, 'Global Optimization of Fixed-Priority Real-Time Systems by RTOS-Aware Control-Flow Analysis', *ACM Transactions on Embedded Computing Systems*, vol. 16, no. 2, pp. 1–25.

Indersain, Neetika Sharma and Dushyant, Singh. 2013, 'Design and Implementation of μc/Os II Based Embedded System Using ARM Controller', *International Journal of Engineering and Technical Research*, vol. 1, no. 2, pp. 1–4.

Jae Hwan, Koh and Byoung, Wook Choi. 2013, 'Real-time Performance of Real-time Mechanisms for RTAI and Xenomai in Various Running Conditions', *International Journal of Control and Automation*, vol. 6, no. 1, pp. 235–246.

Jian, Feng and Hongmei, Jin. 2011, 'μC/OS-II Port for STM32F107VC Processor', *Information Engineering Letters*, vol. 1, no. 1, pp. 1–7.

Jyotirmay, Patel and Solanki, A.K. 2011, CPU Scheduling: A Comparative Study, Proceedings of the 5th National Conference on Computing for Nation Development, INDIACom, March 10–11, 2011: New Delhi, India, pp. 10–11.

Mehdi, Neshat, Mehdi, Sargolzaei, Adel, Najaran and Ali, Adeli. 2012, 'The New Method of Adaptive CPU Scheduling Using Fonseca and Fleming's Genetic Algorithm', *Journal of Theoretical and Applied Information Technology*, vol. 37, no. 1, pp. 1–16.

Padmaja, C., Swathi, N., Sudarshan, E., Srinivas, V. and Anuradha, P. 2022, 'IOT Based Solar Powered Smart Irrigation System', *AIP Conference Proceedings*, vol. 30001, p. 2418.

Reddy, C.V.K., Reddy, I.R., Ahmed, S.Z., Rahul, V., Kumar, K.A., Kumar, M.S., Suman, S. and Reddy, G.V. 2022, 'The Effective Monitoring and Utilization of Parking Area Using IOT Based Technology', *AIP Conference Proceedings*, vol. 40018, p. 2418.

Reddy, S.T., Rajesh, M., Raju, K.G. and Prashanth, B. 2022, 'A Study of Forensic Techniques Used for Bot Nets in IoT', *AIP Conference Proceedings*, vol. 20065, p. 2418.

Sandeep, C.H., Pradeep, C.H., Mendu, M., Thirupathi, V. and Chandhar, K. 2022, 'The Ecosystem of IoT Security Considerations and Generic Framework of IoT', *AIP Conference Proceedings*, vol. 20068, p. 2418.

Sandeep, C.H., Pradeep, C.H., Mendu, M., Thirupathi, V., Pasha, S.N. and Kishan, P.A. 2022, 'Analysis on Various Kinds of Attacks and Appropriate Feasible Solutions Towards IoT', *AIP Conference Proceedings*, vol. 20067, p. 2418.

Singh, A., Goyal, P. and Batra, S. 2010, 'An Optimized Round Robin Scheduling Algorithm for CPU Scheduling', *International Journal on Computer Science and Engineering*, vol. 2, no. 7, pp. 2383–2385.

Srikanth, K. and Samunuri, N. 2013, 'RTOS Based Priority Dynamic Scheduling for Power Applications through DMA Peripherals', *International Journal of Engineering Trends and Technology*, vol. 4, no. 8, pp. 3660–3664.

Sujitha, M. and Kannan, V. 2015, 'RTOS Implementation of Non Linear System Using Multitasking, Scheduling and Critical Section', *Journal of Computer Science*, vol. 10, no. 11, pp. 2349–2357.

Tang, P.L. and De Silva, C.W. 2006, 'Compensation for Transmission Delays in an Ethernet-based Control Network using Variable-Horizon Predictive Control', *IEEE Transactions on Control Systems Technology*, vol. 14, no. 4, pp. 707–718.

Taqwa, Flayyih Hasan. 2014, 'CPU Scheduling Visualization', *Diyala Journal of Engineering Sciences*, vol. 7, no. 1, pp. 16–29.

Thavamani, S., Dandugudum Mahesh, Sinthuja U. and Geetha, Manoharan. 2022, 'Crucial Attacks in Internet of Things via Artificial Intelligence Techniques: The Security Survey', *AIP Conference Proceedings*, vol. 020029, p. 2418.

Vijayakumaran, C. and Janaky, K. 2014, 'An Efficient Multilevel Priority Packet Scheduling for Wireless Sensor Network', *International Journal of Engineering and Computer Science*, vol. 3, no. 5, pp. 5930–5936.

Yu, Anxin Lin, Rui, Kong, Rongbin, She and Shugao, Deng. 2013, 'Design and Implementation of Remote/Short-range Smart Home Monitoring System Based on ZigBee and STM32', *Research Journal of Applied Sciences, Engineering and Technology*, vol. 5, no. 9, pp. 2792–2798.

Yukesh, B. (2016) 'Design of a multi-sensor inertial data acquisition system for patient health monitoring with real time operating system', *International Conference on Advanced Communication Control and Computing Technologies (ICACCCT)*,IEEE, Ramanathapuram, India, 2016, pp. 685–689, doi: 10.1109/ICACCCT.2016.7831727.

Yysookhak, M., Yu, F. R., Khan, M. K., Xiang, Y., & Buyya, R. (2017). 'Attribute-based data access control in mobile cloud computing: Taxonomy and open issues', *Future Generation Computer Systems*, vol. 72, pp. 273–287.

8 Evaluation of Artificial Intelligence Techniques in Disease Diagnosis and Prediction

Harishchander Anandaram,
Neeraj Kumar Mishra, and M. S. Nidhya

INTRODUCTION

The outstanding development of new computer based advances in computation is relentless. Human error can be diminished, clinical outcomes can be improved, and information can be tracked throughout time, all thanks to digital healthcare. Many illnesses, particularly those whose diagnosis relies on imaging or signaling analyses, are being predicted and diagnosed with the use of artificial intelligence (AI) technologies, such as machine learning (ML) and deep learning (DL) algorithms [1, 2]. Artificial intelligence may also be used to identify geographic or environmental hotspots where dangerous behaviors or diseases are common [3, 4]. Because of the modern calculations that consider the programmed extraction of enhanced characteristics [5, 6], ML approaches have found great success in medical picture analysis. Learning is at the heart of ML, and there are three main types of learning methods: unsupervised (association, clustering, and dimensionality), supervised (classification, regression, and composition), and reinforced learning [7]. (Figure 8.1).

ML algorithms conduct a wide variety of calculations and actions on data inputs. Preprocessing data is the first and most important stage in ensuring more accurate forecasts and faster processing times, all of which contribute to better overall data quality. Important characteristics are retrieved and applied in accordance with the chosen ML or DL model for picture classification once the data has been pre-processed. Dimensionality reduction and algorithm performance improvement also helps the computation [8]. In order to arrive at sound judgments and reliable classifications or predictions in the final stage, Figure 8.2, data processing is used for model training and parameter modification depending on the selected method.

With the help of ML, computers can already carry out procedures normally performed by doctors. DL is a popular branch of ML that is used extensively for medical image identification. DL is a strategy for constructing the ML algorithm that

DOI: 10.1201/9781032686714-8

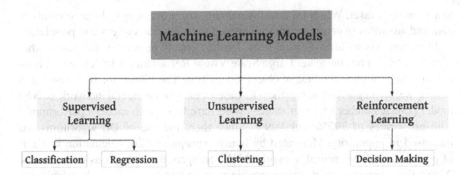

FIGURE 8.1 Algorithms and models used in machine learning

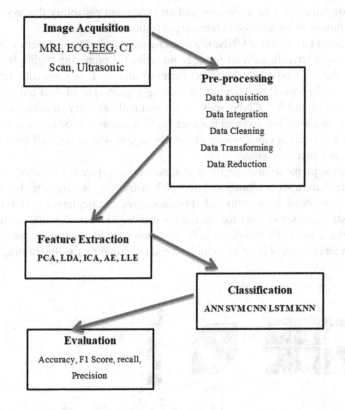

FIGURE 8.2 Medical image categorization involves many stages

involves stacking basic ideas to create a deep structure with many layers of process-ing power. To rephrase, DL is the evolution of ML for dealing with large datasets [9–11]. It is an automated approach to pattern building and extraction for classifica-tion that lets a computer learn which characteristics are most important via exposure to a training dataset. While DL itself is not new, the processing of huge amounts of data and advances in computer power have led to its current success and popularity.

In various visual identification tests, DL has greatly beaten earlier state-of-the-art algorithms. The ImageNet Large Scale Visual Recognition Challenge [12] is an annual competition in the field of object recognition. The 1.3 million high-resolution photos were categorized using the DL-based convolution neural network (CNN) model [13]. A number of better models were introduced with model's performance with error rates of 39.7% and 18.9%. Since then, the use of DL algorithms has increased in popularity. Motivated by human synapses, a DL calculation is a sort of profound artificial neural network (ANN) comprised of numerous more modest ANNs that cooperate to do more troublesome undertakings. In each calculation, these units are coordinated into layers and called nerve cells, where information signals are incorporated and sent on to different cells. In the engineered assortment, they are traded out for an aggregate and an enactment capability that cooperate to show the human mind's unpredictable organization of associations.

A managed DL model, a CNN, is a demonstrated technique for picture investigation and characterization. With fewer boundaries for preparing highlights through backpropagation, CNNs are able to get better results since they use fully linked layers with standard weights. They are made to get geographical data out of pictures. As shown in Figure 8.3, CNNs seek to automatically identify images, extract their properties, and learn hierarchical features via adaptation. The principal advantage of this approach is that it can advance extremely unique highlights with little information and basic preprocessing.

During the preparation stage, the neural network is typically initialized randomly, and the data are also randomly ordered. When training is complete, however, the neural network behaves as predicted. It includes precise calculations, yet its deep and intricate structure sometimes baffles human minds. As a result, while attempting to describe the actions of networks, it is fairly uncommon for the training process to be left out. A trained network often exhibits the same properties as the training data set.

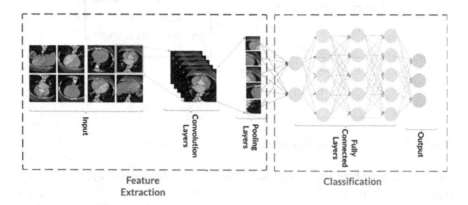

FIGURE 8.3 DL model (CNN) for classifying brain diseases based on CT scan pictures of the brain

The healthcare industry processes a significant volume of data, making it fertile ground for the creation of effective calculations that can be utilized using DL or ML techniques. Medical assessments may employ a wide variety of data types, but photographs are especially useful because of the visual nature of the information they provide. Ultrasound, X-rays, CT scans, MRIs, microscopy, and scintigraphy are only a few of the additional methods used to make medical pictures. Images produced by these methods may vary, however, any picture can be automatically analyzed by AI algorithms for the purpose of illness research and prognosis. Examining the implementation of numerous algorithms is crucial for comprehending how each AI-based model aids in illness diagnosis and prediction [14, 15].

Many studies have been undertaken in many different medical domains to diagnose or forecast different illnesses [16–19], and ML and DL models have recently developed a cutting-edge approach to new techniques in computer-based diagnosis. More precise diagnoses and more productivity have resulted from the use of these techniques [16, 17]. In this study, we take a look at how different illnesses may be diagnosed using ML and DL models.

RELATED WORKS

Recent breakthroughs in AI have garnered the interest of many industries and disciplines of study. The method that is currently most effective is DL; an assortment of techniques and calculations permit PCs to consequently track down confounded designs in massive datasets. Increased data availability ("Big Data"), more intuitive software frameworks, and a proliferation of available processing power have all contributed to these developments. The superior performance of neural networks over other approaches in numerous high-resolution image analysis criteria led to DL's rise to prominence in the image processing field.

In 2012, the second-most noteworthy error rate in picture order work was diminished by half utilizing a CNN model, as estimated by the ImageNet Large Scale Visual Recognition Challenge (ILSVRC) [12]. Until then, it was believed that computers would have a hard time identifying things in photographs of real life. To the point where the ILSVRC classification is practically a solved problem, CNNs have even eclipsed human performance. For many computer vision issues, DL methods are now universally accepted as the gold standard. DL methods have been proposed for application in the diagnosis of acute human illnesses in a number of studies.

Multiple scenarios using models from ML and DL have been utilized by researchers to forecast life-threatening diseases such as liver infection, coronary illness, Alzheimer's disease (AD), and many forms of cancer [20–22]. Some studies [23, 24] have used DL methods for pediatric chest X-rays to identify and discriminate bacterial pneumonia. Extensive work has also been done to characterize the unique aspects of chest computed tomography (CT) imaging for various disorders [25, 26]. Case-based reasoning-based hybrid models were developed for the diagnosis of skin disorders in research [27, 28]. The model's result as an application might be used to identify a variety of skin disorders and provide effective treatment options. It is common practice in the medical field to propose individualized real-time monitoring systems using ANN methods in order to collect crucial data about the body. Patients in critical care situations may benefit greatly from the use of this technology [29]. Scientists in [30] used ANN algorithms to predict the onset of diabetes with 91% accuracy.

Automatic data analysis for illness diagnosis and prognosis is made possible via the use of ML classification algorithms. For vital signs like heart rate and blood pressure, scientists built a custom web-based healthcare monitoring system in real time. Patients in critical care settings may benefit greatly from this [31].

Treatment processes and healthcare technology may benefit from the reconciliation of AI and Internet of Things (IoT) methods in the medical care system. In [32], we suggested the body sensor network (BSN) as a dependable IoT-based systems for healthcare monitoring of human behaviors and the surrounding environment using ML algorithms. Another research [33] proposed a hybrid IoT model for predicting type 2 diabetes based on data from a medical service checking framework and the random forest approach. Likewise, the random forest classifier, which beat past calculations [34], was utilized to evaluate people's chance of type 2 diabetes in light of their lifestyle information. Utilizing the random forest classifier, a versatile based stage was made for continuous identification of tuberculosis illness (TB) antigen-explicit antibodies [35]. An AI system was suggested in a study [14] that classified numerous gastrointestinal disorders using recurrent neural networks (RNN) and long short-term memory (LSTM) networks, with an accuracy of 97.057%.

Strokes and cardiovascular disease may be drastically reduced if hypertension is managed and people are made aware of the risk they are at. Digital healthcare technology and AI were evaluated, and a privacy protection framework for storing peoples' data was proposed [36]. Additionally, several researchers have conducted numerous concentrates on sickness expectations to distinguish and recognise them in their beginning phases. In order to diagnose illnesses at their earliest stages with a precision of 99.50% and an accuracy of 100%, a unique hybrid ML model based on the IoT was presented [37]. Other authors have developed a method for using many factors as predictors of cardiovascular illness. They used a random forest classifier with a mixture model and improved accuracy to 88.7% [38].

An ML method called XGBoost was suggested in a study on the detection of positive urine cultures [39]. The accuracy of this model, which varied from 0.826% to 0.904%, was higher than that of previously produced models. The CNN model was also utilized to extract features from pictures of malaria-infected red blood cells in a separate research [40]. Malaria infection was similarly predicted in another ML-based study [41]. To detect chronic obstructive pulmonary disease (COPD) flare-ups, scientists used a variety of ML techniques. These included logistic regression, support vector machines, naive Bayes, K-nearest neighbor, and random forest They found that the support vector machine (SVM) model performed best [42]. Utilizing information from the 2015–2017 Public Study on Medication Use and Wellbeing, researchers anticipated young narcotic fixation utilizing three unique ML calculations: ANNs, distributed random forests (DRFs), and slope support. Area under the receiver operating characteristic (AUROC) curve scores for prediction performance fall between 0.809 and 0.815 [43]. Others have employed a variety of ML methods to propose a model for distinguishing Coronavirus from an X-beam picture dataset, with results as high as 98.91% accuracy [44]; these include the CNN, random forest (RF), SVM, and boosted decision tree (AdaBoost) classifiers. The utilization of ML and DL strategies makes it possible to gauge an individual's stress levels. Stress may be recognized in various ways, one of which is by means of the utilization of physiological signs. In one study, for instance, researchers looked at several ML models for stress based on heart rate variability (HRV) [45]. ML random forest was shown to be superior to other approaches in this investigation. Others have attempted to foretell the onset of diabetes using a variety of ML models. Logistic regression and support vector machines both did well in their experiments [46]. KNN, SVM, ANN, decision tree, logistic regression, naive Bayes, random forest, and XGBoost were only some of the ML models utilized in a thorough examination to foresee the risks of ongoing type 2 diabetes [47]. The random forest model had the highest overall accuracy (0.91) in this investigation. An improved analysis and separation performance for image-based illnesses has recently been offered using an expanded DL model called 3DCellSeg. Lightweight deep CNNs and a single hyperparameter make this DL technique particularly appealing [48].

AI IN DISEASE DIAGNOSIS

Artificial intelligence encompasses a wide range of topics from mathematics to physics. A machine's "intelligence" would be categorized under AI if it could accomplish anything intelligent automatically [49]. AI systems are trained using data that represents populations [50, 51]. Neural networks and DL are the cornerstones of ML, which is one of the most promising areas of AI. (Figure 8.4).

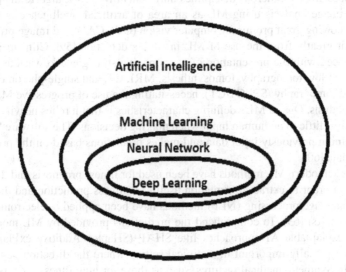

FIGURE 8.4 Connections among various AI, ML, NN, and DL

The objective of ML is to create a self-learning system that can improve with new data, practice and user feedback [52]. In order to do a given job as accurately as possible, the ML algorithm adjusts its parameters depending on the data it receives from the feedback it receives. The end objective is to have it perform as well on previously unknown data sets as it does on the known ones [53].

The use of imaging as a diagnostic tool in medicine has become commonplace. It still has to rely on human interpretation and faces growing difficulties with available resources. Automatic medical imaging diagnosis using AI, particularly in the area of DL, has efficiently addressed issues of human mistake brought on by inaccuracy or a lack of knowledge. Computer-aided diagnosis (CAD), picture disease segmentation, and illness classification based on images all rely heavily on AI. Medical imaging diagnostic activities must be taught via training since realistic simulation of tissues and organ pictures is not possible using simple equations.

AI is the study of how machines learn from data such as images and patterns to solve problems like those faced by humans. It takes constant monitoring and examination of data to detect diseases and stop their spread. Rapid response based on reliable information may work for the personal satisfaction of individuals everywhere [51]. Surgical robots can now be more precisely and efficiently programmed thanks to advancements in AI's application in the healthcare industry [54]. The utilization of AI (artificial intelligence) in medical care has affected information assortment, examination, and advancement [55].

IMAGE-BASED ILLNESS DETECTION USING MACHINE LEARNING

ML is a collection of strategies for automatically recognizing patterns in data, which may subsequently be used to create predictions about new data or judgments under uncertainty. Numerous disciplines may benefit from ML algorithms [56–58]. Medical image analysis using ML as an area of artificial intelligence is expanding and showing great promise. Computer vision (CV), CAD, and image processing all benefit greatly from the use of ML in illness detection [59]. Clinical imaging has advanced with the presentation of new imaging strategies like various cut CT, positron outflow tomography, tomosynthesis, MRI, CT, and single-photon emission computed tomography (SPECT/CT), necessitating the use of progressive ML methods for analysis. One of ML's defining characteristics is that it relies heavily on data and requires little to no human input when making decisions. The software "learns" by examining previously input data and makes predictions based on the previously entered data [60].

Multiple modern ML methods have been used for illness prognosis and diagnosis [61, 62]. In order to extract information useful for illness prediction and diagnosis, natural language processing (NLP) methods have been applied to electronic health records (EHRs) [63]. To comprehend the predictions provided by ML models, we propose explainable AI approaches like SHAP (SHapley Additive exPlanations), which is especially important in a clinical setting where the direction ought to be clear [64]. Synthetic medical pictures, such as those for lung illness, may be generated with the use of generative models like GANs (generative adversarial networks) to improve upon the accuracy of previously collected data [65]. All of these methods are complementary, so you may use as many as you want to get the best results from your model. The choice of method is context- and problem-dependent [66].

IMAGE-BASED ILLNESS DIAGNOSIS USING DEEP LEARNING

Because of its ability to automatically learn a wide variety of characteristics and patterns, DL ranks among the most powerful technologies available today. DL has empowered the improvement of expectation models for use in the precocious detection of illness. DL algorithms outperform conventional ML approaches on account of their superior accuracy, automated feature extraction, and enormous data processing as researchers increasingly rely on tried-and-true pattern analysis techniques. DL algorithms clearly outperform ML ones when it comes to big data. In addition, DL is recommended above other methods for dealing with pictures since its predictive ability is typically superior to that of people [67]. Because medical diagnosis is so heavily dependent on collecting relevant information from pictures, DL has been given particularly clear expression in this area of image processing. The most regularly involved DL calculations in clinical picture-based diagnoses are CNNs, deep neural networks (DNNs), deep belief networks (DBNs), deep automatic encoders (DAEs), deep Boltzmann machines (DBMs), deep convolutional extreme learning machines (DC-ELMs), recurrent neural networks (RNNs), and their variants (bidirectional long short-term memory (BLSTM) networks [68]. Additionally, region aggregation graph convolutional networks (RAGCN) use graph convolutional networks (GCNs) to gather data from different regions of a picture and are, therefore, a DL tool for analyzing medical data. This method was developed for medical pictures like CT and MRI scans, which often include many returns on investment (ROIs) that need to be analyzed independently. To do this, Region Aggregation Graph Convolutional Network (RAGCN) first uses a graph-based technique to divide the picture into sections, from which GCNs may then extract characteristics and generate predictions. An automated technique for estimating bone age using CNNs and GCNs was reported by the authors in [69]. They used CNNs for highlight extraction and GCNs for deduction of bone significant regions. By combining these two organizations, they had the option to make another GCN (RAGCN) that can explore regional characteristics when determining a person's bone age.

Another DL medical data technique, the Lesion Attention Pyramid Network (LAPNet), seeks to identify and categorize lesions in visual diagnostics. To do this, LAPNet employs a pyramidal structure to extract information from the picture at several sizes. Additionally, it makes use of a consideration component to focus on regions of the image that are likely to contain sores; this method was utilized to categorize diabetic retinopathy by the authors of "Lesion-attention pyramid network for diabetic retinopathy grading" [70]. They used a large collection of medical photos to teach a LAPNet how to identify lesion areas.

These are only a few of the many DL methods now in use for illness prognosis and diagnosis. Keep in mind that DL is an ever-evolving area whereby new approaches are continually being created or merged.

DISCUSSIONS

More and more, doctors are turning to AI tools to help them diagnose and treat serious illnesses [71, 72]. With AI's ability to productively engage with medical picture data, it is finding more and more applications in the fields of physical location of diseases and expectations of treatment [73].

To really carry out AI strategies in the medical services framework to improve disease prediction, illness characterization, dynamic cycle exercises, clinical as well as pathological data is required. Also for giving ideal therapy decisions, and assisting individuals with living better and longer, = enormous information is evaluated that is gathered from clinical records or wearable gadgets. Quicker medical analysis and diagnoses are made possible with the use of AI [74, 75]. For instance, this technique may be used to spot potentially malignant tumors in medical photographs, enabling pathologists to make an early diagnosis and begin treatment without having to send tissue or lesion samples to the lab for extensive testing [76]. When it comes to finding people with uncommon or unencoded illnesses, AI-based algorithms are invaluable. Therefore, there is a great deal of potential for early patient diagnosis using AI models for illness diagnosis [76].

There has been a remarkable expansion in the utilization of ML and DL in cardiac illness diagnosis. In cardiology, for example, DL may be employed precisely and advancedly in the analysis and assessment of data thanks to the accessibility of an expansive assortment of clinical imaging innovations including CT, ECG, and echocardiography [77–79]. Atherosclerotic coronary illness is a main cause of death and handicap in the Western world. An excellent effect of therapy is achieved with early detection of this condition. This generation has seen great advances in the recognizable proof of coronary atherosclerotic coronary illness due to ML and DL [80]. CT-fractional flow reserve (CT-FFR) based on ML, for example, may streamline diagnostic processes and cut down on wait times, making it a potent device for anticipating serious antagonistic cardiac events [81, 82]. Additionally, DL-based CT-FFR may minimize computational complexity, shorten runtime, and improve prediction [83, 84]. Early identification of different cardiac problems in 170 patients using SVM and ANN approaches has been reported by the research community [85]. Using SVM and ANN models, they looked at arrhythmia, cardiomyopathy, coronary heart disease (CHD), and CAD. The SVM method achieved an accuracy of 89.1% for arrhythmias, 80.2% for cardiomyopathies, 83.1% for CHD, and 71.2% for CAD. Similarly, the ANN algorithm achieved an accuracy of 85.8% for arrhythmia, 85.6% for cardiomyopathy, 72.7% for CHD, and 69.6% for CAD. Predicting CHD using the South African Heart Disease dataset of 462 examples was the focus of a different research [86]. They used decision tree, support vector machine, and naive Bayes as supervised learning methods to increase the accuracy of CHD diagnoses and predictions. For cardiovascular disorders, they found a library data accuracy of 83.9%, whereas for diabetes, it was 95.7%. Another research comparing ML algorithms for CHD prediction found that the SVM classifier outperformed the competition with a 95% success rate [87]. The prediction ability of SVM, ANN, and decision tree calculations for coronary illness in 502 examples was also studied by other researchers

[88]. SVM's 92.1% accuracy was the highest of the three algorithms tested, easily beating out the other two.

The widespread utilization of AI in medicine has prompted the successful diagnosis and forecasting of neurological disorders. Alzheimer's disease, Parkinson's disease (PD), and brain tumors are only a few of the neurodegenerative disorders that have benefited greatly from the recent ML and DL techniques [89]. By processing and analyzing vast amounts of brain signals and data, AI has made it feasible to uncover insights and connections that are not quickly obvious to the natural eye. DL-based CNN models are the most used method for illness identification currently [90, 91]. Recent research looked at already-trained models for Alzheimer's diagnosis [92]. The EfficientNetB0 model fared best in this evaluation, with an accuracy of 92.98%. Recently, a number of artificial intelligence algorithms were utilized in tandem to detect Parkinson's disease in its earliest stages [93]. The greatest performance in this area in recent studies was achieved by joining the hereditary calculation with random forests, with a precision of 95.58%.

AI's sophisticated algorithms are also very helpful in making accurate early-stage diagnoses and prognoses of breast cancer. Millions of women worldwide lose their lives every year to breast cancer [94]. However, early detection is crucial for treating and managing the disease. Researchers interested in using ML techniques to identify breast cancer often utilize the Wisconsin Breast Cancer Dataset (WBCD). The WBCD was really applied to the breast malignant growth analysis process utilizing the least-squares support vector machine (LSSVM) algorithm, with a 98.53% classification accuracy being reached [95]. Another method for classification accuracy of 99.14% was presented using the WBCD and a hybrid fuzzy-fake framework with a k-nearest neighbor calculation [96]. Involving the WBCD for breast cancer finding and an SVM calculation with highlight determination, scientists had the option to accomplish a 99.51% accuracy [97]. Utilizing the WBCD dataset for prognosis analysis, an optimized SVM was developed in another paper [98]. The two-stage evaluation of this strategy had an overall accuracy of 96.91%. To distinguish between benign and malignant breast tumors, another research [99] used SVM, k-nearest neighbors, and probabilistic neural networks (PNNs) as classifiers. In comparison to two standard datasets, they scored 98.8% and 96.3%accuracy, respectively.

In order to categorize and detect breast cancer, researchers have turned to fuzzy neural approaches, k-nearest neighbor analysis, quadratic classifiers, and combinations thereof [100]. This leads to an accuracy of 94.28% for the fuzzy neural approach, 96.42% for the KNN method, 94.50% for the quadratic classifier method, and 97.14% for the combination of the two. The association technique outperforms all other models in this investigation. For the purpose of breast cancer detection, a machine learning classifier (MLC) was trained using craniocaudal (CC) and/or mediolateral oblique (MLO) mammography image views to categorize segmented areas [101]. When CC and MLO characteristics were combined, the value of the AUROC curve went up to 0.996%. Besides, in another paper [102], the Soothsayer Breast Cancer dataset was separated into two classifications, carcinomas in situ and harmful potential, utilizing the C4.5 calculation. Accuracy of 94% was accomplished during preparation, while 93% was accomplished during testing. classifiers, for example, the decision tree (J48), multilayer perception (MLP), naive Bayes (NB), sequential minimum optimization (SMO), and Occasion Based K-nearest neighbor (IBK) were utilized to analyze and anticipate breast cancer in the Wisconsin Breast Cancer (WBC), Wisconsin Diagnostic Breast Cancer (WDBC), and Wisconsin Prognostic Breast Cancer (WPBC) data sets [103] as a feature of a far-reaching review. Ten times cross-approval was performed to work out the order precision and disarray framework. The WBC, WDBC, and WPBC datasets' accuracy increased to 97.2818%, 97.7153%, and 77.3196%, respectively, after including SMO, J48, NB, and IBK. In another work [104], researchers used the self-supervised (SEER) and University of California, Irvine (UCI) machine learning datasets to suggest various data mining algorithms for breast cancer diagnosis and prediction. The findings showed that the best predictor, a decision tree, obtained an accuracy of 93.62% across both datasets.

It has proven possible to employ ML techniques such as SVMs, random forests, and neural networks for the prediction and classification of genetic illnesses using varying quantities of genomic data. Finding biomarkers for complex genetic illnesses has proven difficult for scientists owing to the wide range of possible genotypes. Predictions of genetic illnesses may be improved by using AI, notably ML and DL approaches. In particular, throughout the training, testing, and validation stages, 85.7%, 84.9%, and 84.3% were obtained by the ANN-based model, respectively [105]. There was a wide range (48–95% accuracy) in ML performance in psychiatry based on genotypes [106]. There is a paucity of gold-standard models and there is a constraint in data sample size, however, AI-based automated predictors are necessary for genetic illnesses [105]. Finding genes or gene clusters that are overexpressed in tumor cells but not in normal cells is a significant difficulty in genetic microarray research [107]. Classifying microarray data from cancer patients has been greatly aided by AI-based approaches. In order to better categorize gene information, three regulated ML techniques were proposed: the C4.5 decision tree, the bagged choice tree, and the helped decision tree. Single decision trees were demonstrated to be second rate than group ML (packed away and developed decision trees) in this study [107]. Autism spectrum disorder (ASD) has been inspected by researchers, who have assessed and decided the best-performing classifier of ASD using datasets of babies, kids, teenagers, and grown-ups due to the condition's hereditary basis. Results showed that MLP outperformed other classification algorithms with a perfect score of 1.000 [108].

The field of dermatology may likewise benefit greatly from the use of AI tools. Data-driven training in ML and DL for skin disease diagnosis, prediction, and classification is amazing. However, the field of dermatological science has been slow to adopt and use these cutting-edge methods. Early detection is crucial for successful treatment of skin cancer. To identify skin cancer, CNN-based algorithms can now analyze photos from skin datasets. Due to inexperience or human error, experts are often inaccurate while trying to diagnose skin cancer. So, the development of AI-based automated systems may greatly improve their ability to identify skin illnesses, which will ultimately save lives and save costs for healthcare providers and their patients [109]. Researchers analyzed skin cancer lesions using two ML-based procedures: group learning and DL [110]. The DL method achieved better results than group learning here, with an accuracy of 91.85% in predictions and 90.1% in skin cancer classifications. Skin cancer has been diagnosed using a hybrid of Bayesian DL and active learning methods [111]. With 75% accuracy, this method outperformed all others in the International Skin Imaging Collaboration (ISIC) 2016 dataset.

Researchers have been able to study datasets more thoroughly and deliver more accurate findings in diagnosing prostate cancer thanks to the integration of digital pathology and AI. Radiotherapy is an effective treatment for prostate cancer, but many men have trouble recognizing the side effects of this treatment [112]. Predicting how a patient would respond to various forms of treatment is a challenging problem, but AI might help. AI-based solutions have also shown promising results in predicting prostate cancer, patient survival, and therapeutic response, and in identifying prostate abnormalities. In aid of prostate cancer early detection, scientists created a supervised AI-based model [113]. Classification accuracy reached 89% using MRI images annotated with histopathological data. Utilizing 3D multiparametric X-ray information given by the PROSTATEx dataset in the undertaking's preparation stage, the researchers of another study [114] suggested a unique DL technique called XmasNet for classifying prostate cancer lesions. XmasNet's AUC of 0.84 is superior to that of traditional ML approaches.

A variety of methods, mostly based on CT scan pictures but also including X-ray images, have been suggested for early recognition of cellular breakdown in the lungs. In spite of the fact that it is challenging to recognize at the beginning phase, it has been shown that doing so builds the endurance of people with cellular breakdown in the lungs [115]. CAD based on a DL-based framework for lung cancer detection was used by 41 out of 1972 teams in the Kaggle Data Science Bowl 2017 competition [116]. Researchers in lung cancer were able to distinguish between adenocarcinoma, squamous cell carcinoma, and small cell carcinoma by employing a deep convolutional neural network (DCNN). [117]. Using three-fold cross-validation, they assessed the likelihood of these three distinct malignancies and found an accuracy of 71%. This model has two fully connected layers in addition to three pooling layers and three convolutional layers. Using the Lung Image Database Consortium (LIDC-IDRI) database, another study examined three profound brain network models (CNN, DNN, and sparse autoencoder (SAE)) for diagnosing and ordering harmless and threatening lung nodes [118]. With a precision of 84%, CNN had the best consequences of the three networks. CNN, deep belief network (DBN), and stacked denoising autoencoder (SDAE) were the three DL-based strategies concentrated on top to bottom to determine lung nodes in CT images in a new exploration [119]. Area under the curve (AUC) execution ideal was accomplished through CNN, which was 0.899 0.018.

In order to effectively treat respiratory infections, early diagnosis and medication prescription are crucial. AI algorithms may aid medical professionals in identifying and understanding lung disorders. In order to identify COPD, a DL-based CNN model was proposed, which analyzed respiratory audio data with 93% accuracy [120]. In addition, a CNN-based framework model was presented for examining

X-ray images to detect Coronavirus. They were able to get an accuracy of 95.7% in their study [121].

CONCLUSION

The area of illness diagnosis and prediction stands to benefit greatly from DL and ML approaches. When it comes to treating sickness, nothing is more important than making an accurate and precise diagnosis. AI has shown amazing precision in distinguishing proof of picture-based ailments and in the prediction of treatment results with respect to endurance rate and treatment reaction. Computing power given by AI techniques must be instantaneous, dependable, and precise in order to perform the processing phases for the large amount of picture data. The outcome of a determination is subject to various variables, including the unwavering quality of the discoveries, the viability of the treatment, and the well-being of the patients. Information, calculations, profound processing techniques, neural networks, and developing methodologies are all part of what makes up AI, which is always growing to fulfill the demands of humans. The motivation behind this exploration is to inspect how well computer-based intelligence strategies can detect and forecast health problems. This study found that SVM performed the best of the available methods for predicting cardiac conditions. In particular, the application of supervised DL networks, such as CNN-based models, has shown substantial advances in the diagnosis of respiratory, lung, skin, and brain illnesses because of their high accuracy and fast picture identification. When diagnosing breast cancer, it is common practice to combine KNN with other networks like SVM to get optimal performance. Because of their promising experimental findings in identifying and categorizing medical pictures, DL and ML have a major bearing on the outcome of many of the illnesses included in this analysis. In other words, medical diagnostic and prognostic systems benefit from AI-based methodologies because of the way they make better use of available resources. The fast progress of AI technology also means that in the not-too-distant future, physicians won't have to struggle as much to make unbiased diagnoses of a wide range of ailments.

REFERENCES

1. Uysal G, Ozturk M. Hippocampal atrophy based Alzheimer's disease diagnosis via machine learning methods. *J Neurosci Methods.* 2020;33:108669. doi: 10.1016/j.jneumeth.2020.108669.
2. Woldaregay AZ, Årsand E, Walderhaug S, Albers D, Mamykina L, Botsis T, et al. Data-driven modeling and prediction of blood glucose dynamics: Machine learning applications in type 1 diabetes. *Artif Intell Med.* 2019;98:109. doi: 10.1016/j.artmed.2019.07.007.
3. Bhatt VK, Pal VK. An intelligent system for diagnosing thyroid disease in pregnant ladies through artificial neural network. International Conference on Advances in Engineering Science Management & Technology (ICAESMT), Uttaranchal University, Dehradun, India, 2019.

4. Yıldırım Ö, Pławiak P, Tan RS, Acharya UR. Arrhythmia detection using deep convolutional neural network with long duration ECG signals. *Comput Biol Med.* 2018;102:411. doi: 10.1016/j.compbiomed.2018.09.009.

5. Kong B, Wang X, Bai J, Lu Y, Gao F, Cao K, et al. Learning tree-structured representation for 3D coronary artery segmentation. *Comput Med Imaging Graph.* 2020;80(101688):2020.

6. Xia C, Li X, Wang X, Kong B, Chen Y, Yin Y, et al. *A multi-modality network for cardiomyopathy death risk prediction with CMR images and clinical information in lecture notes in computer science.* Cham: Springer; 2019.

7. Wu X, Liu X, Zhou Y. *Proceedings of 2021 Chinese intelligent systems conference: Review of unsupervised learning techniques in lecture notes in electrical engineering.* Singapore: Springer; 2022.

8. Dhal P, Azad C. A comprehensive survey on feature selection in the various fields of machine learning. *Appl Intell.* 2022;52(4):4543. doi: 10.1007/s10489-021-02550-9.

9. Xi X, Meng X, Yang L, Nie X, Yang G, Chen H, Fan X, Yin Y, Chen X. Automated segmentation of choroidal neovascularization in optical coherence tomography images using multi-scale convolutional neural networks with structure prior. *Multimed Syst.* 2019;25(2):95. doi: 10.1007/s00530-017-0582-5.

10. Rizk Y, Hajj N, Mitri N, Awad M. Deep belief networks and cortical algorithms: A comparative study for supervised classification. *App Comput Inform.* 2019;15(2):81. doi: 10.1016/j.aci.2018.01.004.

11. Bhatt C, Kumar I, Vijayakumar V, Singh KU, Kumar A. The state of the art of deep learning models in medical science and their challenges. *Multimedia Syst.* 2021;27(4):599. doi: 10.1007/s00530-020-00694-1.

12. Russakovsky O, Deng J, Su H, Krause J, Satheesh S, Ma S, et al. Imagenet large scale visual recognition challenge. *Int J Comput Vision.* 2015;115(3):211. doi: 10.1007/s11263-015-0816-y.

13. Krizhevsky A, Sutskever I, Hinton GE. Imagenet classification with deep convolutional neural networks. *Commun ACM.* 2017;60(6):84. doi: 10.1145/3065386.

14. Owais M, Arsalan M, Choi J, Mahmood T, Park KR. Artificial intelligence-based classification of multiple gastrointestinal diseases using endoscopy videos for clinical diagnosis. *J Clin Med.* 2019;8(7):986. doi: 10.3390/jcm8070986.

15. Nithya A, Appathurai A, Venkatadri N, Ramji DR, Palagan CA. Kidney disease detection and segmentation using artificial neural network and multi-kernel k-means clustering for ultrasound images. *Measurement.* 2020;149:106952. doi: 10.1016/j.measurement.2019.106952.

16. Abedi V, Khan A, Chaudhary D, Misra D, Avula V, Mathrawala D, et al. Using artificial intelligence for improving stroke diagnosis in emergency departments: A practical framework. *Therap Adv Neurol Disorders.* 2020;13:1756286420938962. doi: 10.1177/1756286420938962.

17. Gorris M, Hoogenboom SA, Wallace MB, van Hooft JE. Artificial intelligence for the management of pancreatic diseases. *Digestive Endoscopy.* 2021;33(2):231. doi: 10.1111/den.13875.

18. Sinagra E, Badalamenti M, Maida M, Spadaccini M, Maselli R, Rossi F, et al. Use of artificial intelligence in improving adenoma detection rate during colonoscopy: Might both endoscopists and pathologists be further helped. *World J Gastroenterol.* 2020;26(39):5911. doi: 10.3748/wjg.v26.i39.5911.

19. Xu L, Gao J, Wang Q, Yin J, Yu P, Bai B, et al. Computer-aided diagnosis systems in diagnosing malignant thyroid nodules on ultrasonography: A systematic review and meta-analysis. *European Thyroid J.* 2020;9(4):186. doi: 10.1159/000504390.

20. Bharti R, Khamparia A, Shabaz M, Dhiman G, Pande S, Singh P. Prediction of heart disease using a combination of machine learning and deep learning. *Comput Intell Neurosci.* 2021;2021:8387680. doi: 10.1155/2021/8387680.
21. Suresha PB, Wang Y, Xiao C, Glass L, Yuan Y, Clifford GD. *Explainable AI in healthcare and medicine.* Switzerland: Springer; 2021. A deep learning approach for classifying nonalcoholic steatohepatitis patients from nonalcoholic fatty liver disease patients using electronic medical records.
22. Goenka N, Tiwari S. Deep learning for Alzheimer prediction using brain biomarkers. *Artif Intell Rev.* 2021;54(7):4827. doi: 10.1007/s10462-021-10016-0.
23. Ramani V, Shendure J. Smash and DASH with Cas9. *Genome Biol.* 2016;17:42. doi: 10.1186/s13059-016-0905-4.
24. Rajaraman S, Candemir S, Kim I, Thoma G, Antani S. Visualization and interpretation of convolutional neural network predictions in detecting pneumonia in pediatric chest radiographs. *Appl Sci.* 2018;8(10):1715. doi: 10.3390/app8101715.
25. Depeursinge A, Chin AS, Leung AN, Terrone D, Bristow M, Rosen G, et al. Automated classification of usual interstitial pneumonia using regional volumetric texture analysis in high-resolution computed tomography. *Invest Radiol.* 2015;50(4):261. doi: 10.1097/RLI.0000000000000127.
26. Anthimopoulos M, Christodoulidis S, Ebner L, Christe A, Mougiakakou S. Lung pattern classification for interstitial lung diseases using a deep convolutional neural network. *IEEE Trans Med Imag.* 2016;35(5):1207. doi: 10.1109/TMI.2016.2535865.
27. Jiji GW, Rajesh A, Raj PJD. Decision support techniques for dermatology using case-based reasoning. *Int J Image Graphics.* 2020;20(03):2050024. doi: 10.1142/S0219467820500242.
28. Dabowsa NI, Amaitik NM, Maatuk AM, Aljawarneh SA. *A hybrid intelligent system for skin disease diagnosis in international conference on engineering and technology (ICET),* IEEE Explore, International Conference on Engineering and Technology (ICET,) Antalya Turkey, 2017.
29. Guni A, Normahani P, Davies A, Jaffer U. Harnessing machine learning to personalize web-based health care content, IEEE. *J Med Int Res.* 2021;23(10):25497.
30. Jader R, Aminifar S. Fast and accurate artificial neural network model for diabetes recognition. *NeuroQuantology.* 2022;20(10):2187.
31. Alfian G, Syafrudin M, Ijaz MF, Syaekhoni MA, Fitriyani NL, Rhee J. A personalized healthcare monitoring system for diabetic patients by utilizing BLE-based sensors and real-time data processing. *Sensors.* 2018;18(7):2183. doi: 10.3390/s18072183.
32. Momin MA, Bhagwat NS, Dhiwar AV, Chavhate SB, Devekar NS. Smart body monitoring system using IoT and machine learning. *Int J Adv Res Elect Electron Instrum Eng.* 2019;8(5):1501.
33. Ijaz MF, Alfian G, Syafrudin M, Rhee J. Hybrid prediction model for type 2 diabetes and hypertension using DBSCAN-based outlier detection, synthetic minority over sampling technique (SMOTE), and random forest. *Appl Sci.* 2018;8(8):1325. doi: 10.3390/app8081325.
34. Tigga NP, Garg S. Prediction of type 2 diabetes using machine learning classification methods. *Procedia Comput Sci.* 2020;167:706. doi: 10.1016/j.procs.2020.03.336.
35. Shabut AM, Tania MH, Lwin KT, Evans BA, Yusof NA, Abuhassan K, et al. An intelligent mobile enabled expert system for tuberculosis disease diagnosis in real time. *Expert Syst Appl.* 2018;114:65. doi: 10.1016/j.eswa.2018.07.014.
36. Matsuoka R, Akazawa H, Kodera S, Komuro I. The dawning of the digital era in the management of hypertension. *Hypertens Res.* 2020;43(11):1135. doi: 10.1038/s41440-020-0506-1.

37. Verma A, Agarwal G, Gupta AK, Sain M. Novel hybrid intelligent secure cloud internet of things based disease prediction and diagnosis. *Electronics.* 2021;10(23):3013. doi: 10.3390/electronics10233013.
38. Mohan S, Thirumalai C, Srivastava G. Effective heart disease prediction using hybrid machine learning techniques. *IEEE Access.* 2019;7:81542. doi: 10.1109/ACCESS.2019.2923707.
39. Taylor RA, Moore CL, Cheung KH, Brandt C. Predicting urinary tract infections in the emergency department with machine learning. *PLoS One.* 2018;13:e0194085. doi: 10.1371/journal.pone.0194085.
40. Rajaraman S, Antani SK, Poostchi M, Silamut K, Hossain MA, Maude RJ, et al. Pre-trained convolutional neural networks as feature extractors toward improved malaria parasite detection in thin blood smear images. *PeerJ.* 2018;6:4568. doi: 10.7717/peerj.4568.
41. Won LY, Woo CJ, Eun-Hee S. Machine learning model for predicting malaria using clinical information. *Comput Biol Med.* 2021;129:104151. doi: 10.1016/j.compbiomed.2020.104151.
42. Chenshuo W, Xianxiang C, Lidong D, Qingyuan Z, Ting Y, Zhen F. Comparison of machine learning algorithms for the identification of acute exacerbations in chronic obstructive pulmonary disease. *Comput Methods Programs Biomed.* 2020;188:105267. doi: 10.1016/j.cmpb.2019.105267.
43. Han DH, Lee S, Seo DC. Using machine learning to predict opioid misuse among U.S. adolescents. *Prev Med.* 2020;130:105886. doi: 10.1016/j.ypmed.2019.105886.
44. Saha P, Sadi MS, Islam MM. EMCNet: Automated COVID-19 diagnosis from X-ray images using convolutional neural network and ensemble of machine learning classifiers. *Inform Med Unlocked.* 2021;22:100505. doi: 10.1016/j.imu.2020.100505.
45. Tazarv A, Labbaf S, Reich SM, Dutt N, Rahmani AM, Levorato M. Personalized stress monitoring using wearable sensors in everyday settings. 43rd Annual International Conference of the IEEE Engineering in Medicine & Biology Society (EMBC), Mexico, 2021.
46. Jamal KJ, Simon YF. A comparison of machine learning algorithms for diabetes prediction. *ICT Express.* 2021;7(4):432. doi: 10.1016/j.icte.2021.02.004.
47. Lu H, Uddin S, Hajati F, Moni MA, Khushi M. A patient network-based machine learning model for disease prediction the case of type 2 diabetes mellitus. *Appl Intell.* 2022;52(3):2411. doi: 10.1007/s10489-021-02533-w.
48. Wang A, Zhang Q, Han Y, Megason S, Hormoz S, Mosaliganti KR, et al. A novel deep learning-based 3D cell segmentation framework for future image-based disease detection. *Scientific Rep.* 2022;12(1):342. doi: 10.1038/s41598-021-04048-3.
49. Chen Z, Liu X, Hogan W, Shenkman E, Bian J. Applications of artificial intelligence in drug development using real-world dat. *Drug Discov Today.* 2021;26(5):1256. doi: 10.1016/j.drudis.2020.12.013.
50. Kumar Y, Mahajan M. Recent advancement of machine learning and deep learning in the field of healthcare system. *Comp Intellig Mach Learn Healthcare Informat.* 2020;1:77.
51. Minaee S, Kafieh R, Sonka M, Yazdani S, Soufi GJ. Deep-COVID: Predicting COVID-19 from chest X-ray images using deep transfer learning. *Med Image Anal.* 2020;65:101794. doi: 10.1016/j.media.2020.101794.
52. Nasteski V. An overview of the supervised machine learning methods. *Horizons B.* 2017;4:51. doi: 10.20544/HORIZONS.B.04.1.17.P05.
53. Ghazal TM, Hasan MK, Alshurideh MT, Alzoubi HM, Ahmad M, Akbar SS, et al. IoT for smart cities: Machine learning approaches in smart healthcare—A review. *Future Internet.* 2021;13(8):218. doi: 10.3390/fi13080218.

54. Kumar Y, Singla R. *Federated learning systems for healthcare perspective and recent progress federated learning systems.* Cham: Springer; 2021.
55. Tengnah MAJ, Sooklall R, Nagowah SD. *A predictive model for hypertension diagnosis using machine learning techniques telemedicine technologies.* Coimbatore, INDIA: Elsevier; 2019:139.
56. Ghafari M, Mailman D, Hatami P, Peyton T, Yang L, Dang W, et al. A comparison of YOLO and mask-RCNN for detecting cells from microfluidic images. International Conference on Artificial Intelligence in Information and Communication (ICAIIC), Jeju Island, Republic of Korea, 2022.
57. Tran TV, Khaleghian S, Zhao J, Sartipi M. *SIMCal: A high-performance toolkit for calibrating traffic simulation in IEEE BigData,* Osaka, Japan: ACM, 2022.
58. Sajedian A, Ebrahimi M, Jamialahmadi M. Two-phase Inflow performance relationship prediction using two artificial intelligence techniques: Multi-layer perceptron versus genetic programming. *Pet Sci Technol.* 2012;30(16):1725. doi: 10.1080/10916466.2010.509074.
59. Wolberg WH, Street WN, Mangasarian OL. Machine learning techniques to diagnose breast cancer from image-processed nuclear features of fine needle aspirates. *Cancer Lett.* 1994;77(2–3):163. doi: 10.1016/0304-3835(94)90099-X.
60. Cao R, Bajgiran AM, Mirak SA, Shakeri S, Zhong X, Enzmann D, et al. Joint prostate cancer detection and gleason score prediction in mp-mri via focalnet. *IEEE Trans Med Imaging.* 2019;38(11):2496. doi: 10.1109/TMI.2019.2901928.
61. Marwa E-G, Moustafa HE-D, Khalifa F, Khater H, AbdElhalim E. An MRI-based deep learning approach for accurate detection of Alzheimer's disease. *Alex Eng J.* 2023;63:211. doi: 10.1016/j.aej.2022.07.062.
62. Bhosale YH, Patnaik KS, PulDi-COVID: Chronic obstructive pulmonary (lung) diseases with COVID-19 classification using ensemble deep convolutional neural network from chest X-ray images to minimize severity and mortality rates. *Biomed Signal Process Control.* 2023;81:104445. doi: 10.1016/j.bspc.2022.104445.
63. Li C, Zhang Y, Weng Y, Wang B, Li Z. Natural language processing applications for computer-aided diagnosis in oncology. *Diagnostics.* 2023;13(2):286. doi: 10.3390/diagnostics13020286.
64. Nordin N, Zainol Z, Noor MHM, Chan LF. An explainable predictive model for suicide attempt risk using an ensemble learning and Shapley additive explanations (SHAP) approach. *Asian J Psychiatr.* 2023;79:103316. doi: 10.1016/j.ajp.2022.103316.
65. Chen Y, Lin Y, Xu X, Ding J, Li C, Zeng Y, et al. Multi-domain medical image translation generation for lung image classification based on generative adversarial networks. *Comput Methods Programs Biomed.* 2023;229:107200. doi: 10.1016/j.cmpb.2022.107200.
66. Pagano TP, Loureiro RB, Lisboa FV, Peixoto RM, Guimarães GA, Cruz GO, et al. Bias and unfairness in machine learning models: A systematic review on datasets, tools, fairness metrics, and identification and mitigation methods. *Big Data Cognitive Comput.* 2023;7(1):15. doi: 10.3390/bdcc7010015.
67. Chee CG, Kim Y, Kang Y, Lee KJ, Chae H-D, Cho J, et al. Performance of a deep learning algorithm in detecting osteonecrosis of the femoral head on digital radiography: A comparison with assessments by radiologists. *Am J Roentgenol.* 2019;213(1):155. doi: 10.2214/AJR.18.20817.
68. Aggarwal CC. *Neural networks and deep learning.* Bangalore: Springer; 2018;10: 978.
69. Li X, Jiang Y, Liu Y, Zhang J, Yin S, Luo H. RAGCN: Region aggregation graph convolutional network for bone age assessment from x-ray images. *IEEE Trans Instrum Meas.* 2022;71:1. doi: 10.1109/TIM.2022.3218574.

70. Li X, Jiang Y, Zhang J, Li M, Luo H, Yin S. Lesion-attention pyramid network for diabetic retinopathy grading. *Artif Intell Med.* 2022;126:102259. doi: 10.1016/j. artmed.2022.102259.

71. Zitnik M, Nguyen F, Wang B, Leskovec J, Goldenberg A, Hoffman MM. Machine learning for integrating data in biology and medicine: Principles, practice, and opportunities. *Inf Fusion.* 2019;50:71. doi: 10.1016/j.inffus.2018.09.012.

72. Burgos N, Colliot O. Machine learning for classification and prediction of brain diseases: Recent advances and upcoming challenges. *Curr Opin Neurol.* 2020;33(4):439. doi: 10.1097/WCO.0000000000000838.

73. Moen E, Bannon D, Kudo T, Graf W, Covert M, Valen DV. Deep learning for cellular image analysis. *Nat Methods.* 2019;16(12):1233. doi: 10.1038/s41592-019-0403-1.

74. Chen J, Remulla D, Nguyen JH, Liu Y, Dasgupta P, Hung AJ. Current status of artificial intelligence applications in urology and their potential to influence clinical practice. *BJU Int.* 2019;124(4):567. doi: 10.1111/bju.14852.

75. Chen PHC, Gadepalli K, MacDonald R, Liu Y, Kadowaki S, Nagpal K, et al. An augmented reality microscope with real-time artificial intelligence integration for cancer diagnosis. *Nature Med.* 2019;25(9):1453. doi: 10.1038/s41591-019-0539-7.

76. Keenan TD, Clemons TE, Domalpally A, Elman MJ, Havilio M, Agron E, et al. Retinal specialist versus artificial intelligence detection of retinal fluid from OCT: Age-related eye disease study 2: 10-year follow-on stud. *Ophthalmology.* 2021;128(1):100. doi: 10.1016/j.ophtha.2020.06.038.

77. Johnson KW, Soto JT, Glicksberg BS, Shameer K, Miotto R, Ali M, et al. Artificial intelligence in cardiology. *J Am Coll Cardiol.* 2018;71(23):2668. doi: 10.1016/j. jacc.2018.03.521.

78. Shameer K, Johnson KW, Glicksberg BS, Dudley JT, Sengupta PP. Machine learning in cardiovascular medicine: Are we there yet? *Heart.* 2018;104(14):1156. doi: 10.1136/ heartjnl-2017-311198.

79. Somani S, Russak AJ, Richter F, Zhao S, Vaid A, Chaudhry F, et al. Deep learning and the electrocardiogram: Review of the current state-of-the-art. *EP Europace.* 2021;23(8):1179. doi: 10.1093/europace/euaa377.

80. Lu H, Yao Y, Wang L, Yan J, Tu S, Xie Y, et al. Research progress of machine learning and deep learning in intelligent diagnosis of the coronary atherosclerotic heart disease. *Comp Mathemat Methods Med.* 2022;2022:3016532.

81. Coenen A, Kim YH, Kruk M, et al. Diagnostic accuracy of a machine-learning approach to coronary computed tomographic angiography-based fractional flow reserve: Result from the MACHINE consortium. *Circul Cardiovas Imaging.* 2018;11(6):007217. doi: 10.1161/CIRCIMAGING.117.007217.

82. von Knebel Doeberitz PL, Cecco CND, Schoepf UJ, et al. Impact of coronary computerized tomography angiography-derived plaque quantification and machine-learning computerized tomography fractional flow reserve on adverse cardiac outcome. *Am J Cardiol.* 2019;124(9):134019. doi: 10.1016/j.amjcard.2019.07.061.

83. Li Y, Qiu H, Hou Z, et al. Additional value of deep learning computed tomographic angiography-based fractional flow reserve in detecting coronary stenosis and predicting outcomes. *Acta Radiologica.* 2022;63(1):133. doi: 10.1177/0284185120983977.

84. Zhi-Qiang W, Yu-Jie Z, Ying-Xin Z, et al. Diagnostic accuracy of a deep learning approach to calculate FFR from coronary CT angiography. *J Geriatric Cardiol.* 2019;16(1):42.

85. Faieq AK, Mijwil MM. Prediction of heart diseases utilising support vector machine and artificial neural network. *Indon J of Elect Eng Comp Sci.* 2022;26(1):374.

86. Onsalves AH, Thabtah F, Mohammad RM, Singh G. Prediction of coronary heart disease using machine learning: An experimental analysis. 3rd International Conference on Deep Learning Technologies, Beijing, China, 2019.

87. Ahuja NMS. Prediction of heart diseases using data mining techniques: Application on Framingham heart study. *Int J Big Data Analy Healthcare.* 2018;3(2):1. doi: 10.4018/IJBDAH.2018070101.

88. Xing Y, Wang J, Zhao Z. Combination data mining methods with new medical data to predicting outcome of coronary heart disease. International Conference on Convergence Information Technology (ICCIT 2007). IEEE, Gwangju, Korea (South), 2007:868–872.

89. Khan P, Kader MF, Islam SMR, Rahman AB, Kamal MS, Toha MU, et al. Machine learning and deep learning approaches for brain disease diagnosis: Principles and recent advances. *IEEE Access.* 2021;9:37622. doi: 10.1109/ACCESS.2021.3062484.

90. Wen J, Thibeau-Sutre E, Diaz-Melo M, Samper-González J, Routier A, Bottani S, et al. Convolutional neural networks for classification of Alzheimer's disease: Overview and reproducible evaluation. *Med Image Anal.* 2020;63:101694. doi: 10.1016/j.media.2020.101694.

91. Gautam R, Sharma M. Prevalence and diagnosis of neurological disorders using different deep learning techniques: A meta-analysis. *J Med Syst.* 2020;44(2):49. doi: 10.1007/s10916-019-1519-7.

92. Savaş S. Detecting the stages of Alzheimer's Disease with Pre-trained deep learning architectures. *Arab J Sci Eng.* 2022;47(2):2201. doi: 10.1007/s13369-021-06131-3.

93. Lamba R, Gulati T, Jain A. A hybrid system for Parkinson's disease diagnosis using machine learning techniques. *Int J Speech Technol.* 2022;25(3):583. doi: 10.1007/s10772-021-09837-9.

94. Yu X, Zhou Q, Wang S, Zhang YD. A systematic survey of deep learning in breast cancer. *Int J Intell Syst.* 2022;37(1):152. doi: 10.1002/int.22622.

95. Polat K, Güneş S. Breast cancer diagnosis using least square support vector machine. *Digital Signal Proc.* 2007;17(4):694. doi: 10.1016/j.dsp.2006.10.008.

96. Şahan S, Polat K, Kodaz H, Güneş S. A new hybrid method based on fuzzy-artificial immune system and k-nn algorithm for breast cancer diagnosis. *Comput Biol Med.* 2007;37(3):415. doi: 10.1016/j.compbiomed.2006.05.003.

97. Akay MF. Support vector machines combined with feature selection for breast cancer diagnosis. *Expert Syst Appl.* 2009;36(2):3240. doi: 10.1016/j.eswa.2008.01.009.

98. Maglogiannis I, Zafiropoulos E, Anagnostopoulos I. An intelligent system for automated breast cancer diagnosis and prognosis using SVM based classifiers. *Appl Intell.* 2009;30(1):24. doi: 10.1007/s10489-007-0073-z.

99. Osareh A, Shadgar B. Machine learning techniques to diagnose breast cancer. 5th International Symposium on Health Informatics And Bioinformatics, Ankara, Turkey, 2010.

100. Hsieh S-L, Hsieh S, Cheng P, Chen C, Hsu K, Lee I, et al. Design ensemble machine learning model for breast cancer diagnosis. *J Med Sys.* 2012;36(5):2841. doi: 10.1007/s10916-011-9762-6.

101. Ramos-Pollán R, Guevara-López MA, Suárez-Ortega C, Díaz-Herrero G, Franco-Valiente JM, Rubio-del-Solar M, et al. Discovering mammography-based machine learning classifiers for breast cancer diagnosis. *J Med Sys.* 2012;36(4):2259. doi: 10.1007/s10916-011-9693-2.

102. Rajesh K, Anand S. Analysis of SEER dataset for breast cancer diagnosis using C4 5 classification algorithm. *Int J Adv Res Comp Commun Eng.* 2012;1(2):2278.

103. Salama GI, Abdelhalim M, Zeid MA-E. Breast cancer diagnosis on three different datasets using multi-classifiers. *Breast Cancer.* 2012;32(569):2.

104. Kharya S. Using data mining techniques for diagnosis and prognosis of cancer disease. *Int J Comp Sci Eng Inform Technol.* 2012;2(2):55.

105. Nasir MU, Khan MA, Zubair M, Ghazal TM, Said RA, Hamadi HA. Single and mitochondrial gene inheritance disorder prediction using machine learning. *Computers Mat Continua.* 2022;73(1):953. doi: 10.32604/cmc.2022.028958.

106. Bracher-Smith M, Crawford K, Escott-Price V. Machine learning for genetic prediction of psychiatric disorders: A systematic review. *Mol Psychiatry.* 2021;26(1):70. doi: 10.1038/s41380-020-0825-2.

107. Tan AC, Gilbert D. Ensemble machine learning on gene expression data for cancer classification. Proceedings of New Zealand Bioinformatics Conference Te Papa, Wellington, New Zealand, 2003.

108. Hossain MD, Kabir MA, Anwar A, Islam MZ. Detecting autism spectrum disorder using machine learning techniques. *Health Inform Sci Sys.* 2021;9(1):17. doi: 10.1007/s13755-021-00145-9.

109. Das K, Cockerell CJ, Patil A, Pietkiewicz P, Giulini M, Grabbe S, et al. Machine learning and its application in skin cancer. *Int J Environ Res Pub Health.* 2021;8(24):13409. doi: 10.3390/ijerph182413409.

110. Magalhaes C, Tavares JMR, Mendes J, Vardasca R. Comparison of machine learning strategies for infrared thermography of skin cancer. *Biomed Signal Process Control.* 2021;69:102872. doi: 10.1016/j.bspc.2021.102872.

111. Gal Y, Islam R, Ghahramani Z. Deep bayesian active learning with image data. *Proc Int Conf Mach Learn.* 2017;70:1183.

112. Tătaru OS, Vartolomei MD, Rassweiler JJ, Virgil O, Lucarelli G, Porpiglia F, et al. Artificial intelligence and machine learning in prostate cancer patient management— Current trends and future perspectives. *Diagnostics.* 2021;11(2):354. doi: 10.3390/diagnostics11020354.

113. Khosravi P, Lysandrou M, Eljalby M, Li Q, Kazemi E, Zisimopoulos P, et al. A deep learning approach to diagnostic classification of prostate cancer using pathology-radiology fusion. *J Magn Reson Imaging.* 2021;54(2):462. doi: 10.1002/jmri.27599.

114. Liu S, Zheng H, Feng Y, Li W. Medical imaging: Computer-aided diagnosis - Prostate cancer diagnosis using deep learning with 3D multiparametric MRI. *SPIE Proc.* 2017;10134:581.

115. Xie Y, Meng W-Y, Li R-Z, Wang Y-W, Qian X, Chan C, et al. Early lung cancer diagnostic biomarker discovery by machine learning methods. *Transl Oncol.* 2021;14(1):100907. doi: 10.1016/j.tranon.2020.100907.

116. Kuan K, Ravaut M, Manek G, Chen H, Lin J, Nazir B, et al. Deep learning for lung cancer detection: Tackling the kaggle data science bowl 2017 challenge. *arXiv preprint.* 2017;1705:09435.

117. Teramoto A, Tsukamoto T, Kiriyama Y, Fujita H. Automated classification of lung cancer types from cytological images using deep convolutional neural networks. *Biomed Res Int.* 2017;2017:4067832. doi: 10.1155/2017/4067832.

118. Song Q, Zhao L, Luo X, Dou X. Using deep learning for classification of lung nodules on computed tomography images. *J Healthcare Eng.* 2017;2017:8314740. doi: 10.1155/2017/8314740.

119. Sun W, Zheng B, Qian W. Automatic feature learning using multichannel ROI based on deep structured algorithms for computerized lung cancer diagnosis. *Comput Biol Med.* 2017;89:530. doi: 10.1016/j.compbiomed.2017.04.006.

120. Srivastava A, Jain S, Miranda R, Patil S, Pandya S, Kotecha K. Deep learning based respiratory sound analysis for detection of chronic obstructive pulmonary disease. *PeerJ Comp Sci.* 2021;7:e369. doi: 10.7717/peerj-cs.369.

121. Afshar P, Heidarian S, Naderkhani F, Oikonomou A, Plataniotis KN, Mohammadi A. Covid-caps: A capsule network-based framework for identification of covid-19 cases from x-ray images. *Pattern Recogn Lett.* 2020;138:638. doi: 10.1016/j. patrec.2020.09.010.

9 Exploring COVID-19 Through Intensive Investigation with Supervised Machine Learning Algorithm

R. Dinesh Kumar, G. Prudhviraj, K. Vijay,
P. Senthil Kumar, and Philipp Plugmann

INTRODUCTION

Many statistical, probabilistic, and optimization techniques are used by supervised machine learning (ML) algorithms to learn from the past and find useful models from different datasets [1]. ML algorithms have a wide range of applications, including search optimization, intrusion detection, disease prediction models, spam mail filtering, fraudulent credit card transaction detection, manufacturing system optimization, and the detection of customer purchase behavior. In general, digital data are used in a variety of healthcare research domains, including the study of disease occurrences, the performance evaluation of hospital care, the exploration of prototypes, the prediction of disease risk models, and the analysis of chronic diseases.

In this study, the prediction model for Coronavirus disease is examined in connection to various supervised ML methods. For training and testing purposes, we employed a labeled patient training dataset in this model. Additionally, patients are divided into groups based on their level of risk, including minimal, moderate, and significant risk. The effectiveness of supervised ML algorithms on the depth of disease diagnostic prediction is still being investigated.

In particular, we conducted a brief study that outlines a thorough analysis of articles that have already been published and uses a variety of supervised machine learning algorithms for disease prediction. As a result, the goal of this research is to identify major trends in the performance accuracy of various categories of supervised machine learning algorithms as well as the types of diseases that are being taken into account. Several supervised machine learning algorithms have also been researched for their positives and negatives. The results of this study will improve

DOI: 10.1201/9781032686714-9

our understanding of the hotspots (state of art) and current trends in disease prediction methods utilizing supervised machine learning algorithms, allowing us to better formulate the research objectives. Following a set of recommendations, the results of the analytical investigation that included a thorough comparison of the results of several supervised machine learning algorithms and current studies from the literature have been examined.

Additionally, this work specifically took into account those observations when using multiple supervised machine learning algorithms for a single disease prediction in the same research environment. The comparison together with other supervised machine learning methods produced by this study is a key contribution that is more accurate and effective. Comparison of a single algorithm's effectiveness in several study conditions might be subjective and lead to inaccurate results [2], it is important to avoid this. For diagnosis and risk prediction in diseases, conventional statistical systems and a doctor's perception, knowledge, and experience have typically been used. This practice frequently results in unintended biases, errors, and high costs, which have a negative impact on the standard of care given to patients [3]. Advanced computational technologies, such as machine learning, have emerged as more practical to use and research with the huge availability of medical data.

RELATED WORK

The Top2Vec (for topic modeling) and RoBERTa were used to evaluate a database of more than 100,000 COVID-19 news headlines and articles. The topic modeling findings showed that the most prevalent and widely covered topics in the UK, India, Japan, and South Korea were education, the economy, and sports. The sentiment classification model's validation accuracy was 90%, and the study revealed that the UK, the country in our sample that was the worst hit, also had the largest percentage of negative sentiment [1].

Linear Discriminant Analysis (LDA) is frequently less comprehensible when you look for precise answers to questions regarding an incident for which news pieces have been published all over the world. That is a result of the content's distinct semantics. We pooled news items based on information retrieval using the Term Frequency–Inverse Document Frequency (TF-IDF) score in a data-processing step and topic modeling using LDA with a combination of 1 to 6 n-grams to meet this difficulty. To examine the variations in sentiments in news stories reported across several geographic locations, we used the Valence Aware Dictionary for Sentiment Reasoning (VADER) sentiment analyzer. The findings imply that news coverage by publications with various political alignments supports the reported material [4].

The analysis methodology was used to examine 117 newspaper stories by developing a codebook with 133 codes. The findings demonstrate how the media in all nations over-emphasized public dread of the epidemic while underestimating the risk of the pandemic in its early stages. There is a propensity for criticism of the Chinese government in American and Japanese journals, whereas articles in German newspapers seem to be against the wearing of face masks. The methodology of this study emphasizes the benefits of combining qualitative analysis and quantitative comparison to pinpoint and assess elusive themes in media texts across cultural boundaries [2].

In this study, we seek to quickly identify COVID-19-specific areas that may be present in chest X-ray pictures. In order to segment X-ray images, this research suggests a hybrid COVID-19 detection model based on the improved marine predators algorithm (IMPA). The performance of the IMPA is improved by using the ranking-based diversity reduction (RDR) technique to find better solutions with less iteration. RDR focuses on locating the particles that failed to find better solutions over the course of a certain number of iterations and then directing those particles in the direction of the current best answers. Nine chest X-ray images with threshold levels between ten and one hundred were used to validate the IMPA's performance, and its results were compared with those of five cutting-edge algorithms: equilibrium optimizer (EO), whale optimization algorithm (WOA), sine cosine algorithm (SCA), Harris-hawks algorithm (HHA), and salp swarm algorithms (SSAs). The experimental results show that for a variety of measures the suggested hybrid model beats all other algorithms [5].

Fake information is spread on social media websites about the vaccination. The problem of fake news detection on the Twitter network (now X) is examined. After collecting a dataset and pre-processing, a set of features is extracted from the tweets. This includes the tweet's length and its keywords, number of followers, sentiment, and readability scores. In the next phase, six well-known classifiers are executed on this data, and the best result with the highest accuracy is chosen for the community detection process to study and track the evolution of fake news campaigns. For the analysis, we considered multiple criteria such as the number of communities, their sizes, their leaders, and communities. The results of this research can help decision-makers to understand the underlying and formation of fake news campaigns [3].

By combining the usage of cutting-edge language models to describe the articles with a density-based, streaming, clustering technique that is specifically designed to handle high-dimensional text embeddings, the problem of identifying news clusters at certain periods is addressed. In addition, we offer a technique for automatically assigning semantically appropriate labels to the discovered clusters and present a collection of metrics for monitoring the temporal evolution of clusters In this study, the authors have focused on the city of Rome during COVID-19 and discussed and illustrated to evaluate the proposed system. [6].

The multitask deep learning algorithm in "Covid-19 Outbreak Analysis" [7] illustrated in the following classifies COVID-19 patients and separates COVID-19 abrasion from chest computed tomography (CT) images. The three phases of the proposed architecture are segmentation, reconstruction, and categorization.

The proposal is to do an analysis of COVID-19-affected patients. Machine learning models help in identifying the severity and decision making. Understanding the pandemic will help in implementing proper measures at the right time. Implementing these measures prior to the outbreak will help maintain a country's health and economy. Here we observe and analyze the behavior of data, which helps in providing better results for decision making. The model proposed in this chapter is in adjunct with an interactive application that gives updates regarding the pandemic; indicates the different zones of the local area prescribed by the government, and also serves as a platform for the needy [8].

The Coronavirus (COVID-19) epidemic is spreading worldwide. A database of X-ray, and CT scan images from patients with common bacterial pneumonia, confirmed COVID-19 infection, and common cases, were used to automatically detect Coronavirus infection. The purpose of the study was to evaluate the effectiveness of COVID-19 acquisition. During the COVID-19 scenario, the number of infected cases rose in huge numbers globally. Due to this fact, a vital decision has been taken by medical experts and infected patients to adopt various medical facilities within a reasonable amount of time [9].

First, we looked at how often graphics were used in COVID-19 news reports. The types and frequency of graphs used in COVID-19 news stories were quantitatively analyzed, and the content of the news stories that contained graphs was qualitatively analyzed. Second, we found instances where the incorrect mathematical usage of graphs in news stories about COVID-19 may have influenced readers. These findings' implications for future graph literacy instruction and learning in school mathematics courses are highlighted [10].

The 2019 Coronavirus disease has had a devastating psychosocial impact on the entire human race. Substance use disorders (SUD) are a significant risk factor for infection in marginalized groups, and those with SUD are also more likely to have increased psychosocial load. This chapter examines the complex reciprocal relationship between addiction and COVID-19 [11].

Individuals with SUD are more likely to have poorer COVID-19 side effects. During this time, there is an increase in addictive behaviors (both new and relapsed ones), including behavioral addiction. Emergencies related to withdrawal and deaths are also being recorded more frequently. Those who are addicted, in particular, have trouble getting healthcare, which makes them more likely to buy medicines illegally [12].

The key COVID-19 diagnosis-related methods employed in the study are briefly discussed in this chapter. The task is discussed in relation to various aspects in a number of research works connected to COVID-19 diagnosis. In order to provide some of the essential support for analysis, segmentation methods are used. The main strategies used to improve the effectiveness of COVID-19 disease identification from chest radiograph pictures are image augmentation and transfer learning. F1-score, sensitivity, specificity, and accuracy are the performance indicators used to identify COVID-19.

PROPOSED MODEL

SUPERVISED MACHINE LEARNING MODEL

Through the use of supervised machine learning, pre-programmed algorithms are put into action. These algorithms assess incoming data to create forecasts that fall within a reasonable range. These algorithms tend to produce more accurate predictions as a result of the inclusion of additional data. While there are some discrepancies in how machine learning algorithms are categorized, they can be grouped into three main groups based on their underlying principles and how the key machine is trained. Supervised, unsupervised, and semi-supervised machine learning models are the three main categories [13].

The underlying algorithm is first trained using a labeled dataset in spite of supervised machine learning algorithms. The unlabelled test dataset is then used by the trained algorithm to sort them into similar categories. The supervised machine learning algorithms are well-suited for two different classification and regression issues. The subsequent output variable is a real number in regression issues, such as the likelihood that a person would contract a certain disease. In the following sections, we briefly describe the supervised machine-learning techniques for disease prediction.

Naive Bayes Algorithm

Although the naive Bayes method is a classification system, it is also used in predictive modeling and is key in the early diagnosis of diseases. The family of probabilistic classifiers is entirely based on the Bayes theorem. The naive Bayes method is one type of conditional probability. To categorize a given instance of data values in this model, which is represented by a vector $X=(x_1... x_n)$, where n stands for characteristics of independent variables.

The probability was determined using all class labels and all conditions. The target test data was given to obtain the results. The values of the output data can be utilized to potentially place the patient in a specific class. A person can easily identify if they have COVID-19 or not by determining the source of the probability's value. By using culture, serology, and electron microscopy, viruses can be differentiated and classified using the conventional technique [14]. Using these techniques, coronaviruses are defined as enveloped viruses having a crown nature and a volume between 120 and 160 nm.

Coronavirus diagnosis is difficult to process and the patient's diagnosis procedure can take 48–72 hours to complete. In order to forecast the coronavirus, we can use the naive Bayes classifier, which is a supervised learning method. The diagnosis of each patient will be based on their characteristics and attributes [15, 16]. When a patient exhibits the symptoms listed in Table 9.1, it is possible to determine whether or not they have COVID-19 by using the machine learning methodology.

Support Vector Machine Method

The support vector machine applications include bio-information, hypertext classification of regular text, and face detection [17]. The SVM is accurate for supervised machine learning models that are trained using historical input data and create predictions for the future as an output. SVM is a technique that analyzes data and divides it into two categories. Figure 9.2 represents the SVM supervised learning model and also a regression version applied to classification. Both linear and non-linear forms

TABLE 9.1

Symptoms of COVID-19

Common Symptoms	Other Symptoms	Emergency Symptoms
Fever	Vomiting	High BP
Cough	Headache	Breathing problem
Tiredness	Nausea	Chest pain
Loss of smell or Taste	Throat pain	New confusion

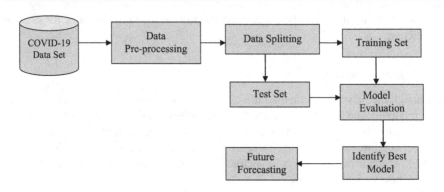

FIGURE 9.1   Architecture of the proposed system

of SVM are made available. There are two data ets that are comparable in SVM model, the trained dataset and test dataset.

If there are two classes, the SVM model divides them completely, or linearly separable. One dividing line is chosen as the best line along with a set of solid lines to divide the dataset on the opposite side. The ideal line is chosen so that it should be as close to the closest function of the two classes in the training dataset as possible. Support vectors serve as the dataset's primary foci during the separation process since the hyperplane should go beyond what many people would consider feasible. The support vector machine model is shown in Figure 9.2. Another parameter is the regularisation parameter. The regularisation parameter is another parameter that determines whether the data will be overfit, or biased to an upper-low value. SVM avoids overfitting. Combining irregular mistakes in the model results in overfitting. Moreover, the model fundamentally constructs additional errors across various datasets. Creating simpler models protects the model from being overfitting.

Classification of SVM Type 1

Classification SVM Type 1 is also called C-SVM classification which uses the training function that minimizes the error function that is defined as follows,

$$\frac{1}{2}w^T w + C\sum_{i=1}^{N} \xi_i \qquad (1)$$

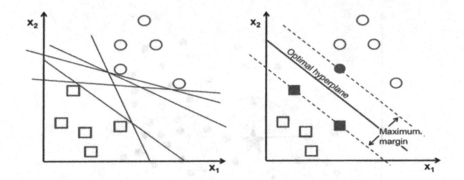

FIGURE 9.2   Support vector machine

Subject to the constraints:

$$Yi(W^T\Phi(x_i) + b) \geq 1 - \xi_i \text{ and } \xi_i \geq 0, i = 1\ldots N \tag{2}$$

Where C is capacity constants, w is the vector of coefficients, b is the constant and ξ_i represents the parameters for handling non-separable data. The index I label the N training cases, ye $\pm$ 1 represents the class label and x_i is the independent variable. The kernel ϕ is used to transform data from the input to the feature space which is larger than c. Thus, C should be chosen with care to avoid overfitting.

K-Nearest Neighbor

One of the straightforward classification techniques is the K-nearest neighbor (KNN) algorithm. It may be possible to take into account a more straightforward naive Bayes classifier. The naive Bayes method requires reflection on probability theory, but the KNN algorithm does not. The number of closest neighbors regarded to take a "vote" from the set is K in the KNN method. For the same sample object, using multiple values for "K" can result in inconsistent categorization outcomes. Figure 9.3 depicts the process the KNN uses to classify a new object. The most recent object is categorized as "black" for K=3; it is categorized as "red" for K=5, at the same moment.

This Euclidean distance metric is applied to real-valued vectors. The distance measure is calculated using the formula below.

$$d(x,y) = \sqrt{\sum_{i=1}^{n}(y_i - x_i)^2} \tag{3}$$

Decision Tree

A decision tree simulates the evaluation logic, such as tests and corresponding results for categorizing data items into a structure resembling a tree. A DT's nodes

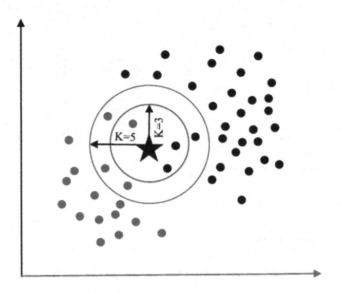

FIGURE 9.3   K-nearest neighbor

typically have different levels, with the major or top-most node being referred to as the root node. All internal nodes are tested on characteristics or input variables. The classification algorithm branches in the direction of the appropriate child node based on the test result and then repeats the test and branching process until it reaches the leaf node. The result outcomes are communicated by the leaf or terminal nodes. The results of all tests at each node along the pathway will provide adequate information to infer the classification of the sample when navigating the classification tree. Figure 9.4 shows an example of a DT along with its components and guidelines.

Entropy for a single attribute is denoted mathematically by

$$E(s) = \sum_{i=1}^{c} - p_i \log_2 p_i \qquad (4)$$

An entropy of zero is designated as a leaf node, whereas an entropy of greater than zero requires further branch splitting.

RANDOM FOREST

A forest is a collection of numerous decision trees, and a random forest (RF) is a collection classifier made up of many DTs. Due to overfitting of the training data, which produces a significant difference in classification effect for a minor alteration in the input data, DTs are generated very frequently. They make mistakes on the experiment dataset because they are very perceptive to their training data. An RF's several DTs are learned using portions of the training dataset that are different from one another. The input vector from that trial must be sent down with every DT of the

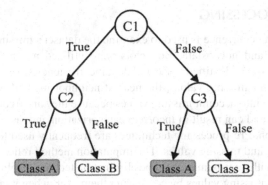

FIGURE 9.4 Decision tree

forest in order to categorize a new sample. Each DT then considers a separate aspect of the input vector and provides a classification result. The forest will then decide which trees are categorized based on which received the most "votes" or based on the forest-wide average. The RF approach reduces the variation produced by a single DT's decision for a similar dataset because it replicates the results from multiple separate DTs. The RF algorithm is depicted in Figure 9.5.

The Gini index can be expressed mathematically as:

$$Gini\ index = 1 - \sum_{i=1}^{n}(p_i)^2 \tag{5}$$

The feature with the lowest Gini index will be selected as the root node by this approach from among all feasible splits. The lowest Gini index gives the lowest impurity.

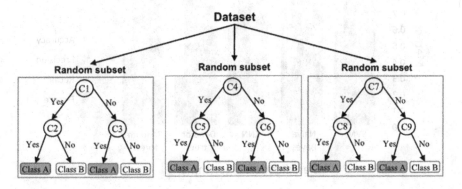

FIGURE 9.5 Random forest

DATA PRE-PROCESSING

Data distribution occurrence is used to examine the dataset's missing values, inconsistent datasets, and noisy data. Box plots were utilized as an outlier detection method. The dataset's identified noise and data inconsistencies were manually rectified. To correct for missing values, the nearest neighbor method was applied. The dataset must take into account missing values because they are frequently present in medical records and can result in incorrect categorization and, as a result, incorrect disease prediction. Pre-processing techniques are frequently used to deal with deletion, imputation, and missing values. The imputation method relies on replacing the missing values with approximations. The elimination technique relies on eradicating the records with missing values by comparing them. Yet, when it comes to medical data, such a strategy is deemed to be unethical. Moreover, when data is deleted, essential information may also be lost. This affects classification and causes medical records to be viewed as connected data [18]. In this investigation, the KNN approach was used as the imputation strategy. To determine the distances between the point of the unknown instance and the points of the other full records in the training dataset, the KNN classifier uses the Euclidean distance formula. The k most comparable scenarios involve the missing value.

PERFORMANCE MEASURES

In this work, the performance of the algorithms was evaluated using four performance metrics – F-score, accuracy, recall, and precision [19].

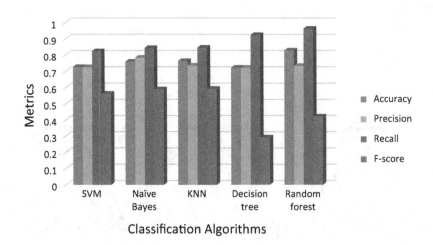

FIGURE 9.6 Classification algorithm accuracy

ACCURACY

The accuracy is the percentage of the total instances evaluated that have correct results, including both true positives and true negatives. The accuracy value is then determined as follows [20]:

$$\text{Accuracy} = \frac{\text{number of true positive} + \text{number of true negative}}{\text{number of true positive} + \text{false positive} + \text{false negative} + \text{true negative}} \tag{6}$$

RECALL

Recall is the proportion of data samples from a class of interest – the "all true positive class." The following equation is used to determine recall:

$$\text{Recall} = \frac{\text{number of true positive}}{\text{number of true positive} + \text{false positive}} \tag{7}$$

PRECISION

The quality of a model's accurate prediction is known as precision. The true positives and false positives are the precision.

$$\text{Precision} = \frac{\text{true positive}}{\text{true positive} + \text{false positive}} \tag{8}$$

F-SCORE

F-score, a single performance indicator, is the mean of recall and precision. Generally, the F-score will be closer to the smaller estimation of precision or recall.

$$\text{F1 score} = \frac{\left(2*\left(\text{Recall}*\text{Precision}\right)\right)}{\left(\text{Recall} + \text{Precision}\right)} \tag{9}$$

EXPERIMENT AND RESULT

The performance evaluation of the Coronavirus detection methods such as decision tree (DT), naive Bayes, random forest, KNN, and SVM with single out cross-validation method. The performance of the proposed methods is evaluated by using different types of metrics such as F-score, accuracy, recall, and precision. The resultant values of various classification algorithms are shown in Table 9.2.

TABLE 9.2
Performance evaluation of different classifier algorithms

Algorithm	Accuracy	Precision	Recall	F-score
SVM	0.825	0.724	0.823	0.562
Naive Bayes	0.758	0.782	0.843	0.589
KNN	0.762	0.734	0.845	0.5921
Decision tree	0.721	0.721	0.923	0.2924
Random forest	0.927	0.832	0.962	0.421

CONCLUSION

Decision tree, naive Bayes, random forest, KNN, and SVM were compared in order to determine which one performed the best in terms of the performance of COVID-19 prediction to determine the most exact prediction and classification technique. The classifiers were implemented and evaluated using the scikit-learn program. Hence, the classifiers have been given access to the COVID-19 patient data collection. The analysis's findings show that every classification technique that has been tried may forecast Coronavirus disease. Regarding the dataset and evaluation dimensions, the random forest technique outperformed other strategies in predicting Coronavirus diseases.

Author Contributions

All authors listed have made a substantial, direct, and intellectual contribution to the work and approved it for publication.

CONFLICT OF INTEREST

The authors declare that there is no conflict of interest with any person(s) or organization(s).

DATASET

The data that support the findings of this study are available upon request.

REFERENCES

1. Ghasiya, P., & Okamura, K. (2021). Investigating COVID-19 news across four nations: A topic modeling and sentiment analysis approach. *IEEE Access*, 9, 36645–36656.
2. Sato, Y. (2022). Cross-cultural analysis of the American, German, and Japanese newspaper coverage on COVID-19. International Electronics Symposium (IES), Surabaya, Indonesia, pp. 595–600.

3. Piedrahita-Valdés, H., Piedrahita-Castillo, D., Bermejo-Higuera, J., Guillem-Saiz, P., Bermejo-Higuera, J. R., Guillem-Saiz, J., & Machío-Regidor, F. (2021). Vaccine hesitancy on social media: Sentiment analysis from June 2011 to April 2019. *Vaccines*, 9(1), 28.

4. Sittar, A., Major, D., Mello, C., Mladenić, D., & Grobelnik, M. (2022). Political and economic patterns in COVID-19 news: From lockdown to vaccination. *IEEE Access*, 10, 40036–40050.

5. Abdel-Basset, M., Mohamed, R., Elhoseny, M., Chakrabortty, R. K., & Ryan, M. (2020). A hybrid COVID-19 detection model using an improved marine predator's algorithm and a ranking-based diversity reduction strategy. *IEEE Access*, 8, 79521–79540.

6. Bogomolov, A., Lepri, B., Staiano, J., Oliver, N., Pianesi, F., & Pentland, A. (2014). Once upon a crime: Towards crime prediction from demographics and mobile data. Conference: ICMI '14: Proceedings of the 16th International Conference on Multimodal InteractionNovember 2014 pp. 427–434.

7. Kanji, J. N., Zelyas, N., MacDonald, C., Pabbaraju, K., Khan, M. N., Prasad, A., & Tipples, G. (2021). False negative rate of COVID-19 PCR testing: A discordant testing analysis. *Virology Journal*, 18(1), 13.

8. Nikil, C. V. S. S., Dalmia, H., & Kumar, G. J. R. P. (2020). Covid-19 outbreak analysis. 2020 International Conference on Smart Technologies in Computing, Electrical and Electronics (ICSTCEE), Bengaluru, pp. 347–350.

9. Siddhu, A. K., Kumar, A., & Kundu, S. (2020). Review paper for detection of COVID-19 from medical images and/ or symptoms of patient using machine learning approaches. 9th International Conference System Modeling and Advancement in Research Trends (SMART), Moradabad, pp. 39–44.

10. Kwon, O. N., Han, C., Lee, C., et al. (2021). Graphs in the COVID-19 news: A mathematics audit of newspapers in Korea. *Educational Studies in Mathematics*, 108, 183–200.

11. Dubey, M. J., Ghosh, R., Chatterjee, S., Biswas, P., Chatterjee, S., & Dubey, S. (2020, September–October). COVID-19 and addiction. *Diabetes & Metabolic Syndrome*, 14(5), 817–823.

12. Loey, M., Manogaran, G., & Khalifa, N. E. M. (2020). A deep transfer learning model with classical data augmentation and cgan to detect Covid-19 from chest ct radiography digital images. *Neural Computing and Applications*, 23, 1–13.

13. Amyar, A., Modzelewski, R., Li, H., & Ruan, S. (2020). Multi-task deep learning based CT imaging analysis for COVID-19 pneumonia: Classification and segmentation. *Computers in Biology and Medicine*, 126, 104037.

14. El-Kenawy, E. S. M., Ibrahim, A., Mirjalili, S., Eid, M. M., & Hussein, S. E. (2020). Novel feature selection and voting classifier algorithms for COVID-19 classification in CT images. *IEEE Access*, 8, 179317–179335.

15. Shah, V., Keniya, R., Shridharani, A., Punjabi, M., Shah, J., & Mehendale, N. (2021). Diagnosis of COVID-19 using CT scan images and deep learning techniques. *Emergency Radiology*, 28(3), 497–505.

16. Wood, A., et al. (2019). Private naive bayes classification of personal biomedical data: Application in cancer data analysis. *Computers in Biology and Medicine*, 105, 144–150.

17. Kumar, R. D., Sridhathan, C., & Kumar, M. S. (2020). Performance evaluation of different neural network classifiers for Sanskrit character recognition. In A. Haldorai, A. Ramu, & S. Khan (Eds.), *Business intelligence for enterprise internet of things. EAI/ springer innovations in communication and computing.* Cham: Springer, pp 185–194.

18. Sujath, R., Chatterjee, J. M., & Hassanien, A. E. (2020). A machine learning forecasting model for COVID-19 pandemic in India. *Stochastic Environmental Research and Risk Assessment*, 34, 959–972.

19. Afshar, P., Heidarian, S., Enshaei, N., Naderkhani, F., Rafiee, M. J., Oikonomou, A., et al. (2020). COVID-CT-MD: COVID-19 computed tomography (CT) scan dataset applicable in machine learning and deep learning. *Scientific Data*, 8.

20. Shakouri, S., Bakhshali, M. A., Layegh, P., Kiani, B., Masoumi, F., Ataei Nakhaei, S., & Mostafavi, S. M. (2021). COVID19-CT-dataset: An open-access chest CT image repository of 1000+ patients with confirmed COVID-19 diagnosis. *BMC Research Notes*, 14(1), 178.

10 Medicine Supply through UAV

Mandala Bhuvana, Kesireddy Rajashekar
Reddy, Varagani Ramu, Bochu Sai Vardhan,
Vijaya Gunturu, and Delukshi Shanmugarajah

INTRODUCTION

Particularly in rural and difficult-to-reach places, air transport may one day transform the delivery of medical supplies. Drone technology shortens the time it takes for patients to get life-saving medical care by enabling faster and more effective delivery of essential medical supplies and equipment. This can be especially useful in emergency situations or places with poor transit facilities.

Drones can help lower the cost of delivering medical supplies in addition to making deliveries quicker. The usage of drones can lower costs for fuel, labor, and vehicle maintenance by minimizing the requirement for land transportation. Particularly in low- and middle-income nations with few resources, this can be crucial.

Drones can be used to carry medical supplies, but there are some drawbacks as well. For instance, drones can have their flight trajectories disrupted and have difficulty reaching their objective due to adverse weather conditions like high winds and rain. Additionally, there are worries regarding the security and safety of drones, especially in crowded regions where they may endanger people or property.

Despite these difficulties, there are a lot of advantages to employing drones to transport medical supplies. Drones are anticipated to play a bigger part in medical supply delivery in the future because of advancements in drone technology and regulations ensuring their safe and effective operation.

LITERATURE REVIEW

According to the authors, De Silvestri et al. [1], "smart cities" must overcome a number of challenges before urban air mobility (UAM) can become a reality. Flying Forward 2020 wants to advance this goal by undertaking trials in five European cities. First presenting the concept of distributing medications, San Raffaele Hospital established an unmanned aerial vehicles service. The use of this service as a platform to investigate its potential scaling to far more complicated situations, namely the transfer of medical supplies between hospitals, is conceivable. The drone service would be most useful in such a setting since the government's CO_2 emissions would

DOI: 10.1201/9781032686714-10

be reduced and urban traffic could be avoided. According to a recent study, the introduction of a quadcopter home delivery that lessens the stress on human-powered logistic procedures could improve the sector of medical logistics. The development of clinic productivity, which would lead to enhanced medical results, represents a crucial advantage for an experienced health establishment. There are enormous scientific and governmental obstacles that must be overcome before this can become a reality. The prototype model program will be made possible by a UAM ecosystem, which incorporates information exchange across a networked collection of online portals that could be controlled by numerous players, in the future scenario envisioned by this research. Unmanned traffic management (UTM) (sometimes referred to as European U-space) solutions were projected to be necessary in this regard to organize UAVs controlled by numerous users in the same environment.

As stated by the authors Tran, Thuy-Hang, and Dinh-Dung Nguyen [2], the writers from this study conducted a thorough investigation of the rules and laws governing drone management. While a few nations regulate drone use, several vehemently forbid use in populated areas. Drone-following models, AirMatrix, pathways and passageways throughout urban dwelling regions, and geofencing are just a few of the solutions for managing drone applications that have been suggested in order to ensure the safety of people, property, and additional users of aircraft. Vietnam is utilized as a case study because neither a license for drone operators nor any limitations on the technical specifications of drones are required there. Therefore, developing Vietnam's aviation safety regulation approach employing cutting-edge Internet of Things (IoT), automation, the 5G network, and machine learning may be a stimulating subject for further study. After reviewing and analyzing the relevant laws and management strategies, the authors provide drone governance and supervision in Vietnam focused on restricted and limited zones. Smaller civilian unmanned aircraft are still be subject to uncertain regulation under this legislation, which raises concerns about potential legal repercussions such as ineffective compensation responsibility, drone law violations, and information security. Content facilitates and aids the flight operations of drones within legal corridors. The reality, however, demonstrates that these circumstances have led to an astounding growth in drone application in Vietnam.

The findings of the writers from Hogan et al. [3], claim that in order for this innovative and growing sector to fulfill its potential, drone regulations must be altered. In an advanced and wealthy nation like Australia, it seems unfathomable that individuals in remote areas will continue to experience inadequate exposure to healthcare systems. In Australia, a lack of appropriate rules aimed at the firm's tolerance for autonomous drone activities has hindered the ability to enhance people's lives. People who reside in rural and suburban areas are currently losing their lives as a result of the nation's strict regulations. All Australians should have access to basic healthcare, regardless of where they live. Today, many Australians must drive a considerable distance to purchase their prescriptions. Two areas in particular that could benefit from improved regulations are confidentiality and loud sound control. Modifications to laws governing the use of aircraft for drug carriers may begin with efforts to ensure effective supervision in the areas of security and sound disturbance.

Taking on the problems that the study's findings suggest might be beneficial to the drone industry's growth and address public health challenges, which may have a good effect on those who do not currently have access to quality healthcare.

Hii et al. [4] claim that the use of drones for the delivery of medication to remote locations has become more widespread recently. The lack of information in relation to the viability and effects of package delivery of medications has caused the require-ment for such an assessment. The effectiveness of metformin was tested in condi-tions similar to 30-minute delivery drones, which included temperatures between 20 and 40°C and vibration frequencies between 0 and 40 Hz. Actrapid experienced only a little bubble that developed after the trip; otherwise, there were no adverse effects. As far as the authors are aware, this marks the first occasion in which a drone has successfully tested the efficacy of insulin, proving that it is possible to deliver insulin via drone while retaining pharmaceutical stability and high-quality medicine. As a result, this study offers a framework for further investigation of other medical proce-dures, environmental factors, or different drones.

The main finding of the study, according to Sham et al. [5], was that opinions on the utilization of aircraft to transport pharmaceuticals and immunizations were still divided, with slightly more than half of the remote healthcare workers (54%) agree-ing that using drones to distribute pharmaceuticals and immunizations is a sound concept, a sensible choice (54%), and a desirable option (52.5%). Their perspective is influenced by a variety of variables, including their gender, the leadership's innova-tion, and the delivery's potential risk. A favorable perspective was also substantially connected with the advantages of using aircraft, including the benefits of speed, renewability, wider interoperability, and lesser complexity. It is crucial to stress that since these inconsistencies might be part of their roles within the healthcare system, it is crucial to better study how employment and department affect attitudes among healthcare workers concerning the use of drones. Understanding the factors influ-encing drone acceptability may make it possible to develop policies and procedures that ease the usage of drones in the supply of medication as well as vaccinations to remote areas.

Sham et al. [6] state that, in the current study, the transportation of intermedi-ate and small containers with various payload options using a moderate drone was studied. The following information were collected from the demographic sheet: age, gender, ethnicity, religion, marital status, occupation, department, hospital/clinic where they are working, and highest education attained. The author's embedded solution, which includes interoperability with medical transporters, features a design that is user centered. The cargo concepts are examined from one number of perspec-tives, starting with package design, which considers the product's material, form, and manufacture, Next, the packages were integrated into the drone's airframe, with a focus on potential rapid system concepts. Special attention was paid to the user's interface with the drone, with a focus on accessibility and security measures. Additionally, a quick-release system idea was set and an industrialized package pro-totype was created. More analyses are now being performed to combine control and surveillance features for the delivery and to assess the delivered item's quality throughout transport.

Innocenti et al. stated in their research that the demonstration and evaluation of a feasible proof of concept for the distribution of medication utilizing a modern city unmanned aerial vehicles and fiducial indicators is the primary contribution of this study. A number of factors were taken into consideration when evaluating the effectiveness of two alternative fiducial marker systems, including object recognition effectiveness using the research's most trustworthy metrics, improved detection distance, pose estimate inaccuracy, and ultimately correct recognition speed. Even though in-depth approaches to teaching and machine knowledge are widely used today, the strength and maintenance requirements of these devices may occasionally outweigh the performance improvements. This study examined the potential for combining aircraft and fiducial indicators that pinpoint the ideal circumstances for the system to operate most efficiently. Future research will focus on performing in-depth evaluations of particular topics, such as security checks against unauthorized assaults, route improvement for increased energy efficiency, and sample restoration in parallel to delivery of drugs.

According to Gunaratne et al. [8], in global healthcare logistics, many developing nations still rely on the same old traditional distribution networks. The growing financial burden, healthcare product waste, and time spent on distribution point to the need for innovative initiatives. It was possible to further enhance the operational and environmental factors by using just UAVs for distribution. Baltic states can decrease logistical issues as well as increase productivity by using drones to get around insufficient transportation networks and infrastructure. For a moderately sized nation, receiving assistance from international donors for the early setup and information exchange is the best execution option. It is essential to carry out research that is pertinent to each nation in order to identify the quadcopter settings that are most advantageous.

Munawar et al. [9], claimed an intelligent method for distributing self-testing kits to patients who may be infected and returning the samples to the labs. It intelligently decides when to distribute the test kits using AI. The benefits of the suggested approach are lower carbon footprints, which are especially beneficial for developing nations. If studies of the effect are the most important consideration, Tabu might not be the greatest option; cross, interpersonal, and micro advantages offer better efficiency benefits. This is primarily because of its insufficient battery power and handling capability for testing kits. Future solutions could involve employing larger, longer-range UAVs that are powered by solar energy.

The authors, Robakowska et al. [10], argued that since drones for private usage have become more affordable in recent years, preemptive efforts to avert disasters and lessen emergency events have become necessary. In the broadest sense, drones are being used more and more to help healthcare. However, in order to deploy these technologies efficiently and ideally, further considerations such as regulating and stabilizing, selecting suitable technology and operational procedures (routing), and fostering collaboration among traders must all be taken into account.. Information and data from ongoing projects must be shared in order to advance the application of aircraft in medicine. There are a number of logistical, scope, and transportability concerns that need to be resolved in order to use the tools and techniques to enhance people's health, especially in dire situations. If drones are to be utilized to save lives worldwide, the system's limitations still need to be removed. In addition, it is necessary to employ additive UAVs to monitor human behavior to avert emergencies and undesired actions in concentrations of people. UAVs could therefore be used in a variety of medical settings. UAVs can transport blood, medicines, automated external defibrillators, vaccines, and other medical supplies. They can also help forensic investigators locate victims and secure hard-to-reach regions. Drone use in medicine has the potential to improve accessibility and quality while also reducing the overall cost of healthcare. Additionally, their urgent use will hasten the time to act, which is essential in a disaster.

The authors, Mohsan et al. [11], Elhaty, I.A., and Jahid, asserted that researchers and medical professionals have studied technological approaches to fighting the virus during the COVID-19 epidemic crisis. Researchers from diverse professions have suggested several approaches and remedies to stop this condition. In this situation, drones have emerged as a powerful tool in the fight against COVID-19 and have also shown great promise by providing quick solutions. In a future of enforced social alienation, aircraft may prove to be one of the most effective partners in the battle over COVID-19-like disorders. Through the use of relevant application scenarios and real-world examples, this research examines drone technology. Several contemporary technologies can be used with drones to enable scattered processing features. Despite this, a few potential problems are also briefly discussed in this presentation. The use of drones for surveillance, inspection, message distribution via barcode scanners and sirens, transportation of essential medicines, antibacterial spraying, patient scanning, and transportation of patient information are some of the COVID-19 risk reduction strategies investigated in this research. The writers also covered other advanced products, such as virtual reality, edge computing, blockchain, machine learning, deep learning, and artificial intelligence, which provide a variety of technical advancements for the minimal to nonexistent involvement of humans in the control of this disease. The study's authors stated that it would benefit academics and professionals by assisting them in overcoming the many difficulties that make using UAVs in these kinds of situations difficult.

The researchers, Quintanilla Garcia et al. [12], claimed that this initiative had shown the viability of the UAV delivering ConOps, which could become rapidly used as supplemental assistance as transportation services in emergency circumstances without the necessity of facilities or support functions continuing to support the procedure, in addition to the ground reality, given that the relatively few executed flight doesn't always start serving a foundation for safeguard the amount of safety associated with such processes. However, since it has not yet attained the necessary maturity level for large-scale implementation, the proposed ConOps has a number of restrictions at this time. There is additional work to be done, particularly in the areas of integration and support for Beyond Visual Line of Sight (BVLOS) operations in inhabited airspace. U-Space/UTM operations, conformity assessment, and methods development that will allow for the widespread deployment of Unmanned Aircraft Systems (UASs) operations will be aided by this project, which acts as an experimental use case.

The research by authors Amicone, D., Cannas, A., Marci, A., and Tortora, G., revealed the following: an AI-enhanced Smart Capsule for the demonstration of the transmission of plasma and serum elements. Also, a flying trial session was conducted to evaluate the advantages of utilizing an intelligence device as just a drone flight controller. The minimal training needed for healthcare professionals to utilize the software that allows them to obtain the information available or, perhaps, take action should there be an inadvertent transport disruption, will be established through further research. The testing revealed that perhaps the Smart Capsule, which acts as a backup device to assure the safety of operations, can initiate and oversee landings while affecting the drone's fundamental safeguards and security measures. The capability of the Smart Capsule to regulate the temperature in a practical environment will be thoroughly tested in future experimental service delivery. In order to determine the drone device's vulnerabilities, an analysis of the system's safety features will be done in the future. The electronics and programming both provide cyberattack issue perspectives will be examined in this activity, and preventative measures will be put in place to prevent malicious users (i.e., hackers) from tampering with the drone system. Using finite element analysis (FEA) simulations, the Smart Capsule will also be examined from a mechanical and thermal point of view, and appropriate crash tests will be carried out for certification. By including AI inside the control loop, it will be possible to execute appropriate risk mitigation methods in the event of an emergency, thus enhancing the Smart Capsule's ability to respond to undesirable situations. Labeling will be added when the future service is made available so that the Smart Capsule can be correctly identified in a medical setting. As a result of our research, we can say that using a Smart Capsule with AI to enable With redundant devices and AI-controlled drones, automated flight will provide a dependable choice for this type of distribution and open up a whole new realm of feasibility.

1. The study [14] by Beck, S. et al. found the following: In the healthcare sector, drones are capable of carrying out a range of useful and interesting tasks. Worldwide adoption of healthcare drones has accelerated due to the COVID-19 pandemic. Due to the emergence of business alliances that have increased on-demand medical shipments, the most underprivileged people in society have maintained their functionality. Due to their qualities of speed, accuracy, and adaptability, drones are particularly appealing for emergency healthcare missions. This study provides critical evidence that backs up the theory that drones might successfully deliver an EpiPen during an allergic reaction. This study demonstrated that formulations following pharmacopeia criteria in a reproducible manner did not cause appreciable changes in epinephrine concentration or chiral proportions during drone delivery or under computer simulation drone conditions. Without a doubt, the rapid advancement of robotics and drone technology will lead to an increase in the use of drones and other similar devices in the future. Clinical staff must undertake drone education to assuage any fears. Additionally, the Civil Aviation Authority (CAA) must first give its

regulatory approval before drone technology can be adopted, integrated, and optimized going forward in healthcare. Future studies on utilizing quadcopters to provide treatment should use this work as a springboard. More work on such applications in medicinal use of UAVs should build on the findings of this study. When creating future missions, the authors suggest conducting an observation survey of the pertinent group of health-care experts due to their viewpoints, acknowledgment, and worries which are crucial to the planning as well as objective review. Furthermore, the pharmacopeia suggestions, pharmaceutical value, and security standards, and most importantly, patient experience before, during, and after the flight must first be taken into account.

2. The research [15] from showed that the vast bulk of the research centered on the strategic level of transportation problems, with vans collaborating with drones to deliver packages. More research is starting to state there is poten-tial for installing fast chargers to expand the drones' field of view. A crucial component of the task's problem, which hasn't accumulated a considerable amount of curiosity yet, is determining the ideal number of drones for each depot or station. Nearly all of those findings concentrated on even a mod-est number of clients and a small network of aerial drones. Only randomly generated data was used to illustrate and apply the entirety of the algebraic calculations and computational efficiency of the suggested models for autono-mous drone challenges. Operations for parcel delivery using drones are still in their infancy. In research studies within theoretical frameworks, determin-ing which of the patient's demands should be prioritized often involves con-sidering various factors The question of "where do aircraft go when there are no more clients to serve?" must be discussed to operate these complex sys-tems. A theoretical foundation that illustrates the essential principles needed to create drone delivery models may be developed using a literature assess-ment of prior investigations and hypotheses on autonomous drone systems in logistics. The survey's strategic conclusions may be supported by a number of research projects, both on the planning and engineering sides, such as the cost control of drones, storage areas, batteries, and battery charging stations.

3. Borghetti et al. [16] reported that logistics for the last mile are essential for getting goods to customers. For last-mile delivery of goods, vans, the major part that are powered by internal combustion engines, are commonly used. Drones might reduce the amount of vans on the route in Milan, reduc-ing traffic jams and harmful emissions. Results indicated that after a few years of operation, this approach brings in money for the logistics opera-tor. Drones might lessen interactions between the courier and the final cus-tomer, which would make it less likely that individuals would be in close proximity to one another. The following are the research's shortcomings. 1) Particular restrictions and laws apply to the use of drones in metropolitan areas, particularly those that are heavily populated. 2) Because there are so many dense, towering buildings in a metropolis, the last stage of delivery can be crucial.

DISCUSSIONS

Drones, or unmanned aerial vehicles, have been a topic of interest for many years now, with advancements in technology allowing for more practical and versatile uses in various industries. One of the most significant industries where drones have made a considerable impact is the medical sector. The delivery of medical supplies, the management of healthcare, and the conduct of medical research have all been changed by the usage of drones in the medical industry.

ADVANCEMENTS IN DRONE TECHNOLOGY

Since the inception of drones, there have been numerous advancements in their design and technology. These advancements have allowed drones to be more versatile and efficient in their applications, especially in the medical sector.

One of the significant advancements has been the development of autonomous drones, which can fly without the need for human intervention. These drones can fly to remote locations, delivering essential medical supplies and equipment, without putting human lives at risk.

Another advancement in drone technology is the integration of sensors and cameras, which enable monitoring in real time and tracking of medical deliveries. The use of sensors and cameras also enables drones to fly in challenging terrains, ensuring that essential medical supplies reach remote and underserved areas.

The integration of artificial intelligence and machine learning algorithms in drones has allowed for more accurate and efficient medical deliveries. These algorithms can track the drone's location, monitor the environment, and adjust the drone's flight path to avoid obstacles, thereby reducing the chances of accidents and damage to medical supplies.

MEDICAL APPLICATIONS OF DRONES

The delivery of medical supplies, the management of healthcare, and the conduct of medical research have all been changed by the usage of drones in the medical industry. Drones can be used in medicine for the following purposes:

Drones are being used to transport vaccines, medications, and medical supplies to remote and underserved areas for use in medical deliveries. As a result, there is a lower chance of medical supplies being harmed or lost in transit. This has increased the speed and efficiency of medical deliveries.

Telemedicine: Patients in remote locations can receive remote medical consultations from doctors using drones that are equipped with telemedicine technology. For people who live in remote areas with few healthcare facilities, this has improved access to healthcare.

Medical Research: Drones equipped with cameras and sensors can be used to collect data for medical research purposes. For example, drones may be utilized to keep an eye on and track the transmission of infectious illnesses, such as malaria, in far-off places.

CONCLUSION

In conclusion, technology has the potential to completely transform how medical supplies are delivered to remote areas. With the ability to fly over challenging terrain and reach areas that are difficult to access by road, drones can bring life-saving medicines and equipment to those in need. The speed and efficiency of drones mean that vital supplies can reach their destination much faster than by traditional means, reducing the amount of time people have to wait for medical assistance. This can save countless lives and improve the overall health and well-being of people in remote areas.

However, there are some challenges that need to be overcome before drone technology can be fully embraced for medical supply delivery. One of the biggest challenges is the issue of security. Drones can be vulnerable to hacking, which could put medical supplies at risk. Additionally, drones can also be vulnerable to environmental conditions, such as strong winds, which can disrupt flights. To overcome these challenges, it is essential to have robust security protocols in place and to have well-designed drones that are able to withstand harsh conditions.

Another challenge is cost. Drone technology can be expensive, and the cost of setting up a drone delivery system for medical supplies can be prohibitive for some countries. To overcome this challenge, governments and private organizations will need to work together to ensure that the necessary resources are made available to make drone technology a reality for medical supply delivery. This could include funding for research and development, as well as investment in infrastructure, such as airfields and landing zones.

REFERENCES

De Silvestri, S.; Pagliarani, M.; Tomasello, F.; Trojaniello, D.; Sanna, A. Design of a Service for Hospital Internal Transport of Urgent Pharmaceuticals via Drones. *Drones* 2022, *6*, 70. https://doi.org/10.3390/drones6030070

Tran, T.-H.; Dinh-Dung, N. Management and Regulation of Drone Operation in Urban Environment: A Case Study. *Social Sciences* 2022, *11*, 474. https://doi.org/10.3390/socsci11100474

Hogan, W.; Harris, M.; Brock, A.; Rodwell, J. What Is Holding Back The Use of Drones for Medication Delivery in Rural Australia? *Sustainability* 2022, *14*, 15778. https://doi.org/10.3390/su142315778

Hii, M.S.Y.; Courtney, P.; Royall, P.G. An Evaluation of the Delivery of Medicines Using Drones. *Drones* 2019, *3*, 52. https://doi.org/10.3390/drones3030052

Sham, R.; Siau, C.S.; Tan, S.; Kiu, D.C.; Sabhi, H.; Thew, H.Z.; Selvachandran, G.; Quek, S.G.; Ahmad, N.; Ramli, M.H.M. Drone Usage for Medicine and Vaccine Delivery during the COVID-19 Pandemic: Attitude of Health Care Workers in Rural Medical Centres. *Drones* 2022, *6*, 109. https://doi.org/10.3390/drones6050109

Saponi, M.; Borboni, A.; Adamini, R.; Faglia, R.; Amici, C. Embedded Payload Solutions in UAVs for Medium and Small Package Delivery. *Machines* 2022, *10*, 737. https://doi.org/10.3390/machines10090737

Innocenti, E.; Agostini, G.; Giuliano, R. UAVs for Medicine Delivery in a Smart City Using Fiducial Markers. *Information* 2022, *13*, 501. https://doi.org/10.3390/info13100501

Gunaratne, K.; Thibbotuwawa, A.; Vasegaard, A.E.; Nielsen, P.; Perera, H.N. Unmanned Aerial Vehicle Adaptation to Facilitate Healthcare Supply Chains in Low-Income Countries. *Drones* 2022, *6*, 321. https://doi.org/10.3390/drones6110321

Munawar, H.S.; Inam, H.; Ullah, F.; Qayyum, S.; Kouzani, A.Z.; Mahmud, M.A.P. Towards Smart Healthcare: UAV-Based Optimized Path Planning for Delivering COVID-19 Self-Testing Kits Using Cutting Edge Technologies. *Sustainability* 2021, *13*, 10426. https://doi.org/10.3390/su131810426

Robakowska, M.; Ślęzak, D.; Żuratyński, P.; Tyrańska-Fobke, A.; Robakowski, P.; Prędkiewicz, P.; Zorena, K. Possibilities of Using UAVs in Pre-Hospital Security for Medical Emergencies. *International Journal of Environmental Research and Public Health* 2022, *19*, 10754. https://doi.org/10.3390/ijerph191710754

Mohsan, S.A.H.; Zahra, Q.U.A.; Khan, M.A.; Alsharif, M.H.; Elhaty, I.A.; Jahid, A. Role of Drone Technology Helping in Alleviating the COVID-19 Pandemic. *Micromachines* 2022, *13*, 1593. https://doi.org/10.3390/mi13101593

Quintanilla García, I.; Vera Vélez, N.; Alcaraz Martínez, P.; Vidal Ull, J.; Fernández Gallo, B. A Quickly Deployed and UAS-Based Logistics Network for Delivery of Critical Medical Goods during Healthcare System Stress Periods: A Real Use Case in Valencia (Spain). *Drones* 2021, *5*, 13. https://doi.org/10.3390/drones5010013

Amicone, D.; Cannas, A.; Marci, A.; Tortora, G.; Smart, A. Capsule Equipped with Artificial Intelligence for Autonomous Delivery of Medical Material through Drones. *Applied Sciences* 2021, *11*, 7976. https://doi.org/10.3390/app11177976

Beck, S.; Bui, T.T.; Davies, A.; Courtney, P.; Brown, A.; Geudens, J.; Royall, P.G. An Evaluation of the Drone Delivery of Adrenaline Auto-Injectors for Anaphylaxis: Pharmacists' Perceptions, Acceptance, and Concerns. *Drones* 2020, *4*, 66. https://doi.org/10.3390/drones4040066

Benarbia, T.; Kyamakya, K. A Literature Review of Drone-Based Package Delivery Logistics Systems and Their Implementation Feasibility. *Sustainability* 2022, *14*, 360. https://doi.org/10.3390/su14010360

Borghetti, F.; Caballini, C.; Carboni, A.; Grossato, G.; Maja, R.; Barabino, B. The Use of Drones for Last-Mile Delivery: A Numerical Case Study in Milan, Italy. *Sustainability* 2022, *14*, 1766. https://doi.org/10.3390/su14031766

11 A Novel Method of Human Activity Recognition from Sensor Data Using Artificial Intelligence

N. Badrinath, Jeevana Jyothi Pujari,
D. Jagadeesan, G. Asha, and Konatham Sumalatha

INTRODUCTION

Monitoring daily activities has received great attention due to the rapid spread and cost reduction of sensing hardware, which led to an extraordinary surge (Katz and Chimn (1962), Khan and Ghani (2021)). Recognizing human activity has become a crucial technique in several real-world applications like healthcare, gait analysis, assisted living, security, and smart homes, and it's because of the widespread use of wearable sensors such as mobile phones, smartwatches, and sports bracelets Teng et al. (2020), Wang et al. (2019), Janidarmian et al. (2017). The wearable (such as mobile phones, smartwatches, and sports bracelets) Qian, Pan, and Miao (2021) sensor method describes the use of intelligent electronic devices integrated into wearable objects or in direct contact with the body to measure biological and physiological sensor signals like heart rate, blood pressure, body temperature, accelerometers, or other interesting characteristics like motion and location. These sensors are connected to a communication tool such as a laptop, smartphone, or specially designed embedded system Nguyen et al. (2019). An application server receives raw signals to monitor, visualize, and analyze them in real-time. Human activity recognition can be achieved through wearable sensors that capture time series data from accelerometers Khaertdinov, Ghaleb, and Asteriadis (2021a), gyroscopes, magnetometers, etc. Kwapisz, Weiss, and Moore (2011). These sensors are simple to use and can be easily accessed through our smart devices. They effectively recognize human activity in real time from a distance due to their small data volume and easy sharing capabilities through the internet (Mahmud et al. (2020)).

DOI: 10.1201/9781032686714-11

Mobile devices provide a chance to gather vast amounts of data on human behavior, such as physical activity, sleep, nutrition, medication adherence, and social interactions. The use of smartphones and mobile devices, referred to as "mHealth" (mobile health) is a growing field in healthcare which is still discussing whether or not mHealth (Dicianno et al. (2015) Chiarini et al. (2013)Kumaresan et al. (2017)) is a specialized field. Through sensors, apps, and connectivity, mHealth platforms can collect this data continuously and passively, leading to a better understanding of behavior patterns and their impact on mental health (Faiola, Papautsky, and Isola (2019)). Still, it is clear that many people worldwide use mobile technology to access health services and information (Mechael (2009) Schnall et al. (2018)). have great potential for improving healthcare in developing and remote areas of developed countries (Hossain (2016)). Mobile digital devices, such as smartphones, tablets, and iPods, have gained popularity due to their portability and ease of use for internet connectivity, which enables remote internet access from almost any location (Lupton (2013)). Nowadays, smartphones are becoming smarter and can assist individuals with their everyday activities, and many smart healthcare systems utilize body sensor apps (Li et al. (2021b)). It uses sensors and advanced smartphone features to distinguish between different types of movements, such as walking or running (Raza et al. (2023)).

Analyzing human activity goes beyond data collection; advanced methods like ML, data mining, and artificial intelligence (AI) can uncover insights to improve health outcomes (Miah et al. (2022)). AI is widely used in various fields for predictive analysis purposes, and this technology can learn from vast amounts of data and provide more accurate predictions than humans. AI is particularly useful in classifying activities based on visual and sensor data. Different ML (Li et al. (2021a)) algorithms, like support vector machines, RF, decision tree, and hidden Markov models, can be utilized, but deep learning (DL) techniques (Hamper et al. (2016)) are now the most popular and effective for many applications (Raza, Munir, and Almutairi (2022) Mukhopadhyay et al. (2021)). Deep neural networks (DNNs) are most commonly used in contemporary sensor-based Human Activity Recognition (HAR) research. CNNs, recurrent neural networks (RNNs), or hybrid models that combine the two designs are the most common architectures used to solve this challenge (Khaertdinov, Ghaleb, and Asteriadis (2021b)).

Our main research contributions focus on recognizing human activities and the significance of our study:

- A novel approach presents for detecting human activity by analyzing sensor data and utilizing AI methods; along this technique, we can accurately predict various types of human activities, including walking, running, jogging, standing, and jumping through sensor data. This methodology can potentially revolutionize the tracking and monitoring of human movement in numerous applications, from sports performance to healthcare.
- The other goal of this research is to analyze the performance of the ML classifier and proposed CNN-LSTM model.

- To check the model's ability, we implement our model on another real-world dataset.

The arrangement of this piece of writing is explained below. We used a specific methodology to conduct the research, which we shall describe in Methodology. A thorough literature assessment will be found in Related Word to identify any knowledge gaps. We shall review the experimental outcomes of our utilized models in Experimental and Evaluation. We shall then present our findings.

RELATED WORK

Based on recent research, scholars Mahmud et al. (2020) suggested a multi-stage training approach using DNNs and wearable sensor data for proper human action recognition. The approach optimizes feature extraction and performs exceptionally by merging deep CNN feature extractors from diverse transformed spaces. Experiments performed on public datasets demonstrated an average five-fold cross-validation accuracy of 99.29% on the UCI-HAR database, 99.02% on the USC-HAR database, and 97.21% on the SKODA database. Another study, Chung et al. (2019), explored on-body sensor positioning and data acquisition for HAR systems. Using eight body-worn inertial measurement unit (IMU) sensors and an Android mobile device, the study discovered that a low sampling rate from specific wrists, right ankle, and waist sensors was sufficient for recognizing Activities of Daily livings (ADLs). Also, utilizing a two-level ensemble model and classifier-level sensor fusion enhanced classification performance, with custom weights derived for multimodal sensor fusion based on activity-specific characteristics for human action recognition.

Scholars Ashry, Ogawa, and Gomaa (2020) introduced a high-performance continuous human activity recognition (CHAR) system utilizing deep neural networks. It concentrates on recognizing consecutive activities from IMU sensors on smartwatches. Dataset CHAR-SW is collected and cascading bidirectional long short-term memory (Bi-LSTM) with featured data performs well in accuracy, computational time, and storage space. The introduced technique for online operation proves effective, achieving real-time activity recognition with 91% accuracy. Experimental results demonstrate improved accuracy, reduced processing time by 86%, and reduced storage space by 97.77% compared to using raw data or existing approaches. Dua, Singh, and Semwal (2021) proposed an end-to-end deep neural network model for HAR that performs automatic feature extraction and classification without requiring extensive feature engineering or handcrafted techniques. The model utilizes CNN and gated recurrent unit components. Experiments conducted on raw data from wearable sensors achieve high accuracies of 96.20% on UCI-HAR, 97.21% on WISDM, and 95.27% on Physical Activity Monitoring Dataset (PAMAP2).

The classification of body gestures and motion to predict states of action or behavior during physical activity, known as HAR (Gumaei et al. (2019)), is an important field of study. To measure parameters such as range of motion, speed, velocity, and magnetic field orientation, inertial measurement units (IMUs) are commonly used. ML techniques (Kutlay and Gagula-Palalic (2016)), including extreme gradient boosting (XGBoost), multilayer perceptron (MLP), CNN (Challa, Kumar, and Semwal (2022)), LSTM, CNN and LSTM hybrid (ConvLSTM) and autoencoder by RF (AE w/ RF) are compared using raw data from on-body inertial sensors to classify human activities. The mHealth dataset, which includes 12 physical activities recorded from 10 subjects using 4 inertial sensors, was used in the study. The study results show that MLP and XGBoost achieved the highest accuracy (90.55%, 89.97%), precision, recall, and F1 score. Hamper et al. (2016) and Miah et al. (2022) obtained a dataset from a similar mHealth study that recorded the vital signs of 10 volunteers with different backgrounds, and these volunteers wore body sensors while doing different physical activities. They used five ML algorithms to analyze and predict human health behavior (XGBoost, naive Bayes, DT, RF, and logistic regression). XGBoost performed best with a 95.2% accuracy rate, 99.5% sensitivity rate, 99.5% specificity rate, and 99.66% F-1 score. This research shows a promising future for mHealth in predicting human behavior.

After reviewing the entire literature, we've discovered limitations, challenges, and deficits. Let's discuss this now:

- Traditional approaches depend substantially on practitioners' experience to extract interesting features; the deep features are difficult to learn, and a considerable amount of well-labeled, annotated training data (Khaertdinov, Asteriadis, and Ghaleb (2022)) needs to be included (Nguyen et al. (2019)).
- Unfortunately, trustworthy datasets are limited (Li et al. (2021a)), and research on wearable-sensor-based data is currently limited (Qian, Pan, and Miao (2021)), also, extracting useful features becomes increasingly complex with more sensors (Mahmud et al. (2020)).
- The camera-based sensors have various limitations, such as the inability to detect activity if the user is not within the camera's field of view, which creates privacy concerns for the user (Yadav et al. (2021)).
- The accuracy of sensors still needs to be stable, so there are several challenges in improving signal processing and transmission.
- The available evidence on the efficacy of technological devices for promoting healthy behavior change is currently little (Walsh and Groarke (2019)).

METHODOLOGY

We conducted our research through a series of steps, each with a specific task to complete. The initial step involved gathering data, followed by preprocessing the dataset, extracting features, and analyzing it. We then constructed our model and assessed its effectiveness. Below is a detailed description of each stage. Figure 11.1 depicts the entire methodology of our study.

DATASET

The mHealth dataset, fetched from the UCI Machine Learning Repository, is a large compilation of recordings that capture body movements and vital signs (Banos and

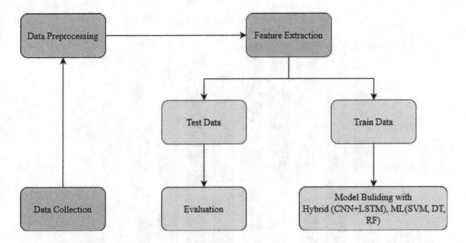

FIGURE 11.1 Overall steps of our research

Saez (2014)). This dataset includes recordings of the body's movements and vital signs from 10 volunteers who participated in 12 physical activities: standing still, Sitting and relaxing, lying down, walking, climbing stairs, waist bends forward, frontal elevation of arms, knees bending (crouching), cycling, Jogging, running, and jump front and back. The dataset records data collected by sensors placed on the person's chest, right-wrist, and left-ankle, measuring the movement of various body parts, including acceleration, rate of turn, and magnetic field orientation. At a sampling rate of 50 Hz, the sensors recorded the acceleration (speed of motion) and turn rate of those body parts. Furthermore, the sensor placed on the chest:

- Provides 2-lead ECG measures
- Provides possible information on fundamental heart monitoring
- Checks for different Arrhythmias or the impacts of exercise on the ECG

The mHealth dataset is a useful resource for ML research and experimentation. It allows the exploration of activity recognition algorithms, sensor fusion techniques, and the development of reliable models for predicting human activity.

The dataset has 1,215,745 rows in total, spread across 14 columns (1 target and 13 features). The total numbers of each class are shown in Figure 11.2. The columns of the dataset are as follows:

- Our features are: alx, aly, alz, glx, gly, glz, arx, ary, arz, grx, gry, grz. (In Table 11.1, we discussed these features).
- Our target variable is referred to as "Activity" and is located in the remaining column. (In Table 11.2, we discussed the activities and their corresponding labels).

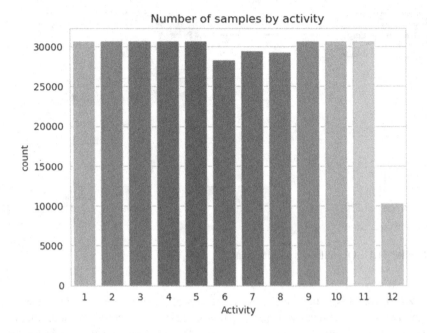

FIGURE 11.2   Number of samples by class

TABLE 11.1

The accelerometer captured 3-axis acceleration data, and the gyroscope captured 3-axis angular velocity data through the sensors

Prefix	Description
al	Acceleration from left-ankle sensor
ar	Acceleration from right-wrist sensor
gl	Gyro from left-ankle sensor
gr	Gyro from right-wrist sensor
Suffix	**Description**
x	X-axis
y	Y-axis
z	Z-axis

DATA PREPROCESSING

We have carefully examined the dataset for any zero, null, or missing values and checked the data types we deal with. We have found no null values in the dataset. The data collected by the sensors is stored as floats, whereas the activities are labeled

TABLE 11.2

Labels and descriptions of each class

Label	Activity Description
1	Standing still (1 min)
2	Sitting and relaxing (1 min)
3	Lying down (1 min)
4	Walking (1 min)
5	Climbing stairs (1 min)
6	Waist bends forward (20x)
7	Frontal elevation of arms (20x)
8	Knees bending (crouching) (20x)
9	Cycling (1 min)
10	Jogging (1 min)
11	Running (1 min)
12	Jump front & back (20x)

using integers. To serve our operational needs, we have expanded the subject value for operational purposes. The correlation matrix of selected columns is shown in Figure 11.3.

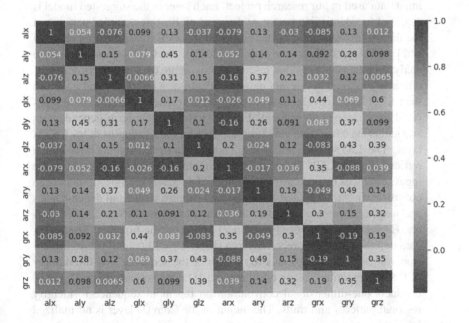

FIGURE 11.3 Correlation matrix for the selected columns

FEATURE ENGINEERING

Given that this was the sampling frequency specified in the dataset, we chose a window size of 1 second and a frequency of 50 Hz for our analysis. The window size is calculated by multiplying the window dimension and frequency settings. A dictionary maps the distinct activity classifications in the dataset to numerical labels. This procedure is carried out to simplify additional processing and analysis. By majority vote, the activity labels are determined for each formed window. The activity labels for the retrieved features are connected to them. A new data frame is created to hold the extracted features and activity labels for future modeling and analysis. We scaled the dataset using standard scaler.

METHOD

Our study evaluated the effectiveness of ML and a CNN-LSTM model combination. The integrated architecture can efficiently identify human activities from sensor data by capturing the spatial and temporal properties of the input signals by utilizing the advantages of both the CNN and LSTM layers. This method has been effective in several applications, including activity tracking, wearable technologies, and human behavior analysis. The purpose of utilizing Ml is to demonstrate how well the classifier works using activity recognition.

- **Hybrid CNN-LSTM model:** We provide a brand-new model architecture designed specifically to capture and handle the unique characteristics of the input data used in our research project. Each layer in the suggested model is essential for identifying pertinent data and replicating complex patterns in the data. The foundation of our proposed model is a combination of CNN and LSTM layers (Kosar and Barshan (2023)). The CNN layers are made to analyze input sequences that resemble geographic data in a particular way, efficiently extracting relevant data and identifying regional trends. We use the LSTM layer, a potent part of recurrent neural networks, to capture temporal dependencies in sequential data. Modeling long-term dependencies and gathering contextual data from earlier states are strengths of the LSTM layer. The LSTM layer efficiently captures both the spatial information retrieved by the CNN layers and the temporal relationships present in the input sequences by processing the output of the CNN layers. Our model can concurrently extract spatial attributes and capture sequential dependencies and long-term context thanks to the combination of CNN and LSTM layers. Each layer in our suggested model has a specific function: The sensor data that the input layer receives has a shape [100, 13 (x- and y-axes dimension values)], which indicates the length of the sequence and the number of features recorded. Using 64 filters, the 1D convolution n layer (Conv1D) conducts one-dimensional convolution on the input sequence to identify regional patterns and traits. The output of the Conv1D layer is normalized by the batch normalization layer, ensuring uniform scaling of features.

By choosing the highest value inside each local zone, the MaxPooling1D layer performs down sampling, capturing key elements while shortening sequence length. With 64 units, the LSTM layer simulates the temporal and long-term dependencies in the sensor data and the dynamics of human behavior. The dense layer with 128 units linearly transforms the output of the LSTM layer, and nonlinearity is introduced by the rectified linear unit (ReLU) activation function, allowing for the extraction of more abstract information and the learning of higher-level representations. We can forecast the work at hand accurately using this model architecture, and we can also spot regional and global trends in the incoming data. The layered method guarantees that each element contributes to the overall model performance without duplication. Figure 11.4 displays a graphical illustration of our suggested model. The hyperparameters during the training process of our proposed model are shown in Table 11.3.

- **Machine learning model**: In our research, we also used several machine classifiers to identify human activity from mobile phone sensor data. The classifiers are created through several processes. First, we did not categorize the action according to custom or normal practice. To determine the ideal parameters for our classifier, we used GridSearchCV (Chandrasekhar and Peddakrishna (2023)). The RF (Putri (2023)), DT (Gong et al. (2023)), and SVM classifiers (Choudhary, Pathak, and Chaubey (2023)) are employed in our basic model. RF is a flexible machine learning technique that accurately and robustly classifies human behavior from sensor data using a group of decision trees. This classifier uses several parameters to produce the best results. The following settings were used: n estimators (50, 100, 150), max depth (10, 20, None), min samples split (2,5,10), min samples leaf [1, 2, 4], and max features: (auto, sqrt). In our research project, SVM has been used to categorize human actions from sensor data precisely. SVM effectively differentiates activity patterns by locating an ideal hyperplane, making it suitable for handling complex decision boundaries and high-dimensional

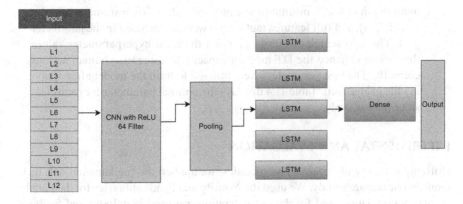

FIGURE 11.4 Proposed CNN-LSTM architecture

TABLE 11.3

hyperparameters of our proposed model during training

Parameter	Value
monitor	val loss
patience	50
verbose	1
optimizer	adam
loss	sparse categorical crossentropy
metrics	sparse categorical accuracy
epochs	10

TABLE 11.4

Best parameters for different models

Model	Best Parameters
DT	Criterion gini, Max Depth: 10, Max Features: auto Min Samples Leaf: 4, Min Samples Split: 10
SVM	C: 0.1, Gamma: scale, Kernel: linear
RF	Max Depth: None, Max Features: auto, Min Samples Leaf: 1 Min Samples Split: 2, N Estimators: 150

data common in human activity detection. GridSearch values for the SVM classifier were C (0.1, 1, 10), kernel ("linear," "rbf," "sigmoid"), and gamma ("scale," "auto"). Our research used the DT algorithm to identify human activity from sensor data. A grid search examined various hyperparameter combinations to improve the DT model. Criterion (gini, entropy), maximum depth (10, 20), minimum samples split (2, 5, 10), minimum samples leaf (1, 2, 4), and full features (auto, sqrt) were all included in the parameter grid. The grid search sought to pinpoint the ideal hyperparameter values that would enhance the DT model's capacity to categorize human actions correctly. The best parameters were utilized to train the model after applying the grid search. Table 11.4 displays the optimal parameter for each classifier for our study.

EXPERIMENTAL AND EVALUATION

Utilizing a variety of libraries and modules, we applied the fundamental ML technique in our research study. We used the NumPy and Pandas libraries for data analysis and manipulation, and for data visualization, we used Matplotlib and Seaborn.

Our approach relied heavily on scikit-learn, a large ML library that gave us access to many different algorithms and tools. Using the train test split function, we divided our dataset into training and testing sets. Model performance was then evaluated using cross-validation with StratifiedKFold. Using scikit-learn's metrics module, several performance metrics, including accuracy, precision, recall, and F1-score, were calculated. The categorization outcomes were examined using confusion matrices. GridSearchCV was used for hyperparameter tuning to identify the ideal parameters for enhancing model performance. Our research used the Keras library with TensorFlow as the backend to create the CNN-LSTM model. We used a computer system with an Intel Core i5 8th generation processor, 12 GB of RAM (random access memory), and a central processing unit (CPU) clocked at 3.6 GHz to implement and experiment for our research work. It's important to note that we lacked access to a dedicated graphics processing unit (GPU) for faster computations. The execution of our experiments took a bit longer as a result. But despite the lack of a GPU, we could still run our code and perform the necessary computations. For our investigation, we employed the Python programming language and Google Coolab.

Our model is evaluated using four research questions (RQ):

RQ1: How well does the ML model perform?
RQ2: How does the CNN-LSTM model perform?
RQ3: How well does our suggested model perform compared to the ML model?
RQ4: How does the proposed model perform on another real-world dataset?

RESULT ANALYSIS

RQ1: How Well Does the ML Model Perform?

Table 11.5 overviews how well our baseline models identified human activity. The RF, SVM, and DT models were assessed using F-score, precision, and recall metrics. The RF model successfully classified human actions with a precision of 95.67%, which is an amazing result. It also showed a recall of 95.21%, capturing a large percentage of real positive occurrences. The harmonic mean of precision and recall, known as the F-score, was calculated to be 95.23%, suggesting a performance that balanced precision and recall. The SVM model also performed well, with a precision and recall of 94.76% and 93.54%, respectively. The precision and recall of the

TABLE 11.5
Performance of baseline model

Model	Precision	Recall	F-score
RF	95.67%	95.21%	95.23%
SVM	94.76%	93.54%	94.60%
DT	94.31%	94.72%	94.08%

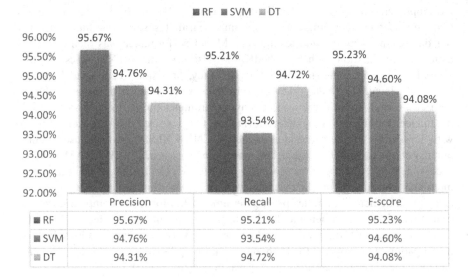

FIGURE 11.5 The performance of precision, recall, and F-score of machine learning classifier

SVM were calculated to be in a strong balance with an F-score of 94.60%. Precision and recall for the DT model were 94.31% and 94.72%, respectively. However, it provided a significantly lower F-score of 94.08%, showing a compromise between recall and precision. Overall, the ability of our baseline models to identify human activity from the sensor data showed promise. The best precision, recall, and F-score results came from the RF model, demonstrating how well it can categorize human activities. While the DT model displayed competitive results, the SVM model also performed well. Figure 11.5 displays the performance of ML classifiers graphically.

RQ2: How Does the CNN-LSTM Model Perform?

From Table 11.6, with a precision of 98.01%, our CNN-LSTM model demonstrated its ability to categorize human activities correctly. The high precision score shows

TABLE 11.6

The performance of precision, recall, and F-score of proposed CNN-LSTM model

Evaluation Metrics	Value
Precision	98.01%
Recall	98.10%
F-score	98%

the model's ability to reduce false positive predictions and deliver accurate outcomes. The model also showed a recall of 98.10%, demonstrating its ability to catch a substantial fraction of true positive events. According to the high recall score, the model appears to have successfully detected the bulk of human activities in the dataset. The F-score was determined to be 98%. This balanced score, which considers precision and recall, represents the model's overall performance. A high F-score indicates that the model successfully balances precision and recall, demonstrating its efficacy in correctly identifying human actions. The performance of our model during the training and testing phases is shown by the training loss value of 0.022% and the test loss value of 0.063%, respectively. The loss value represents the difference between the model's projected outputs and the actual labels. The model has been successfully trained to reduce the errors in its predictions when the loss value is smaller. Since the model has learned the patterns and characteristics in the training data and is generalizing effectively to unknown test data, our situation's training and test losses may be reasonably small. This shows that the model works well to minimize errors and has attained good accuracy. It suggests that the model can make predictions with high dependability and confidence. Overall, the modest training and test losses show how well our model captures the fundamental patterns in the data and produces reliable forecasts. The training and validation loss of the proposed model is shown in Figure 11.7. Figure 11.6 shows the result graphically.

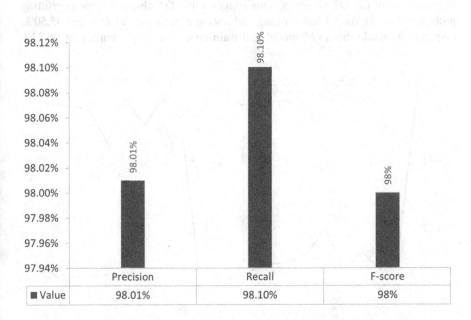

FIGURE 11.6 The performance of precision, recall, and F-score of proposed CNN-LSTM model

TABLE 11.7

Accuracy of our Proposed CNN-LSTM model and machine learning classifier

Model	Training Accuracy	Testing Accuracy
RF	97.25%	95.50%
SVM	96.82%	95.00%
DT	99.16%	94.89%
CNN-LSTM	99.35%	98.29 %

RQ3: How Well Does Our Suggested Model Perform Compared to the ML Model

Table 11.7 compares the performance of our proposed CNN-LSTM model with that of the RF SVM and DT ML classifiers. A phenomenal training accuracy of 99.35% and a testing accuracy of 98.29% were attained by our CNN-LSTM model. These findings show that the model can accurately and successfully learn to classify human actions. In terms of accuracy, the CNN-LSTM model surpassed all other ML classifiers, proving its supremacy in identifying intricate spatial and temporal patterns in sensor data. The DT model had the greatest training accuracy of 99.16% among the ML classifiers. However, compared to the CNN-LSTM model, its testing accuracy of 94.89% was lower. There is a significant gap in the training accuracy and testing accuracy of the DT classifier. This suggests that DT also has some overfitting problems. The RF model had training and testing accuracy of 97.25% and 95.50%, respectively, while the SVM model had training and testing accuracy of 96.82%

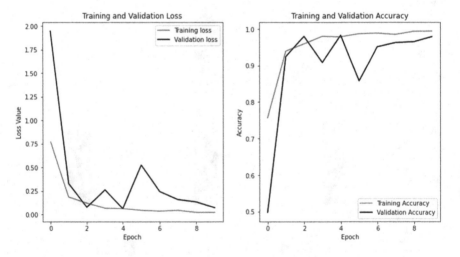

FIGURE 11.7   training and validation loss and accuracy of the proposed CNN-LSTM model

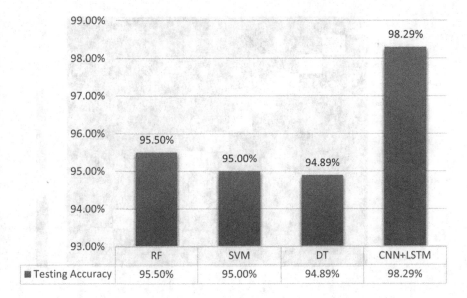

FIGURE 11.8   Testing accuracy of all models

and 95.00%. Therefore, it can be seen that the suggested model increases accuracy by about 2.1% when compared to RF and SVM. Comparing Tables 11.5 and 11.6 further demonstrates that our suggested model achieves the best results from the ML classifier in all evaluation criteria. The accuracy of all models is shown graphically in Figure 11.8. Training and validation accuracy of the proposed model is shown in Figure 11.7.

DISCUSSION

HAR is essential to track and evaluate people's movements for medical applications. It can help identify and treat several medical disorders, including the prevention of general well-being loss in the elderly, tracking physical therapy exercises, identifying gait anomalies, and fall prevention. It can also be used for sports and fitness, enabling tracking of physical activities like swimming, cycling, and running while offering in-sightful information on performance indicators, training intensity, and advancement. It can also be utilized for behavior analysis.

This demonstrates how crucial it is to create an accurate system. As previously mentioned, two models are chosen for our investigation, and both exhibit promising performance. However, there are certain restrictions in the dataset. Figure 2 shows that there is a slight imbalance in the data. This is also evident in the confusion matrix in Figure 11.9. Therefore, additional studies can be done to address these constraints. We test our model with another dataset (Davis and Owusu (2016)). Six

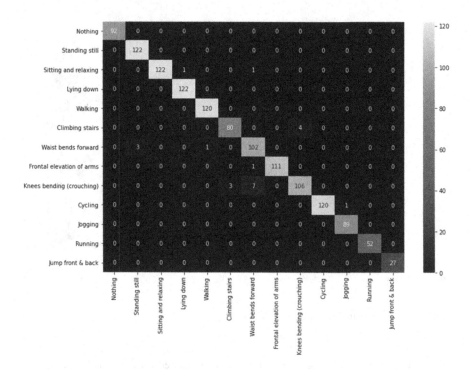

FIGURE 11.9   Confusion matrix of proposed CNN-LSTM model

TABLE 11.8
The performance of the proposed model on another dataset

Evaluation Metrics	Value
Precision	91.29%
Recall	92.43%
F-score	91.65%
Accuracy	93.42%

classes make up the dataset. In this dataset our proposed model performs admirably. The performance of our suggested model on this dataset is shown in Table 11.8.

Results from Table 11.8 indicate that our suggested model on a different dataset were encouraging. The model successfully identified positive events with a precision of 91.29% and recall of 92.43%. The balanced performance of the model was further supported by the F-score of 91.65%. The model's overall accuracy of 93.42% demonstrates its potent capacity for prediction. These findings demonstrate how well our suggested model can identify and foretell human activity patterns.

CONCLUSION

Nowadays, activity recognition has become a crucial part of various practical applications, particularly in the field of healthcare, and researchers are highly concerned about improving the performance of recognition algorithms. This chapter aims to present a new method for recognizing human activities using data from wearable sensor devices and the acceleration and rate of turn of various body parts using ML and a hybrid CNN-LSTM Model. The proposed human activities, such as walking, jogging, standing, jumping, and lying down, etc. identification methodology, which uses recording data of body movement and vital signs and artificial intelligence, is a potential way to precisely identify and categorize different human motions. The techniques rely on data collected from sensor devices on the accelerometers, gyroscopes, and magnetometers, which have been developed and assessed. How we interact with smartphones, smartwatches, and embedded microprocessors may change if ML and DL are implemented on these devices. We can apply AI directly for activity recognition because of the growing availability of powerful CPUs and sensors on mobile devices. We compared the effectiveness of using ML and a hybrid CNN-LSTM model in identifying human activities from sensor data. By combining the strengths of both CNN and LSTM layers, the integrated architecture efficiently captures the spatial and temporal properties of the input signals. Our CNN-LSTM model showed the highest accuracy among the applied AI-based techniques and performed exceptionally well in identifying human motion.

Furthermore, the method could be tested on a larger scale and on various datasets to evaluate its performance in different scenarios.

REFERENCES

Ashry, Sara, Tetsuji Ogawa, and Walid Gomaa. 2020. "CHARM-deep: Continuous human activity recognition model based on deep neural network using IMU sensors of smart watch." *IEEE Sensors Journal* 20 (15): 8757–8770.

Banos, Garcia Rafael Oresti, and Alejandro Saez. 2014. "m-Health dataset." *UCI Machine Learning Repository.* https://doi.org/10.24432/C5TW22.

Challa, Sravan Kumar, Akhilesh Kumar, and Vijay Bhaskar Semwal. 2022. "A multi-branch CNN-BiLSTM model for human activity recognition using wearable sensor data." *The Visual Computer* 38 (12): 4095–4109.

Chandrasekhar, Nadikatla, and Samineni Peddakrishna. 2023. "Enhancing heart disease prediction accuracy through machine learning techniques and optimization." *Processes* 11 (4): 1210.

Chiarini, Giovanni, Pradeep Ray, Shahriar Akter, Cristina Masella, and Aura Ganz. 2013. "m-Health technologies for chronic diseases and elders: A systematic review." *IEEE Journal on Selected Areas in Communications* 31 (9): 6–18.

Choudhary, Parul, Pooja Pathak, and Abhishek Chaubey. 2023. "Activity recognition system via unification of CNN and SVM in complex domain." 2023 6th International Conference on Information Systems and Computer Networks (ISCON). Mathura, India: IEEE, 1–6.

Chung, Seungeun, Jiyoun Lim, Kyoung Ju Noh, Gague Kim, and Hyuntae Jeong. 2019. "Sensor data acquisition and multimodal sensor fusion for human activity recognition using deep learning." *Sensors* 19 (7): 1716.

Davis, Kadian, and Evans Owusu. 2016. "Smartphone dataset for human activity recognition (HAR) in ambient assisted living (AAL)." *UCI Machine Learning Repository.* https://doi.org/10.24432/C5P597.

Dicianno, Brad E, Bambang Parmanto, Andrea D Fairman, Theresa M Crytzer, Daihua X Yu, Gede Pramana, Derek Coughenour, and Alan A Petrazzi. 2015. "Perspectives on the evolution of mobile (m-Health) technologies and application to rehabilitation." *Physical Therapy* 95 (3): 397–405.

Dua, Nidhi, Shiva Nand Singh, and Vijay Bhaskar Semwal. 2021. "Multi-input CNN-GRU based human activity recognition using wearable sensors." *Computing* 103: 1461–1478.

Faiola, Anthony, Elizabeth Lerner Papautsky, and Miriam Isola. 2019. "Empowering the aging with mobile health: A m-Health framework for supporting sustainable healthy lifestyle behavior." *Current Problems in Cardiology* 44 (8): 232–266.

Gong, Zheng, Yahui Ding, Yifan Chen, and Michael J Cree. 2023. "Wearable microwave medical sensing for stroke classification and localization: A space-division-based decision tree learning method." *IEEE Transactions on Antennas and Propagation*, 1–1.

Gumaei, Abdu, Mohammad Mehedi Hassan, Abdulhameed Alelaiwi, and Hussain Alsalman. 2019. "A hybrid deep learning model for human activity recognition using multimodal body sensing data." *IEEE Access* 7: 99152–99160.

Hamper, Andreas, Jonathan Wendt, Christian Zagel, and Freimut Bodendorf. 2016. "Behavior change support for physical activity promotion: A theoretical view on mobile health and fitness applications." 2016 49th Hawaii International Conference on System Sciences (HICSS). Koloa, HI, USA: IEEE, 3349–3358.

Hossain, Mohammad Alamgir. 2016. "Assessing m-Health success in Bangladesh: An empirical investigation using IS success models." *Journal of Enterprise Information Management* 29 (5): 774–796.

Janidarmian, Majid, Atena Roshan Fekr, Katarzyna Radecka, and Zeljko Zilic. 2017. "A comprehensive analysis on wearable acceleration sensors in human activity recognition." *Sensors* 17 (3): 529.

Katz, Sidney, and Md Chimn. 1962. "Multidisciplinary studies of illness in aged persons: II." *A New Classification of Functional.*

Khaertdinov, Bulat, Stylianos Asteriadis, and Esam Ghaleb. 2022. "Dynamic temperature scaling in contrastive self-supervised learning for sensor-based human activity recognition." *IEEE Transactions on Biometrics, Behavior, and Identity Science* 4 (4): 498–507.

Khaertdinov, Bulat, Esam Ghaleb, and Stylianos Asteriadis. 2021a. "Contrastive selfsupervised learning for sensor-based human activity recognition." 2021 IEEE International Joint Conference on Biometrics (IJCB). Shenzhen, China: IEEE, 1–8.

Khaertdinov, Bulat, Esam Ghaleb, and Stylianos Asteriadis. 2021b. "Deep triplet networks with attention for sensor-based human activity recognition." 2021 IEEE International Conference on Pervasive Computing and Communications (PerCom). Kassel, Germany: IEEE, 1–10.

Khan, Nida Saddaf, and Muhammad Sayeed Ghani. 2021. "A survey of deep learning based models for human activity recognition." *Wireless Personal Communications* 120 (2): 1593–1635.

Kosar, Enes, and Billur Barshan. 2023. "A new CNN-LSTM architecture for activity recognition employing wearable motion sensor data: Enabling diverse feature extraction." *Engineering Applications of Artificial Intelligence* 124: 106529.

Kumaresan, Senthil, Satya Sai Srinivas, Anutosh Maitra, and Nataraj Kuntagod. 2017. "Rapid m-Health-a mobile healthcare application development framework." *International Journal of Telemedicine and Clinical Practices* 2 (1): 42–62.

Kutlay, Muhammed Ali, and Sadina Gagula-Palalic. 2016. "Application of machine learning in healthcare: Analysis on m-Health dataset." *Southeast Europe Journal of Soft Computing* 4 (2): 2233–1859.

Kwapisz, Jennifer R, Gary M Weiss, and Samuel A Moore. 2011. "Activity recognition using cell phone accelerometers." *ACM SigKDD Explorations Newsletter* 12 (2): 74–82.

Li, Kathy, In~igo Urteaga, Amanda Shea, Virginia J Vitzthum, Chris H Wiggins, and No'emie Elhadad. 2021a. "A generative, predictive model for menstrual cycle lengths that accounts for potential self-tracking artifacts in mobile health data." *arXiv preprint arXiv:2102.12439.*

Li, Wei, Yuanbo Chai, Fazlullah Khan, Syed Rooh Ullah Jan, Sahil Verma, Varun G Menon, Kavita, and Xingwang Li. 2021b. "A comprehensive survey on machine learning-based big data analytics for IoT-enabled smart healthcare system." *Mobile Networks and Applications* 26: 234–252.

Lupton, Deborah. 2013. "Quantifying the body: Monitoring and measuring health in the age of m-Health technologies." *Critical Public Health* 23 (4): 393–403.

Mahmud, Tanvir, AQM Sazzad Sayyed, Shaikh Anowarul Fattah, and Sun-Yuan Kung. 2020. "A novel multi-stage training approach for human activity recognition from multimodal wearable sensor data using deep neural network." *IEEE Sensors Journal* 21 (2): 1715–1726.

Mechael, Patricia N. 2009. "The case for m-Health in developing countries." *Innovations: Technology, Governance, Globalization* 4 (1): 103–118.

Miah, Jonayet, Muntasir Mamun, Md Minhazur Rahman, Md Ishtyaq Mahmud, AM Islam, and S Ahmad. 2022. "MHfit: Mobile health data for predicting athletics fitness using machine learning models." Proceedings of the International Seminar on Machine Learning, Optimization, and Data Science, Jakarta, Indonesia.

Mukhopadhyay, Subhas Chandra, Sumarga Kumar Sah Tyagi, Nagender Kumar Suryadevara, Vincenzo Piuri, Fabio Scotti, and Sherali Zeadally. 2021. "Artificial intelligence-based sensors for next generation IoT applications: A review." *IEEE Sensors Journal* 21 (22): 24920–24932.

Nguyen, HD, Kim Phuc Tran, Xianyi Zeng, Ludovic Koehl, and Guillaume Tartare. 2019. "Wearable sensor data based human activity recognition using machine learning: a new approach." *arXiv preprint arXiv:1905.03809.*

Putri, Klarissa Ardilia. 2023. "Analysis of land cover classification results using ANN, SVM, and RF methods with r programming language (case research: Surabaya, Indonesia)." In Prof. Dr. Khan B. Marwat (ed.) *IOP conference series: Earth and environmental science,* Vol. 1127, 012030. IOP Publishing.

Qian, Hangwei, Sinno Jialin Pan, and Chunyan Miao. 2021. "Latent independent excitation for generalizable sensor-based cross-person activity recognition." In Prof. Dr. Khan B. Marwat (ed.) *Proceedings of the AAAI conference on artificial intelligence,* Vol. 35, 11921–11929.

Raza, Ali, Kashif Munir, and Mubarak Almutairi. 2022. "A novel deep learning approach for deepfake image detection." *Applied Sciences* 12 (19): 9820.

Raza, Ali, Mohammad Rustom Al Nasar, Essam Said Hanandeh, Raed Abu Zitar, Ahmad Yacoub Nasereddin, and Laith Abualigah. 2023. "A novel methodology for human kinematics motion detection based on smartphones sensor data using artificial intelligence." *Technologies* 11 (2): 55.

Schnall, Rebecca, Hwayoung Cho, Alexander Mangone, Adrienne Pichon, and Haomiao Jia. 2018. "Mobile health technology for improving symptom management in low income persons living with HIV." *AIDS and Behavior* 22: 3373–3383.

Teng, Qi, Kun Wang, Lei Zhang, and Jun He. 2020. "The layer-wise training convolutional neural networks using local loss for sensor-based human activity recognition." *IEEE Sensors Journal* 20 (13): 7265–7274.

Walsh, Jane C, and Jenny M Groarke. 2019. "Integrating behavioral science with mobile (m-Health) technology to optimize health behavior change interventions." *European Psychologist* 24(1): 38–48.

Wang, Jindong, Yiqiang Chen, Shuji Hao, Xiaohui Peng, and Lisha Hu. 2019. "Deep learning for sensor-based activity recognition: A survey." *Pattern Recognition Letters* 119: 3–11.

Yadav, Santosh Kumar, Kamlesh Tiwari, Hari Mohan Pandey, and Shaik Ali Akbar. 2021. "A review of multimodal human activity recognition with special emphasis on classification, applications, challenges and future directions." *Knowledge-Based Systems* 223: 106970.

12 Robotics and its Navigation Techniques
The Present and Future Revelations

K. Rajchandar, D. Kothandaraman,
Geetha Manoharan, and Gabriel Kabanda

INTRODUCTION

The growth of artificial intelligence is mainly related to the proliferation of software-based robotic systems. Some examples of these systems are mobile robots, unmanned aerial vehicles, and, increasingly, semiautonomous autos. Developing intelligent, human-friendly robots capable of interacting with and regulating an environment focused on humans is still a long way off using the technologies now available. This is because of the significant gap between the physical and algorithmic worlds. The creation of a trustworthy artificial intelligence that is embodiment-aware, conscious not only of itself but also its surroundings, and able to adapt its operating systems to the interactive body it is controlling is the end goal of the newly emerging field of study known as machine intelligence (MI), which brings together robotics and artificial intelligence (AI). Combining control, perception, and machine learning (ML) systems with artificial intelligence and robotics is necessary if completely autonomous intelligent systems are ever going to become a part of everyday life. This chapter begins with a discussion of machine intelligence's history, beginning in the twelfth century. After that, it describes the present status of robotics and artificial intelligence, evaluates major systems and recent research paths, explains the remaining problems, and imagines a future of man and machine that has not yet been made.

PATH PLANNING: HISTORY IN A NUTSHELL

Mobile robots that operate on their own are referred to as being autonomous. Active route planning is required to accomplish autonomous navigation between the multiple waypoints located in the surrounding area. The fundamental issue continues to be the search for an ideal route that is both free of collisions and adjacent positions. The

DOI: 10.1201/9781032686714-12

challenges described above were tackled in the past using breadth-first and depth-first search techniques, whereby a map was used to provide a potential route across the environment. The Bellman-Ford algorithm [1] is an example of an advanced search strategy that reduces the time complexity of a problem by offering an optimum route. Because prior approaches demand a significant amount of computing work, solving the challenge of finding the best route across an environment is a difficult problem to solve. A* [2] and Dijkstra's algorithm [3] are the two algorithms that are most often used for minimizing the amount of computing work required and determining the best route between the beginning and ending points by using neighborhood exploration strategies. At the moment, heuristic-modified A*, and Dijkstra's algorithms are used in conjunction with each other to determine the best route. However, some of the methods can only be used after the data has been processed.

Several different algorithms were proposed by academics as potential solutions to enhance route planning. Grid environment adaptive rapidly exploring random tree (RRT) algorithm [4] has been one of the ways that was created to build a dynamic route with post-processing feedback components. The grid served as the inspiration for the development of this program. It was feasible, via the use of genetic algorithms, to develop a cooperative co-evolutionary adaptive genetic algorithm [5], which is a kind of parallel computing that deals with routes. In the context of multi-agent systems, this was done. A technique that was created using the knowledge-based genetic algorithm (GA) [6], which provides route planning by combining local search methods with standard genetic algorithms, was also utilized to build an approach that was pretty similar to the one that was mentioned in the previous paragraph. In addition, sampling and distance metrics are utilized in the growth of 3D inflexible body route planning procedures [7].

The multi-objective genetic algorithm (MOGA) [8] is a point-to-point self-directed mobile robot path-planning algorithm that compares several dominant and non-dominant solutions, each of which requires additional computational time to compute an optimal solution. Despite this, there is not a visible improvement in the quantity of computational effort that is required for post-processing as a consequence of this utilization. The trajectory of an autonomous mobile robot may be planned with the help of this method. To circumvent the issue of repeatedly comparing the solutions that were acquired, an auxiliary method referred to as the multi-objective shortest path evolutionary algorithm (MOSPEA) [9] uses an outside set to exclude non-dominated results. This method was developed to circumvent the issue. The methodology that was just discussed, which works toward identifying the best possible fitness function but can do so only via the application of certain post-processing techniques, may result in a significant increase in the total amount of time that must be spent calculating. Researchers have developed a modified version of the A* algorithm to enhance the method for discovering the fitness function without the need for post-processing. This improvement was made possible by the A* algorithm.

A collision-free path-planning technique for several mobile robots traveling concurrently on a 2D gridded map is what the spatio-temporal A* system [10] is developed for. It is an improvement on the strategy that the genetic algorithm uses to calculate the fitness function. The technique was developed to be used as a

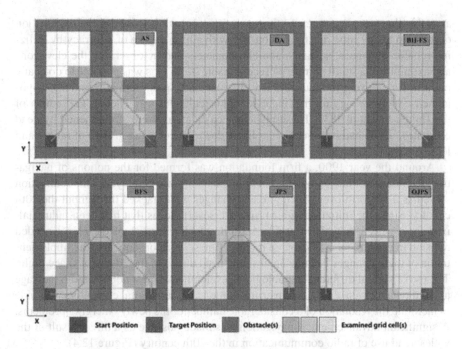

FIGURE 12.1   AS: A* star Algorithm (AS), DA: Dijkstra Algorithm, BH-FS: Breadth first search algorithm, BFS: Best first search algorithm, JPS: Jump point search algorithm, OJPS: Orthogonal jump point search algorithm.

collision-free path-planning procedure for several mobile robots that travel together on a gridded two-dimensional map [11]. This was the motivation for the development of the technology (Figure 12.1).

However, more memory space is necessary to create a considerable quantity of data without producing duplicates of the data. Hierarchical Task Network (HTN) [12] path-planning method for multiple robots makes use of the process of conflict resolution to discover optimal routes for the robots.

The aforementioned methods need a significant amount of processing time and take up a larger amount of space as they search for the best route [13]. They post-process many data consistently, which increases the difficulty of managing dynamic and large-scale settings.

MODERN ERA OF NAVIGATIONS

Marine navigation had developed into an entirely methodical technique by the time the 19th century came to a close. This method integrated the need for ease of use and reliability among its practitioners with the need for rigor and precision found in the expertise of astronomers, mathematicians, cartographers, and equipment designers. This allowed marine navigation to become increasingly accurate and reliable. To prepare for any potential journey, volumes of sailing instructions and sea maps were available that were accurate and thorough. It was possible to determine the location of the ship at any time during the journey with a high degree of accuracy by utilizing reliable almanacs, sextants, and chronometers and by observing the elevations and azimuths of a small number of recognizable stars. This was possible to do at any point in time during the journey. The first person to propose a conventional trigonometric methodology for making essential calculations was Thomas H. Sumner of the United States in 1837. These planetary calculations were strengthened by dead estimates, which had become more reliable due to the ongoing growth of ranges and logs (Figure 12.2, 12.3).

Around the year 1900, a firm foundation was formed for the notions of navigation, techniques of navigation, and instruments that were employed. This foundation served as a platform over which massive changes were stacked throughout the 20th century. Since the initiation of air transport, several ideas that had been industrialized for a period in which travel was limited to the surface of the planet needed revision in order to be applicable in modern times. As a consequence, the implementation of these changes became an absolute need due to of the advent of space flight. The use of several different technological developments, most notably radio message and radio navigation, electronic instrumentation, and fast digital computers, was able to facilitate the resolution of a considerable number of the newly emerging problems. A significant advancement in navigation was made possible as a direct result of the widespread use of radio communication in the 20th century (Figure 12.4).

During World War II, the advent of new navigational technology, such as radar, Loran, and radio direction finding, were responsible for the shifts that occurred in the field of navigation. These days, a mariner or pilot may simply flick a switch to turn on a Loran or global positioning system (GPS) receiver and immediately determine their position and path within a few yards of accuracy. The progress that has been made in technology makes this occurrence conceivable. It is most common for inertial guidance systems to be used in the navigation of aircraft, spacecraft, and

FIGURE 12.2   Amazon delivery robot

FIGURE 12.3   Waymo autonomous car

FIGURE 12.4   Essential components of self-driving cars

submarines; these systems make it feasible to travel without ever establishing contact with a terrestrial base. In these kinds of systems, the navigation of the vehicle is taken care of by a computer with the help of a device called an inertial navigator.

This apparatus consists of a gyroscope to show the direction in which the vehicle is moving and an accelerometer to measure variations in both speed and heading. Cruise missiles include inertial guidance systems and terrain-following radar that allow them to travel a thousand miles and hit their intended target. This makes it possible for cruise missiles to engage in long-range combat. In the 1960s, work on the construction of navigation satellites was initiated, which eventually led to the establishment of the GPS in the United States in the 1990s. Because it collects and analyses signals received from satellites, GPS can offer information about the position in addition to other types of information. This is possible because GPS uses satellites. Using this procedure, the GPS can offer the aforementioned information. During this intervening period, both Russia and China have developed navigation systems that are functionally comparable to GPS. Because mobile phones and other types of electronic devices increasingly come equipped with GPS receivers, it is now feasible to build navigation systems that are compatible with automobiles as well as other modes of transportation.

ARTIFICIAL INTELLIGENCE IN NAVIGATIONS

Over the past few hundred years, the transportation sector has gone through several upheavals and rebellions. However, we are now at the stage where advances are being achieved in the form of artificial intelligence in the transport area. There is a solid reason why artificial intelligence is attracting the attention of transportation officials all across the world. AI is proving its worth, whether it be in the method of self-driving cars, which provide higher dependability, road state monitoring, which offers better protection, or traffic flow examination, which offers improved efficiency. Several people now employed in the transport commercial have realized the mammoth potential of AI, and it is expected that the worldwide market will reach enormous profit by the year 2026. These expenditures can help businesses take advantage of developing technologies such as computer vision and ML to shape the future of transport in a way that improves the safety of travelers, diminishes the number of accidents that happen on the highways, and lessens the burden of traffic congestion. To achieve this goal, it is necessary to design the future of transportation in a manner that will influence the future of transportation. The application of deep learning and machine learning (ML) [14] in the transportation sector may also contribute to the development of "smart cities", akin to what we have witnessed in Glasgow. Here, this technology analyzes vehicle parking durations, monitors parking fines, and observes trends in traffic density (Figure 12.5).

The following topics will be discussed:

- Transportation
- Parking
- Navigation
- Self-driving cars
- Traffic recognition
- Pedestrian discovery
- Traffic movement examination

FIGURE 12.5   Pioneer 3-AT mobile robot

- Computer vision-powered car parks managing
- Road state monitoring
- Automatic traffic event finding
- Computerized license plate recognition
- Motorist monitoring

KEY TECHNOLOGIES IN THE LATEST ROBOTS

People's need for intelligent robots is growing as a result of the growing number of tasks that may be performed by robots and the necessity of societal advancement. The surrounding environment of intelligent robots is often uncharted territory and fraught with unpredictability. The following essential technologies play a significant role in the research process about the development of such robots:

MULTI-SENSOR EVIDENCE FUSION

Students' interest in the field of multi-sensor evidence fusion technology has grown significantly over the last several years, making it one of the disciplines that have received the most amount of focus from this demographic. It provides a practical solution for robots to carry out tasks in settings that are complicated, dynamic, uncertain, and unknown by combining control systems, signal dispensation, artificial intelligence, possibility, and statistics. These fields are all brought together in one solution. Robots will now be able to complete jobs in environments that are foreign to them because of this advancement. Robots make use of a wide variety of different sorts of sensors. Following the various functions that they serve, we categorize them as either internal measurement sensors or exterior measurement sensors. Figure 12.6 presents the characteristics of the robot various functions that they serve, we categorize them as either internal measurement sensors or exterior measurement sensors. Reference: https://techvidvan.com/tutorials/robotics-and-artificial-intelligence/Core

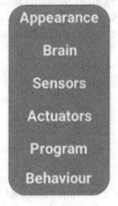

Robot Characteristics

Appearance

Brain

Sensors

Actuators

Program

Behaviour

FIGURE 12.6   Basic characteristics of robots

FIGURE 12.7   ROS bot

measurement sensors are used to ascertain the current internal status of the following robot apparatuses:

- The sensor of a certain location and angle
- Position and angle sensor with arbitrary values
- The sensor of speed and angle
- The sensor for acceleration
- The tilt angle sensor
- Sensor for azimuth

Auxiliary sensors comprise measurement, sensor recognition, contact, proximity sensor, distance sensor pressure, sliding sensor, force, torque sensor, tilt, direction, and attitude sensors.

The process of fusing information from numerous sensors to obtain information that is more dependable, accurate, or complete is known as multi-sensor information fusion. The combined multi-sensor system has the potential to more accurately represent the features of the item that has been identified, remove any confusion about the information, and increase the information's trustworthiness. The information that was obtained from several sensors all at the same time possesses the characteristics of duplication, complementary, legitimate updates, and low cost. The following can now be considered to be the most important multi-sensor data aggregation methods:

1. The Bayesian method of estimating
2. Theory of Dempster and Schafer
3. Kalman filter
4. Neural networks
5. Wavelet transform

NAVIGATION AND POSITIONING

In robotics research, one of the most important and difficult problems to solve is autonomous navigation, which is a basic technology in the robot system. The basic responsibilities of navigation may be broken down into three points:

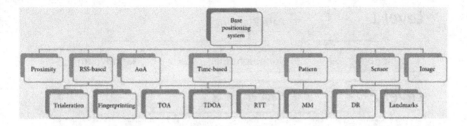

FIGURE 12.8 Classification of a positioning system

1. **Positioning on a global scale based on our comprehension of the environment:**
2. Location of man-made road signs or specified items to finish situating the robot and supply resources for route planning
3. **Recognizing targets and identifying potential obstructions:** Detection and identification of particular impediments or targets in real-time, which will help to enhance the control system's overall stability
4. **Safety and soundness of mind:** Analysis of the robot's working environment, identification of impediments and moving objects, and prevention of harm to the robot (Figure 12.8).

PLANNING OF PATH

Path forecasting is an essential component of robotics investigation, hence it is unavoidable that path forecasting will use this technology. One or more optimization criteria will decide which path will be the best possible one to take. Find the route through the robot workplace that is the most time- and resource-effective, one that will get you from the initial state to the end state while also enabling you to navigate around any challenges you encounter along the way. The process of creating the route may, in a general sense, be divided into two distinct approaches: the methods that are considered traditional and the methods that are considered intelligent.

The old-style path-planning approaches mainly comprise the following: free space, graph search method, grid decoupling method, and artificial potential field (APF) method. Even though the procedures mentioned above are employed for the majority of the global planning that is involved in robot path planning, these methods still have the potential for advancement in terms of their capability for efficient route search and path optimization. Within the realm of classical algorithms, the APF approach is a well-developed and active planning technique. When designing the route, it employs the environmental potential field model, however, it does not take into account whether or not the path is the best option. The intelligent route planning method incorporates many approaches used in the field of artificial intelligence into the overall process of path planning. These techniques include genetic algorithms, fuzzy logic, and neural networks. The goal of this method is to improve the precision with which robot path planning accounts for potential obstacles, accelerate the planning process, and satisfy the requirements of practical applications (Figure 12.9).

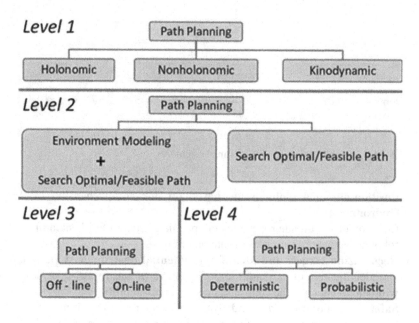

FIGURE 12.9 Classification of path planning levels

The following are some of the algorithms that are utilized the most: fuzzy method, neural networks, genetic algorithm, qlearning, and hybrid algorithm. These approaches are effective in reaching certain research objectives, even though the potentially dangerous setting was either known or unknown.

ROBOT VISUALIZATION

The graphic processing scheme is a crucial component of autonomous robots. A computer, an image capture card, and a camera are all components of this system.

The following is a list of some of the duties that are carried out by the robot vision system:

Image acquisition
Altering images and doing statistical research
Both input and display are available (Figure 12.10).

The key duties are the recognition of faces, the extraction of characteristics, and the segmentation of pictures. The primary challenge faced by the visual system is determining how to do information processing in a precise and effective manner. At the moment, the processing of visual information is progressively being improved, including the following:

Compression and filtering of images
Obstacle avoidance

Robot Application

FIGURE 12.10   Image processing in robotics

Identification of specific signs
3D perception and dispensation

The most important and difficult phase in visual information processing is the one that identifies the surrounding environment and any potential obstacles. In the field of visual information processing, one frequent technique is known as edge extraction. Mobile robots that need to interpret pictures while they are moving have a difficult time meeting real-time requirements for traditional image edge abstraction approaches such as the incline approach utilizing local information and the second-order variance method. This is because these methods require a significant amount of processing power and time. These techniques are among the most often used ones. It is proposed that a method based on computational intelligence should be used for the extraction of image edges to achieve this objective. This method employs several different methodologies, some of which are predicated on neural networks and others of which make use of fuzzy inference rules. Particularly noteworthy are the contributions of Professor Bezdek J.C. Recently, an in-depth investigation of the relevance of applying fuzzy logic interpretation for image edge extraction was carried out. This strategy can only be used for visual navigation. It does this by including the road information essential for the robot to go out-of-doors into the fuzzy rule base. This knowledge includes things like white highway lines and information about road borders. This helps to boost both the efficiency and robustness of road identification. In addition to this, the combination of genetic algorithms with fuzzy logic is another possibility that has been proposed.

INTELLIGENT PROGRAMMING

As a consequence of the development of robotics, it has been shown that conventional control theory has several shortcomings. Among these issues is the incapacity to accurately resolve the actual objects that are being represented, as well as the ill-conditioned processes that contain insufficient information. Over the last several years, several researchers have created a variety of intelligent control strategies for robotic systems.

The robotic system's intelligent control mechanisms are as follows: fuzzy based control, neural network control algorithm, and fusion of intelligent control skills.

Aspects such as these are included in the incorporation of intelligent control technology:

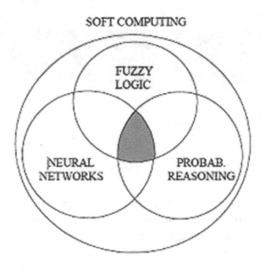

FIGURE 12.11   Soft computing system

Combination of variable structure control with fuzzy control
Combining neural networks with control over variable structures
Combining fuzzy control with control provided by neural networks
Fuzzy control techniques based on genetic algorithms are also included in
  intelligent fusion technology (Figure 12.11).

HUMAN–MACHINE INTERFACE TECHNOLOGY

The goal of the technology known as the human–machine interface is to investigate
ways to simplify the process by which humans may connect with computers. To attain
this goal, the robot controller must meet not only the most basic requirements but also
the requirements for a human–machine interface that is user-friendly, adaptable, and
convenient. In addition to this, it is important for the computer to grasp both the material and the language being used, to be able to communicate, and even to translate or
converse across many languages. Before these functions can be put into use, research
into the many different ways that knowledge might be represented is necessary. As a
result, the study of technology that facilitates human–machine interaction has both
significant practical use and significant theoretical import. The technology that allows
for communication between humans and machines has made significant strides in
recent years. A few examples of technologies that have lately made their way into
practical applications are text recognition, audio combination and recognition, photo
identification and processing, and machine translation. In addition, human–machine
interface technologies include crucial components such as monitoring technologies,
remote operation technologies, communication technologies, and human–machine
interface devices and communicating technologies [15]. The development of knowledge for remote operations is a significant area of study (Figure 12.12).

What is a Human-Machine Interface?

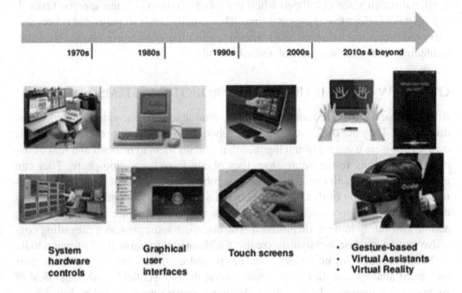

FIGURE 12.12 Human–machine interface

LEARNING IN INTELLIGENCE USING ALGORITHMS

Computer vision, an AI component, enables robots to move, assess their environment, and determine how they should behave. Machine learning, which is, once again, an element of both computer programming and artificial intelligence, enables robots to learn how to accomplish their duties from people.

It was in 1956 that John McCarthy first used the phrase "artificial intelligence," and ever since then, it has been a subject of considerable discussion and debate. This is because AI can endow inanimate objects, such as robots, with consciousness and the ability to reason and choose for themselves. Numerous AI varieties are used for robots, each corresponding to a particular application and set of responsibilities. They are as follows:

WEAK AND STRONG ARTIFICIAL INTELLIGENCE

Simulation of human cognition and behavior is the goal of this form of artificial intelligence. The orders and replies for the robots have been specified. Nevertheless, the robots are incapable of comprehending directions. They are solely responsible for carrying out the work necessary to get the relevant answer when a valid command is sent. Siri and Alexa are the best models to use for illustrating this point. The artificial intelligence included in these gadgets will only carry out the actions requested by the device's owner.

SPECIALIZED ARTIFICIAL INTELLIGENCE

Artificial intelligence is utilized when the robot needs to execute specific tasks. It can only do a select few things at a time. This usually refers to industrial robots that carry out preset tasks that are labor-intensive and time-consuming, such as painting, tightening, and other activities of a similar kind.

OVERVIEW OF AVAILABLE MOBILE ROBOTIC SYSTEMS

Robotic systems include humanoid robots, unmanned rovers, entertainment dogs, drones, etc. They are different from other robots in that they can walk about on their own and possess a degree of intelligence that enables them to respond and make conclusions according to the impressions they obtain from their atmosphere. They can also communicate with one another. Mobile robots need a basis of input data, a technique for decryption of that information, and a system for responding to an altering environment by performance activities. Mobile robots also require a way of executing actions (including moving themselves). The organism must possess a prevailing cognitive system to sense and familiarize itself with an environment that is foreign to it.

Mobile robots can now walk, run, jump, and do other motions much as their biological analogs can. In recent years, several distinct subfields within the field of robotics have emerged. These sub-disciplines comprise maneuvered mobile robots, legged robots, airborne robots, robot visualization, artificial intelligence, and a great deal of other specialized types of robots. These sub-disciplines need skills in a variety of technological domains, including mechanism, microelectronics, and computer science, among others. The expression and perception of emotions, artificial intelligence, autonomous driving, network communication, cooperative work, nanorobotics, pleasant human–robot interfaces, safe human–robot interaction, and nanorobotics are the developing themes that are now leading the pack. Furthermore, these new advances are being implemented in a variety of areas, including the health field, healthcare, informatics, comfort design, engineering, distribution of goods, and amenity robots, to mention just a few of these fields' applications. These inclinations will continue to grow more and further as time goes on.

When it comes to the creation of a mobile robot, the most important aspect is the ability to navigate. The objective is for the robot to successfully navigate through an environment, either one that it is familiar with or one that it is not familiar with, moving from one location to another while taking into consideration the readings from the sensors to achieve the goals that are intended for it.

The following activities are grouped as types of mobile robot navigation:

Producing a model of the planet in the shape of a map
The process of computing a route from a starting place to a target location that avoids collisions
Moving in the direction of the computed trajectory while avoiding running into any obstacles

CONCLUSION

The present technical state of robotic navigation and AI has been shown throughout this chapter. It has investigated the issues occurring right now and shed light on the potential directions these technologies might go in the era of artificial intelligence. Our day-to-day lives, as well as our society, will be affected by MI exploring the potential for artificial intelligence in digital marketing [16]. It presents a wide range of possibilities for addressing the current issues and those that society may already predict. The responsibility that comes along with the use of this technology is not something that should be taken lightly. The development of this technology needs to emphasize dependability, safety, putting people first, and enhancing technological advances in business management techniques [17]. It is necessary, for instance, to bring into existence the circumstances of a framework that forbids this technology's use in a way that would be harmful to individuals and humanity as a whole.

REFERENCES

1. Hu, Dichao. "An introductory survey on attention mechanisms in NLP problems." Proceedings of SAI Intelligent Systems Conference. Springer, Cham, 2019.
2. Yao, Junfeng, et al. "Path planning for virtual human motion using improved A* star algorithm." 2010 Seventh International Conference on Information Technology: New Generations. Las Vegas, NV, USA: IEEE, 2010.
3. Biswas, Siddhartha Sankar. "Fuzzy real-time Dijkstra's algorithm." *International Journal of Computational Intelligence Research* 13.4 (2017): 631–640.
4. Lin, Na, and Yalun Zhang. "An adaptive RRT based on the dynamic step for UAV path planning." *Journal of Information &Computational Science* 12.5 (2015): 1975–1983.
5. Cai, Zixing, and Zhihong Peng. "Cooperative coevolutionary adaptive genetic algorithm in path planning of cooperative multi-mobile robot systems." *Journal of Intelligent and Robotic Systems* 33.1 (2002): 61–71.
6. Hu, Yanrong, et al. "A knowledge-based genetic algorithm for path planning in unstructured mobile robot environments." 2004 IEEE International Conference on Robotics and Biomimetics. Shenyang, China: IEEE, 2004.
7. Kuffner, James J. "Effective sampling and distance metrics for 3D rigid body path planning." IEEE International Conference on Robotics and Automation. Proceedings. ICRA'04. 2004, Vol. 4. New Orleans, LA, USA: IEEE, 2004.
8. Castillo, Oscar, Leonardo Trujillo, and Patricia Melin. "Multiple objective genetic algorithms for path-planning optimization in autonomous mobile robots." *Soft Computing* 11.3 (2007): 269–279.
9. Pangilinan, Jose Maria A., and Gerrit K. Janssens. "Evolutionary algorithms for the multiobjective shortest path problem." *World Academy of Science, Engineering, and Technology* 25 (2007): 205–210.
10. Wang, Wenjie, and Wooi-Boon Goh. "Multi-robot path planning with the spatiotemporal A* algorithm and its variants." International Conference on Autonomous Agents and Multiagent Systems. Springer, Berlin, Heidelberg, 2011.
11. Rajchandar, K., et al. "An approach to improve multi-objective path planning for mobile robot navigation using the novel quadrant selection method." *Defence Science Journal* 71.6 (2021): 748.

12. Zeng, Suying, Yuancheng Zhu, and Chao Qi. "Htn-based multi-robot path planning." 2016 Chinese Control and Decision Conference (CCDC). Yinchuan, China: IEEE, 2016.

13. Zuo, Lei, et al. "A hierarchical path planning approach based on A_* and least-squares policy iteration for mobile robots." *Neurocomputing* 170 (2015): 257–266.

14. Singh, Barinderjit, Dr. Rachana Jaiswal Habibulla Palagiri, Dr. Shweta Mogre, Dr. Rashmi Badjatya Rawat, and Geetha Manoharan. "An investigation on the use of machine learning methods for predicting employee performance, manager." *The British Journal of Administrative Management* (2022, September): 15–22, ISSN: 1746–1278. https://tbjam.org/vol58-special-issue-06/

15. Manoharan, Geetha, Dr. Barinderjit Singh, Dr. Y. Saritha Kumari, Dr. Aarti Dawra, Dr. Rachana Jaiswal, and Aijaj Ahmad Raj. "An evaluation of machine learning techniques and how they affect human resource management and sustainable development, manager." *The British Journal of Administrative Management* (2022, September): 4–14, ISSN: 1746–1278. https://tbjam.org/vol58-special-issue-06/

16. Manoharan, Geetha, Dr. Harish Purohit, Dr. Somanchi Hari Krishna, Dr. Neelam Sheoliha, Shivangi Ghildiyal, and Dr. R. Krishna Vardhan Reddy. "An overview of exploring the potential of artificial intelligence approaches in digital marketing, manager." *The British Journal of Administrative Management* (2022, September): 48–59, ISSN: 1746–1278. https://tbjam.org/vol58-special-issue-06/

17. Ramachandran, Dr. K. K., Dr. S. Shyni Carmel Mary, Dr. Arun Kant Painoli, Harish Satyala, Barinderjit Singh, and Geetha Manoharan. "Assessing the full impact of technological advances on business management techniques, manager." *The British Journal of Administrative Management* (2022, September), ISSN: 1746–1278. https://tbjam.org/vol58-special-issue-06/

13 Powering the Future
How AI is Advancing the Field of Robotics

Mantosh Mahapatra, Parag Raut,
Aksh Dharaskar, Rohit Naluri, and Saliha Bathool

INTRODUCTION

Automation has substantially enhanced as a result of the combination of artificial intelligence (AI) and robotics. The area of robotics is concerned with the design, manufacturing, and operation of automated systems capable of performing duties that are normally performed by humans. AI is the creation of systems that are capable of doing actions that usually necessitate human intellect, such as learning knowledge, deciding on actions, and resolving issues.

Combining robots with AI has acquired traction in recent years, with many sectors implementing such innovations to improve performance, productivity, and safety precautions. The present study's goal is to investigate the good and bad elements of bringing the use of AI towards the discipline of robotics, as well as the potential social repercussions [1].

The integration of artificial intelligence into robotics yields a notable advantage in executing intricate tasks with heightened precision and efficacy. AI-enabled robots possess the ability to traverse and engage with their surroundings, render judgments founded on data scrutiny, and assimilate knowledge from past encounters to enhance their efficacy.

The utilization of AI in robotics presents a plausible advantage of enhancing safety in environments that pose potential risks. In particular industries, including mining, oil and gas, and nuclear power, the use of robots for dangerous jobs that might otherwise endanger human workers is viable. Artificial intelligence offers an opportunity to improve robot control as well as danger identification and response abilities [2].

AI offers the ability to improve the effectiveness of resource allocation while reducing waste. Robotic technology offers the ability to improve agricultural practices by making tracking crops easier and optimizing irrigation purposes and fertilization processes, reducing resource and chemical use.

DOI: 10.1201/9781032686714-13

AI AND ROBOTICS: A SYNERGISTIC RELATIONSHIP

There is an evident connection between robotics and AI in the context of medical methods. Artificial intelligence-enhanced robotics can assist in surgical operations by performing actions such as binding and cutting with more accuracy and precision than human surgeons. Robots powered by AI have the possibility of being employed in a variety of healthcare circumstances, such as patient monitoring, prescription dispensed medication, and rehabilitation.

Robotic capabilities in the setting of production-related applications may be enhanced with artificial intelligence. Robots having AI features might execute and conduct an assortment of tasks involving, but not confined to, installation, the painting process, and soldering. Robots possess the ability to acquire knowledge from their surroundings and adapt their actions to enhance their efficacy, resulting in heightened efficiency and productivity [3].

The merging of robots and AI has been having an important effect on the service economy. The concept of artificial intelligence algorithms connected to automated machinery has the capacity to perform a range of jobs, including, but not limited to, housekeeping, cooking, and assisting customers. The aforementioned robots possess the ability to engage in customer interaction, address their requirements, and assimilate knowledge from their engagements to enhance their operational efficacy.

The amalgamation of artificial intelligence and robotics presents a plethora of advantages; however, it is imperative to acknowledge and tackle the associated obstacles and constraints. One of the primary obstacles pertains to the requirement for human supervision. Although AI has the potential to automate numerous tasks, certain tasks necessitate human intervention, particularly in scenarios that involve ethical or moral judgments. In healthcare settings, robots may be required to make decisions pertaining to patient care. However, it is imperative that these decisions are subject to the final approval of human healthcare professionals.

Another barrier is the likelihood of AI systems demonstrating discrimination or prejudice, especially when trained using data deemed intrinsically biased. To avoid the perpetuation of already present prejudices and discriminatory practices, it is critical to ensure that AI systems are designed and trained in an unbiased and egalitarian manner.

There exist apprehensions regarding the possibility of exploiting AI systems for malevolent objectives, such as the creation of self-governing armaments. The regulation of AI-powered robotics is imperative to mitigate potential societal harm resulting from their development and utilization.

The future of robots and AI will continue to be emphasized by continued technological developments and increased integration among other developing sectors, such as the Internet of Things (IoT) along with the usage of cloud-based computing. As AI-powered robots evolve to acquire knowledge and adapt to new surroundings and conditions, their autonomy and flexibility are predicted to increase [4].

CHALLENGES AND LIMITATIONS OF AI IN ROBOTICS

While the use of AI in robots has numerous potential advantages, there are additional problems and constraints to consider. Cost, ethical considerations, and the necessity for human monitoring are among the difficulties.

The expense of technology is one of the primary barriers to incorporating artificial intelligence in robotics. The expense of developing and maintaining AI-powered robots might discourage smaller businesses and individuals from employing them. This cost may also limit the potential for widespread adoption of AI-powered robotics in industries where labor costs are already low.

The ethical dilemmas raised by the utilization of AI in robotics provide an important hurdle. The prospect that AI-powered robots could replace human employment opportunities is one ethical concern. Although this could increase efficiency and productivity in the immediate future, it could result in adverse long-term consequences for society. It is essential to consider how AI-powered robots can be used to complement human workers rather than replace them.

The possibility for robots powered by AI to be taken advantage of for nefarious ends, such as the manufacturing of autonomous weapons, is another ethical concern. An array of ethical difficulties, such as the likelihood of unintentional harm and the absence of culpability for the robots' conduct, are raised by the creation and implementation of autonomous weapons. It is essential to regulate the development and use of AI-powered robotics to prevent them from being used in ways that could be harmful to society.

Another limitation of AI in robotics is the need for human oversight. While AI can automate many tasks, there are some tasks that require human intervention, particularly in situations that require ethical or moral judgments. For example, in healthcare applications, robots may need to make decisions about patient care, but these decisions must ultimately be approved by human healthcare professionals.

Another restriction of AI in robots is the possibility of bias and discrimination. Biased data can be used to train AI systems, resulting in biased outcomes. To prevent preserving earlier preconceptions and inequality, it is critical that AI systems are designed and schooled to be impartial and comparable [5].

Furthermore, the combination of AI and robots may result in a concentration of power in the hands of only a few persons or institutions. As a result, the creation of AI-powered robots necessitates substantial skill and resources, thereby limiting the possibility of mass adoption by smaller organizations or individuals [6]. This concentration of power may result in unequal distribution of the advantages and disadvantages connected to AI-powered robots.

While incorporating AI into robots has numerous potential advantages, there are also some hurdles and constraints to consider. Cost, ethical considerations, the necessity for human monitoring, and the possibility of prejudice and discrimination are among the obstacles. In order to be sure that these innovations are put to use for the greater good of society, it is of the utmost importance for researchers to carry out the study and development of them in an ethical and accountable way. It is also vital to take into consideration any dangers that can arise from using AI-powered robotics, such as the prospect of autonomous ammunition and the effect on employment. By addressing these challenges and limitations, we can maximize the potential benefits of AI-powered robotics while minimizing their negative impacts.

ETHICS IN AI

The science-fiction writer Isaac Asimov predicted the potential perils of unsupervised artificial intelligence units before they were developed. Hence, Asimov's Three Laws of Robotics were formulated to reduce the risks. Artificial intelligence ethics are an assortment of concepts and practices aimed at directing the development and accountability of AI technology. The first law of Asimov's code of ethics restricts robots from performing any form of assault on humans or allowing humankind to be injured by refusing to act. Second, robotic agents are intended to follow human commands unless those instructions violate the previous law. Only the third principle pleads in the name of the robots, which are expected to safeguard themselves by complying with both the first and second rules. Robot ethics, often known as roboethics, deals with the likelihood that someday robots might be an imminent danger to humans, or even killer robots used in conflict might grow into an obstacle for human society. The basic objective of the code of principles is to advise participants when presented with a moral choice regarding the application of AI.

Amazon, Microsoft, Elon Musk, and the International Business Machines Corporation (IBM) agree that the potentially limitless future of computational intelligence merits immediate investigation, while also acknowledging that major ethical challenges must be resolved. Artificial intelligence image and speech recognition systems are now in use. Among these systems are enterprises, which have an immediate influence on humans. The information assigned through such systems has prejudices by the system itself. Face + + by Microsoft had been a predicted facial recognition system that had to find out the gender. After all, the method had been developed using 10-year-old data which contained a greater number of male candidates.

Models based on machine learning may yield outcomes if they utilize associated specifications on high parameters. A reason for worry has come to light regarding injustice and an absence of transparency. Some open-source programs created by grassroots organizations aim to raise consciousness regarding this prejudicial system. Dr David Leslie, a member of the CAHAI Bureau (the Council of Europe's Ad hoc Committee on Artificial Intelligence), highlighted the importance of establishing standards for the development and application of such systems, stating that expanding in performing activities that demand intelligence has ushered in rebellion to an extensive spectrum of cognitive tasks in computation procedures, so the systems themselves are not able to be held either explicitly or straightaway responsible for such conduct. Humans are now accountable for whatever prejudices exist in these program-based innovations, but this may change in the future.

COGNITIVE COMPUTING

When carrying out the process of calculating, the brain of a person includes a system that brings in entirely novel and established information. KGS, known as a knowledge-growing system, is a computerized system with intelligence built around an aspect of a particularly contemporary method of computational customizing, that obtains fresh knowledge that develops in the brain of humans throughout the time being and enhances cognitive potential. This serves as the basis of creating the "cognitive agent," that behaves and considers rationally as an individual. This expands and broadens the doors of cognitive artificial intelligence techniques, and this may be submitted to various extraction of knowledge, coordination in self-sufficient intermediary systems, recognition of human illness, as well as advancing cognitive technology, that results in the creation of artificially intelligent machines that utilize the human spectrum of intelligence.

Cognitive artificial intelligence progresses to further advanced abilities. Memorizing is a basic procedure that occurs on an internet connection that has been filled with already remembered information. When a user encounters unexpected chest discomfort in the middle of the night, an electronic medical aid (humanoid) can be used at the runtime while making note of the user and what is happening. The humanoid or robot must be competent in making sound decisions while concurrently analyzing the individual's health records and making a recommendation only after that. IBM Watson, the supercomputer, may have an enhanced decision-making feature, with reduced expenses and improved results, by using extensive analysis of content and argumentation based on evidence. There's also a cost to everything. Cognition is unable to assess threats, enabling random information such as interpersonal, economic, and political considerations to slip through the cracks. For instance, if a country's economy is shifting and an analogy indicates an optimal spot for mining or drilling for oil, the approach of cognition ought to take this into account, and consequently human contact remains critical for comprehending the dangers entailed and arriving at the final decision.

Human awareness of substances is incomprehensible. Is it feasible for machines to behave rationally with no human intervention? Therefore, enhanced statistics is an advantageous step forward.

APPLICATIONS OF AI IN ROBOTICS

The merging of artificial intelligence with robotics presents a plethora of potential advantages; however, it is imperative to acknowledge the various obstacles and constraints that accompany this integration. Several obstacles that arise in this context encompass financial implications, ethical considerations, and the necessity for human supervision.

The cost of technology remains a significant obstacle for the successful integration of AI with that of robotics. Plus, the development and maintenance costs of AI-powered robots can be significant, potentially constraining their availability to smaller entities and individuals. The cost factor could potentially impede the extensive implementation of AI-driven robotics in sectors where labor expenses are already minimal.

The utilization of artificial intelligence in robotics poses a noteworthy obstacle due to the ethical considerations involved. An ethical issue that arises with the implementation of AI-powered robots is the possibility of human workforce replacement, which may result in job displacement and unemployment [7]. Although the immediate effect of this approach may be a boost in efficiency and productivity, its potential long-term repercussions on society could be detrimental. It is imperative to contemplate the potential for AI-driven robotics to serve as a supplementary resource to human labor, rather than supplanting it entirely [8].

An additional ethical consideration pertains to the potential deployment of AI-driven robots for malevolent objectives, such as the creation of self-governing armaments. The utilization and advancement of autonomous weaponry give rise to various ethical apprehensions, such as the possibility of inadvertent damage and the absence of responsibility for the actions executed by the machines. The regulation of AI-powered robotics is imperative to mitigate potential societal harm resulting from their development and utilization [9].

The constraint of artificial intelligence in the field of robotics pertains to the requirement for human supervision. Although AI has the ability to automate numerous tasks, certain tasks necessitate human intervention, particularly in scenarios that demand ethical or moral judgments. In healthcare settings, robots may be required to make decisions pertaining to patient care. However, it is imperative that such decisions are subject to the final approval of human healthcare experts.

The incorporation of AI into robotics poses a potential hazard of consolidating power among a limited number of individuals or entities. The creation of robots that are powered by artificial intelligence necessitates considerable proficiency and substantial resources, potentially constraining the possibility of broad-based implementation by minor entities or individuals. The centralization of power in the context of AI-driven robotics has the potential to result in disparities in the allocation of the pros and cons that tag along with it.

AI in the field of robotics presents numerous advantageous prospects, yet it is imperative to acknowledge and address various obstacles and constraints. The obstacles encompass financial implications, ethical considerations, the necessity for human supervision, and the possibility of partiality and prejudice. It is imperative to persist in the investigation and advancement of these technologies in a conscientious and moral fashion to guarantee their utilization for the betterment of humanity. It is vital to contemplate the prospective hazards linked with the utilization of AI-fueled robotics, including the plausibility of self-governing armaments and the repercussions on the workforce. Through the mitigation of challenges and limitations, it is possible to optimize the advantageous potential of robotics powered by artificial intelligence, while simultaneously reducing their adverse effects [10].

The healthcare sector has incorporated AI-powered robots to perform various functions such as aiding in surgical procedures, patient monitoring, and dispensing medication. Surgical robots have the potential to enhance surgical precision and control, thereby resulting in better patient outcomes [11]. AI-enabled robots have the capability to continuously observe patients, identifying alterations in their physical state and notifying medical practitioners of possible concerns prior to their escalation to a critical level.

The utilization of AI-powered robots in the manufacturing sector is prevalent for various tasks including but not limited to assembly, painting, and quality control. The implementation of robots in the workforce has resulted in heightened efficiency and productivity due to their superior speed and precision in comparison to human labor. Deep learning and machine learning approaches are used in AI to explain to robots how to do various jobs accurately and efficiently. The use of AI-powered robots for agricultural purposes has been reported for a variety of activities such as planting crops, gathering, and spraying. The use of self-driving robots has the potential to reduce human labor while increasing efficiency in operations. Autonomous cars, which are outfitted with powerful sensors and artificial intelligence computations, can detect crop health and apply the proper quantity of fertilizer or pesticide.

Machine learning is an AI algorithm that enables automated systems to learn from data and subsequently make predictions or decisions based on the acquired knowledge. The algorithm in question is frequently employed in the field of robotics to accomplish tasks such as object recognition and navigation.

Deep learning is a form of machine learning technology that uses neural networks to learn from massive amounts of data. The method in issue is used in the domain of robotics, notably for image identification, audio recognition, and handling natural language jobs.

Reinforcement learning is an approach to machine learning in which robotic agents are instructed to gain information through iterative exploration. The method in issue is used in the field of robotics, especially for task execution such as gameplay and environmental navigation [12].

To summarize, the inclusion of AI in the realm of robotics has culminated in the establishment of a wide range of applications in a variety of industries, including medical care, production, and agriculture. AI methods, such as deep learning, machine learning, and reinforcement training, are used to teach robots methods to do tasks with accuracy and efficiency. It is possible to uncover new AI applications in robotics through ongoing research and development efforts, ultimately improving the efficiency and production of many businesses [13].

FUTURE OF AI IN ROBOTICS

The field of AI in robotics is poised for significant growth and development as technological advancements continue to progress. The future of this field is characterized by a multitude of possibilities. The topic of modern technology and possible advancements in the disciplines of robotics and artificial intelligence will be covered in this section.

A new technology called quantum computer technology has an opportunity to transform robotics and the field of artificial intelligence. Quantum computers possess the ability to execute computations at a significantly accelerated pace in comparison to classical computers. This technological advancement has the potential to facilitate the creation of more sophisticated artificial intelligence algorithms and robotic systems. The utilization of quantum computers has the potential to enhance the real-time data processing capabilities of robots, thereby augmenting their efficiency and decision-making prowess.

Neuromorphic computing is a computational paradigm that emulates the architecture and operation of the human brain. The technological advancement under consideration possesses the capability to bring about a paradigm shift in the field of robotics by facilitating robots to execute tasks in a manner that closely resembles human performance. The implementation of neuromorphic computing has the potential to facilitate the creation of sophisticated artificial intelligence algorithms for robotics, which can facilitate the acquisition of knowledge through experience and the ability to adjust to novel surroundings [14].

Swarm robotics is a specialized area of robotics that concentrates on the synchronization and organization of extensive ensembles of robots to execute various tasks. The implementation of this technology holds the promise of transforming various sectors, including agriculture and manufacturing, through the facilitation of coordinated efforts among numerous robots to execute intricate operations. The utilization of swarm robotics may potentially facilitate the emergence of novel artificial intelligence algorithms that facilitate inter-robot coordination and adaptability in response to dynamic environmental conditions.

One of the most important areas of research in robotics and artificial intelligence is the development of systems that can adapt to new situations and tasks. Modern robotic systems are usually adapted to specific jobs and environments. However, the development of advanced AI algorithms and robotic equipment has the potential to promote the development of flexible robots capable of more easily adjusting to unexpected conditions and tasks. The situation has the potential to enable the development of robotic systems with greater flexibility and multifunctionality, allowing them to perform a wide range of operations in a variety of environments [15].

The use of artificial intelligence in robotics has huge and exciting potential. Emerging technologies like quantum computing, brain computing, and swarm automation have an opportunity to revolutionize the area of robotics by boosting robotics' efficiency and efficacy in job execution. With the ongoing progress made by researchers, it is anticipated that there will be a proliferation of sophisticated artificial intelligence algorithms and robotic systems that possess the ability to adjust to novel environments and tasks. The realization of a more efficient, productive, and sustainable future can be achieved through the optimization of AI and robotics.

IMPLICATIONS FOR SOCIETY

The implementation of AI-powered robotics has the capacity to transform various sectors and enhance efficacy. However, it is imperative to take into account the societal ramifications that may arise. This segment will delve into the ethical and societal implications that arise from the integration of AI in robotics.

Concerns about loss of employment are substantial when it comes to the utilization of machine learning in the robotics industry. With the increasing advancement and capability of robots, there is a possibility that they could undertake tasks that were traditionally executed by humans. The possibility of substantial job displacement in specific sectors may result in economic upheaval and societal turmoil. Notwithstanding, there is a contention that the advancement of robotics propelled by AI could potentially generate novel employment prospects in domains such as robotics engineering, programming, and upkeep.

Privacy and security are among the concerns that are associated with AI-driven robotics. With the increasing integration of robots into our daily routines, there is a possibility that they may accumulate and retain substantial volumes of data concerning individuals, which may comprise personal and sensitive details. It is imperative to safeguard this information in order to avert any unauthorized utilization or access. Moreover, the possibility of cyber-attacks and hacking on robotics systems poses a significant threat that could result in severe repercussions for both individuals and the wider community [16].

The impartiality and lack of bias in AI algorithms utilized in robotics systems are contingent upon the quality of the data on which they are trained. In the event that the data utilized for algorithmic training incorporates prejudiced or biased elements, such partialities may be sustained in the system's determinations and operations. This prompts significant inquiries regarding the attribution of accountability and responsibility in the event of an adverse occurrence. The question of accountability arises in cases where an AI-powered robotic system renders a decision that results in harm to an individual. As the prevalence of AI-driven robotics increases in society, it will be necessary to address these questions.

Transparency is a crucial ethical consideration that warrants attention. The intricacy of AI-powered robotic systems can pose a challenge to human comprehension of their decision-making processes and operational behaviors. Transparency is a crucial aspect in the design of these systems, necessitating lucid explications of their mechanisms and rationales for decision-making. The aforementioned capability will empower both individuals and organizations to make well-informed decisions regarding their interactions with said systems and to hold them accountable in instances where it is deemed necessary.

In light of the rapid progress of AI-driven robotics, it is imperative to deliberate upon the ethical and societal ramifications of this technology. Several concerns need to be addressed, including job displacement, privacy and security, bias, accountability, and transparency. Through proactive measures, it is possible to ensure that the implementation of AI-driven robotics is conducive to the betterment of society, rather than engendering adverse effects or introducing novel challenges. The responsible development and utilization of AI-driven robotics necessitate the collaborative efforts of researchers, policymakers, and industry leaders in establishing a comprehensive framework [17].

The healthcare sector has incorporated AI-powered robots to perform various functions such as aiding in surgical procedures, patient monitoring, and dispensing medication. Surgical robots have the potential to enhance surgical precision and control, thereby resulting in better patient outcomes [18].

VIEWPOINT

The brain might be divided into multiple "principal tiers" which are then further divided into a variety of categories, each of these critical strata functioning as an instance of a brain network linked to other networks that exist at multiple levels. Because of its pulsing properties, each major layer may absorb data independently of the other layers; nevertheless, when many layers are stimulated at the same time to address distinct problems, the interaction interrupts each core activity. The link between it and disruption would result in cognitive evolution across all disciplines. Awareness may be defined as a set of mechanisms that limit brain connection. These systems are an essential component of the mind, and awareness manifests as physiological and electromagnetic waves as a result of their connection/interruption.

This includes and organizes a significant trend-setting creation of businesses of applications based on artificial intelligence and the field of machine learning, having an independent prioritize on company structures and AI as a computer service, along with the basic technology of AI in an array of different sectors. Such notions could assist in clarifying and/or indicate the beginnings of consciousness, whereas they do not completely clarify or provide a solution. In accordance with Chalmers [19], the task is to characterize simultaneously first-person data about personal experiences and third-party data regarding brain behavior and processing. With little space for individual experiences, most modern explanations of awareness are based on third-party facts and the physiological nature underpinnings of awareness. In addition, some of the issues, such as phasing dispersion, transitory motion, the drop-in peak electroencephalogram (EEG) activity produced by transcranial magnetic stimulation (TMS), and each of the phases, including the fact that the two systems split, are unresolved or not well-defined subjects that models of awareness must address. Finally, different hypotheses, especially genetics, seek to clarify awareness and conscious perception through processes that are neither reproducible nor workable [19].

CONCLUSION

The current study investigated the relationship that exists between artificial intelligence and automation, as well as the possible benefits and drawbacks of using the technology. The incorporation of AI technology appears to be improving robots' functional capabilities, allowing them to do complex tasks with more accuracy and efficacy. The expansion of AI has resulted in the rise of a wide range of robotics applications that are driven by this technology and are used in a variety of industries.

Despite the potential benefits, incorporating AI into robots comes with a number of obstacles and constraints. Cost issues, ethical considerations, and the need for human oversight are among the essential aspects to be considered. The risks associated with the use of artificial intelligence in robots have been discussed, including the possibility of self-governing armaments and worker displacement.

Machine learning, deep learning, and reinforcement learning are three forms of artificial intelligence algorithms used in robotics that have recently been investigated. Furthermore, the research addressed the possible impact of emerging technologies like quantum information processing and neuromorphic programming on this sector. Furthermore, we have discussed the societal implications of artificial intelligence-powered robots, including topics such as employment displacement, security and privacy concerns, partiality, accountability, and transparency.

To conclude, the implementation of AI-powered robotics exhibits the capability to transform various sectors and enhance productivity. It is imperative to take into account the ethical and societal ramifications of this technology and strive towards its conscientious advancement and utilization. Potential avenues for additional investigation in this domain encompass enhancing the transparency and accountability of robotics systems propelled by AI, investigating nascent technologies, and contemplating novel ethical frameworks to tackle the distinctive predicaments presented by AI in robotics.

The progression of AI-powered robotics necessitates collaboration among researchers, policymakers, and industry leaders to ensure that its implementation yields societal advantages. The complete utilization of AI in robotics and the realization of its actual potential can only be achieved under such circumstances.

REFERENCES

1. Rajan, Kanna, and Alessandro Saffiotti. "Towards a science of integrated AI and robotics." *Artificial Intelligence* 247 (2017): 1–9.
2. Brady, Michael. "Artificial intelligence and robotics." *Artificial Intelligence* 26.1 (1985): 79–121.
3. Kitano, Hiroaki, et al. "Robocup: A challenge problem for AI and robotics." In Hiroaki Kitano (Ed.) *RoboCup-97: Robot Soccer World Cup I 1*, (pp. 1–19). Springer, Berlin Heidelberg, 1998.
4. Mir, Umar Bashir, et al. "Critical success factors for integrating artificial intelligence and robotics." *Digital Policy, Regulation and Governance* 22.4 (2020): 307–331.
5. Bhaumik, Arkapravo. *From AI to Robotics: Mobile, Social, and Sentient Robots*. New Delhi: CRC Press, 2018.
6. Ashrafian, Hutan. "AIonAI: A humanitarian law of artificial intelligence and robotics." *Science and Engineering Ethics* 21 (2015): 29–40.
7. Müller, Vincent C. "Ethics of artificial intelligence and robotics." *Stanford Encyclopedia of Philosophy* (2020).
8. Bogue, Robert. "The role of artificial intelligence in robotics." *Industrial Robot: An International Journal* 41.2 (2014): 119–123.
9. Froese, Tom, and Shigeru Taguchi. "The problem of meaning in AI and robotics: Still with us after all these years." *Philosophies* 4.2 (2019): 14.
10. Shaukat, K., et al. "The impact of artificial intelligence and robotics on the future employment opportunities." *Trends Comput. Sci. Inf. Technol* 5 (2020): 50–54.
11. Andreu-Perez, Javier, et al. "Artificial intelligence and robotics." *arXiv preprint arXiv:1803.10813* (2018).
12. Smith, Aaron, and Janna Anderson. "*AI, robotics, and the future of jobs*." PewResearch Center (2014).
13. Rajan, Kanna, and Alessandro Saffiotti. "Towards a science of integrated AI and robotics." *Artificial Intelligence* 247 (2017): 1–9.
14. Andras, Iulia, et al. "Artificial intelligence and robotics: A combination that is changing the operating room." *World Journal of Urology* 38 (2020): 2359–2366.

15. Ebers, Martin. "Regulating AI and robotics: Ethical and legal challenges." *Algorithms and Law* (2019): 37–99.
16. Mir, Umar Bashir, et al. "Critical success factors for integrating artificial intelligence and robotics." *Digital Policy, Regulation and Governance* 22.4 (2020): 307–331.
17. Rosenschein, Stanley J. "Formal theories of knowledge in AI and robotics." *New Generation Computing* (1985): 345–357.
18. Wisskirchen, Gerlind, et al. "Artificial intelligence and robotics and their impact on the workplace." *IBA Global Employment Institute* 11.5 (2017): 49–67.
19. Chalmers, D. (1995). "The puzzle of conscious experience." *Scientific American* 273 (2013): 80–86. "How can we construct a science of consciousness?" *Annals of the New York Academy of Sciences* 1303: 25–35. https://nyaspubs.onlinelibrary.wiley.com/doi /10.1111/nyas.12166

14 Smart Rings
Redefining Wearable Technology

Sushanth Kumar Patnaik, Neha Medimi, and Ezendu Ariwa

INTRODUCTION

Modern wearable gadgets have transformed how we connect with our environment and one another in recent years, causing wearable technology to see a meteoric rise in popularity [1]. Among these ground-breaking gadgets, the smart ring has emerged as a small, multipurpose item that expertly combines style and cutting-edge technology. In this chapter, we delve into the definition of smart rings, their historical development, and their numerous uses in a variety of industries.

DEFINITION AND OVERVIEW OF SMART RINGS

In essence, a smart ring is a wearable electronic device that is worn on the finger like a typical ring. Despite its unassuming exterior, a smart ring is packed with networking capabilities, sensors, and processors that enable it to do an incredible array of functions. These rings frequently connect with other smart devices, such as PCs, smartwatches, or smartphones, boosting user experiences and extending their functionality [2].

Smart rings have become effective instruments that perfectly fit into our daily lives, including features such as fitness tracking and remote-control capabilities. The goal of every smart ring is to provide customers with greater accessibility and convenience, even though each one may have a different design and feature set to fit different needs and preferences.

HISTORICAL BACKGROUND AND THE EVOLUTION OF SMART RINGS

The development of wearable technology and the miniaturization of electrical components are at the foundation of the idea of smart rings. Wearable technology was first discussed in science fiction literature and movies in the middle of the 20th century. This was a device that could be worn on a finger and do various tasks. But smart rings didn't really start to take off until much later [3].

DOI: 10.1201/9781032686714-14

Historical Background

Early digital wearables: The concept of wearable technology was initially introduced with the introduction of the first digital watches in the 1970s. Some of the first wearable electronics that were mass-produced were wrist-worn devices, and their potential sparked interest in other wearable technologies.

The rise of smartphones: Thanks to the smartphones' swift development in the late 2000s and early 2010s, consumers now have access to a higher level of connectivity and capability. As smartphones became more prevalent, there was an increase in demand for wearable technology to enhance and extend the functionality of these pocket-sized computers.

The Fitbit and Fitness Trackers: Wearable technology gained great popularity thanks to the fitness tracker market. Companies like Fitbit have created wearable devices for tracking steps, heart rate, sleep patterns, and other health-related indications. The way was now open for later, more sophisticated wearable electronics [4].

Evolution of Smart Rings

- **Early attempts:** In the early 2010s, there were a number of smart ring trials, but their functionality was limited, and they lacked the technology to be used broadly. Primitive features like call and message notifications or gesture controls predominated in the early iterations of smart rings.
- **Miniaturization advances:** More complicated sensors and components may be fitted into smaller devices as low-power electronics and miniaturization progressed. This innovation made it feasible to design smart rings with additional functions and a higher level of intelligence.
- **Expanded functionality:** Smart rings began to include additional features in the mid-2010s, such as heart rate monitoring, sleep tracking, NFC (near field communication) for contactless payments, and voice assistant connectivity with Siri and Google Assistant.
- **Integration with IoT:** Smart rings began interacting with other smart gadgets, such as smartphones, smartwatches, and smart home systems, as the Internet of Things (IoT) expanded. They might act as controllers, opening doors, changing smart home settings, or even acting as a way to authenticate devices.
- **Focus on health and wellness:** Smart rings emerged to satisfy the needs of customers who worried about their health as fitness and wellness tracking gained popularity. They were equipped with cutting-edge health monitoring tools, including the ability to measure body temperature, stress levels, and blood oxygen levels.
- **Fashion and personalization:** As smart rings advanced, design and aesthetics became increasingly important. Manufacturers started working on creating appealing, distinctive, and fashionable smart rings in order to appeal to a wider audience than just tech enthusiasts.
- **Ecosystem integration:** In recent years, smart rings have been integrated into more expansive technology ecosystems, supporting a range of platforms

and operating systems. This integration has improved both compatibility and user experience.

Importance and Applications of Smart Rings in Various Fields

Smart rings have become increasingly popular due to their adaptability, which has a positive effect on both our personal and professional lives [5].

- **Health and fitness:** Smart rings have become indispensable tools for both health enthusiasts and medical professionals. These rings provide users with real-time health information so they may make informed decisions about their health. They include heart rate monitors, sleep trackers, and activity sensors. Additionally, certain smart rings have the capability to recognize early symptoms of particular diseases, which may revolutionize preventative healthcare practices.
- **Communication and connectivity:** Thanks to smart rings, we can now maintain connections in whole new ways. Thanks to connectivity with smartphones, users may receive notifications, read messages, and even accept calls without having to pull out their phones. This constant connectivity supports enhanced productivity and convenience in our fast-paced digital world.
- **Access control and security:** Access control and security now function more effectively and simply thanks to smart rings. Utilizing Bluetooth or NFC technology, these rings can be used as keys to facilitate contactless transactions, verify user identities, and open doors.
- **Personalization and style:** One of the most enticing aspects of smart rings is the way they merge fashion and technology. Customizable designs and interchangeable bands allow users to express their individuality while utilizing cutting-edge technology.
- **Gaming and entertainment:** Smart rings have also been utilized by the gaming and entertainment industries. Some rings allow users to employ gesture controls to engage with interactive games or smart devices, further integrating the real and virtual worlds.

DESIGN AND TECHNOLOGY OF SMART RINGS

Overview of Physical Design and Form Factors

Compact and lightweight, smart rings are made to be pleasantly worn on a finger. They are usually subtle and unobtrusive in comparison to larger wearables like smartwatches. Although smart rings come in a range of designs, most of them have a strong emphasis on elegance and simplicity to guarantee that they complement a wide range of outfits and occasions [6].

Depending on the manufacturer and target market, smart rings can come in a range of designs, including minimalist bands, traditional ring forms, or even programmable modules that slide into different ring shells. Some smart rings have changeable sizes to fit a range of finger sizes.

Materials Used in Smart Rings

The comfort, longevity, and aesthetic appeal of smart rings are significantly influenced by the material selection. Common materials used in the construction of smart rings include:

- **Metals:** High-quality metals like titanium, stainless steel, or gold are often used for their tensile strength and elegance.
- **Ceramics:** Ceramic materials offer a sleek and scratch-resistant finish, making them popular choices for smart ring manufacturers.
- **Composite materials:** Smart rings may incorporate composite materials, combining the advantages of multiple materials, such as lightweight alloys or ceramic-infused coatings.

Sensor Technology and Components Integrated into Smart Rings

Smart rings contain a range of sensors and elements to enable their capabilities. Some common sensors and components found in smart rings include [7]:

- **Heart rate monitor (HRM):** HRM sensors continuously track the wearer's heart rate using optical or electrical techniques.
- **Accelerometer and gyroscope:** These sensors can sense motion and keep track of activities, enabling gesture control.
- **NFC:** NFC technology offers contactless communication, which enables features like mobile payments and information exchange with other nearby devices.
- **Infrared sensors:** Infrared sensors could be utilized as proximity sensors for gesture-based interactions or for temperature monitoring.
- **Vibration motors:** Haptic feedback vibration motors vibrate to notify the wearer of notifications or other events.
- **Microprocessors:** Smart rings govern the device's general functionality and process data using tiny but potent microprocessors.

Connectivity Options and Compatibility

Smart rings frequently use wireless connectivity features to connect to other gadgets or the internet. The most common connectivity options include:

- **Bluetooth:** Bluetooth enables smart rings to be connected to computers, tablets, and mobile devices. It makes it possible to synchronize and transmit information such as settings, activity data, and notification data.
- **Wi-Fi:** A few high-end smart rings may have Wi-Fi connectivity, enabling instant internet access and other functions.

Device compatibility is crucial to achieving a smooth integration into users' daily lives. Most smart rings are designed to work with both iOS and Android devices, making them useful for a variety of users. In order to increase interoperability, manufacturers may create companion apps that give additional capabilities and settings for the smart ring. Additionally, smart rings might be able to establish connections with other Internet of Things (IoT) platforms and gadgets, enabling users to control their smart homes or access certain services using straightforward gestures or commands through their rings.

FUNCTIONALITIES AND FEATURES OF SMART RINGS

Fitness and Health Tracking Capabilities

Smart rings have gained popularity as a popular wearable technology that provides users with insightful information about their physical well-being and a variety of fitness and health tracking capabilities. These features aim to promote a healthy way of life by enticing users to maintain an active lifestyle, monitor their vital signs, and make knowledgeable decisions on their overall health. Here are some key fitness and health tracking capabilities commonly found in smart rings:

- **Heart rate monitoring:** Heart rate monitoring is one of the key features of smart rings. Using built-in optical sensors, smart rings may continuously track the wearer's heart rate throughout the day. Using this real-time data, users can keep track of their heart rate trends throughout various activities, such as exercise, relaxation, or stressful situations. By keeping an eye on their heart rates, users can gain a better understanding of their cardiovascular health and alter the intensity of their workouts for maximum effectiveness and safety [8].
- **Sleep tracking:** mart rings with sleep tracking capabilities monitor the wearer's sleeping habits and provide illuminating information on the caliber of their sleep. The ring can monitor various phases of sleep, such as rapid eye movement (REM) sleep, deep sleep, and light sleep, as well as the duration and number of interruptions. Using sleep tracking data, users can identify any sleep-related issues, such as insomnia or unpredictable sleeping habits, and subsequently alter their lives to improve their general sleep hygiene.
- **Activity tracking:** Users may monitor their daily physical activity and progress toward fitness goals thanks to the crucial activity-tracking feature of smart rings. The rings typically count steps, calculate distance traveled, and calculate calories burned based on the wearer's activity levels. This information encourages users to achieve their fitness goals and encourages them to be more conscious of their daily physical exercise.
- **Exercise recognition:** Some state-of-the-art smart rings can automatically detect and track particular workouts or physical activities, such as walking, jogging, cycling, or swimming. This automatic exercise identification tool ensures that customers receive correct information about their activities by doing away with human entry.
- **Stress and well-being monitoring:** The wearer's stress levels are measured via smart rings with stress-monitoring features using a variety of sensors.

These rings can detect changes in skin conductance and heart rate variability, two physiological indicators of stress. By exercising relaxation techniques or engaging in mindfulness exercises and monitoring their stress levels, users can learn how to effectively manage their stress.

- **Blood oxygen saturation (SpO2) tracking:** Some smart rings have SpO2 sensors built in to measure the wearer's blood oxygen levels. In order to detect potential health issues like sleep apnea or respiratory diseases, continuous monitoring is helpful. A vital indicator of respiratory and circulatory health is blood oxygen saturation.
- **Body temperature monitoring:** Some smart rings come equipped with temperature sensors to monitor the wearer's body temperature. Monitoring temperature changes can help identify early signs of sickness or changes in overall health.
- **Sedentary reminders** In order to promote a more active lifestyle, smart rings can send sedentary reminders when the user has been inactive for a while. These notifications encourage users to get some exercise during the day and take little breaks.
- **Data visualization and insights:** Smart rings usually come with companion apps that provide clients with data visualization and insights based on their fitness and health measurements. Thanks to these visual representations, users can establish reasonable fitness objectives, track their development, and identify trends.

It's important to keep in mind that while smart rings are useful tools for tracking fitness and health, they are not medical devices and cannot be used in place of professional medical advice or diagnosis. Users who have particular health issues should consult with medical professionals for guidance and any necessary treatment.

NOTIFICATION AND COMMUNICATION FEATURES OF SMART RINGS

Users may stay connected and informed without constantly checking their devices thanks to the useful communication and notification features that smart rings offer. Thanks to these capabilities, wearers can receive important notifications, alerts, and messages immediately on their fingers. Here are the key notification and communication features commonly found in smart rings:

- **Call alerts:** Smart rings can vibrate, provide visual clues, or show consumers a small display to alert them to incoming calls. The smart ring notifies the wearer as soon as they receive a call on their connected smartphone. This feature makes it simpler to prevent missing critical calls when the user's phone is on silent or is in a bag or pocket.
- **Message notifications:** Smart rings have tiny screens or vibration patterns that can provide notifications for text messages, emails, and other app alerts. Wearers can decide if a response is required or if they may attend

to the message later on their connected device by having quick access to message previews.

- **Social media alerts:** Users can receive alerts for social media interactions including likes, comments, and direct messages thanks to some smart rings that integrate social media notifications. Wearers are kept up to date on their social activities thanks to this feature, especially when they can't frequently check their phones.
- **Customizable notification settings:** Smart rings frequently include companion apps that let users set specific preferences for notifications. Users can select the apps or notification kinds they want to receive on their smart rings, ensuring that only the most pertinent and significant information is alerted to.
- **Do-not-disturb mode:** Smart rings may provide a "do-not-disturb" mode that temporarily mutes all alerts during particular periods or activities to prevent pointless interruptions. When wearers prefer uninterrupted concentration, such as during meetings, workouts, or other circumstances, this feature is useful.
- **Silent alarms:** By sending out moderate vibrations to wake the wearer without disturbing others, smart rings can act as discrete alarm clocks. For individuals who prefer a quiet wake-up call or who require gentle reminders, silent alarms are perfect.
- **Rejected call and message notifications:** Smart rings can warn wearers when a call is declined, or a message is left unanswered in addition to alerting users to incoming calls and messages. It is reassuring to know that the call or message has been acknowledged using this feature.
- **Quick response and actions:** Some smart rings enable users to respond to calls or messages with prepared short responses right from their ring. When a complete answer is not yet possible, these customizable responses offer a convenient method to acknowledge messages or calls.
- **Women's safety using IoT:** Every woman, from all backgrounds, fights every day to be secure and defend herself from the roving gaze of brutally insensitive men who routinely violate the dignity of women by molesting and abusing them. The hunters' territory has grown to include the streets, public transportation, and open spaces in particular. A smart security wearable device for women based on the Internet of Things is suggested as a result of the atrocities that women currently endure. With a Raspberry Pi Zero, a Raspberry Pi camera, a buzzer, and a button to activate the services, it is implemented as a smart ring (named SMARISA). This device is highly portable and can be activated by the victim after an assault by just pressing a button. This will retrieve the victim's current location and use the Raspberry Pi camera to capture an image of the attacker. The victim's smartphone will be used to transfer the position and URL of the captured image to specified emergency contact numbers or the police, eliminating the need for additional hardware components and making the device more portable [9].

Notification and communication functions in smart rings increase connectivity and productivity by keeping users informed with only a quick glance at their finger. These features enable smart rings to fit in with users' daily activities, offering a discrete and useful way to stay connected while travelling.

SMART RING PAYMENT FUNCTIONALITY AND CONTACTLESS PAYMENTS

Smart rings have advanced beyond conventional wearables and now handle contactless payments, enabling users to purchase conveniently and safely without having to remove their wallets or phones from their pockets. By combining NFC technology with current safe payment options, this capability is made available. Here's an overview of payment functionality and contactless payments in smart rings:

- **Contactless payment technology:**
  With appropriate payment terminals in stores, restaurants, and other locations, NFC-enabled smart rings can safely exchange data. A user can make a wireless payment by simply tapping their smart ring on an NFC-enabled payment terminal.
- **Digital wallet integration:**
  To enable contactless payments, users must set up their digital wallets on their smart rings. The digital wallet securely stores the user's credit card or payment information, allowing for quick and secure financial transactions. Some smart rings may additionally enable other payment methods like mobile wallets, prepaid cards, or loyalty cards.
- **Security and authentication:**
  Security comes first with smart rings that enable contactless payments. To prevent theft or unauthorized access to the user's financial information, the rings are manufactured with multiple layers of encryption and authentication. By requiring the wearer to enter a PIN or use biometric verification, such as fingerprint identification, to authorize payments, many smart rings provide an additional layer of security.
- **Wide acceptance:**
  NFC-enabled payment terminals are being used by more merchants and businesses as contactless payments continue to gain popularity worldwide. Due to the widespread popularity of contactless payments, smart rings are a useful and practical alternative to traditional payment methods.
- **Daily convenience:**
  Users can take advantage of the ease of single-tap transactions thanks to the payment functionality integrated into their rings. Smart rings make paying for goods like groceries, coffee, or public transport easier and do away with the need to carry physical cards or cash.
- **Transaction history and monitoring:**
  Users usually receive full transaction histories and expense details via smart ring companion apps. Users may assess their purchasing patterns, better manage their budgets, and keep track of their expenses.

- **Limits and security features:**
  Smart rings may contain pre-set transaction constraints or let users customize them in order to offer even more control over their spending. In addition, a few smart rings have built-in theft prevention measures that let users remotely turn off the capacity to make payments in case a ring is lost or stolen.
- **Cross-platform compatibility:**
  Smart rings that include payment capabilities are designed to work with various payment methods and financial institutions, ensuring interoperability with a range of merchants and banks.

Smart rings with contactless payment capabilities help to make online purchases more secure and safe in addition to being efficient and convenient. As contactless payments continue to improve in technology, smart rings are likely to gain importance.

ARTIFICIAL INTELLIGENCE IN SMART RINGS

AI has greatly increased the capabilities of smart rings and fundamentally altered how we interact with wearable electronics. This section will analyze how AI works in smart rings and provide real-world examples to highlight our points.

Machine learning-based health monitoring is one of the primary applications of AI in smart rings. These devices could collect a range of physiological data, such as heart rate, sleep patterns, and levels of exercise. Smart rings can utilize machine learning algorithms to analyze this data and give vital details about the user's health and well-being. For instance, the Oura Ring uses AI algorithms to find patterns in sleep data, allowing users to improve their sleeping habits and overall sleep quality.

Fitness and activity tracking that is personalized is made possible by smart rings powered by AI. These gadgets may be learned from and adjusted to the user's activity patterns by utilizing machine learning and offering specialized recommendations and objectives. For instance, the Motiv Ring makes use of AI algorithms to comprehend a user's heart rate, sleep schedule, and level of activity. It then offers tailored exercise targets and feedback to assist users in successfully achieving their fitness objectives.

Smart rings can send context-sensitive notifications thanks to AI. By analyzing a range of contextual factors, such as location, time, and user behavior, smart rings may give pertinent information at the appropriate time. For example, an AI-powered smart ring may be able to detect when a user is participating in a conference and automatically silence non-urgent notifications to minimize disruptions.

Voice interaction and NLP capabilities are included in AI-powered smart rings, allowing users to interact with the device using voice commands. In order to interpret user queries and provide applicable responses, these rings can use speech recognition and language-understanding algorithms. For instance, the voice-activated interactions offered by the Amazon Echo Loop smart ring, which incorporates AI, enable hands-free access to a variety of services and features.

Smart rings with AI algorithms may be able to aid with fall detection and emergency response. By examining sensor data and movement patterns, these rings are able to recognize approaching falls and activate automatic alerts or emergency messages. The Apple Watch, for instance, utilizes AI algorithms to determine whether a user has fallen seriously. The choice to alert predetermined emergency contacts or dial emergency services is then presented.

AI-driven smart rings with gesture support can enhance user interfaces. With the help of AI algorithms, these rings can recognize and comprehend hand motions, enabling users to interact with their devices more intuitively. AI is used, for instance, by the Nod smart ring to identify and interpret hand gestures.

AI in smart rings can provide behavioral insights and predictive analytics based on user data. By analyzing past trends and user behavior, these rings can forecast user desires and preferences. For instance, a smart ring can learn the user's daily habits and provide proactive reminders or suggestions based on their habits and preferences.

AI enhances the capabilities of smart rings, turning them into intelligent wearable computers. AI algorithms enhance the user experience and provide meaningful data, whether it be through context awareness, gesture control, personalized fitness tracking, or health monitoring. Several examples of real-world products, such as the Oura Ring, Motiv Ring, Amazon Echo Loop, and Apple Watch, demonstrate how AI enables smart rings to provide personalized, context-aware, and proactive features, thereby improving users' health, convenience, and overall well-being.

SMART RINGS' MACHINE LEARNING ALGORITHMS FOR DATA ANALYSIS AND INTERPRETATION

Machine-learning algorithms play a significant role in the ability of smart rings to analyze and interpret data obtained from numerous sensors. These algorithms can produce and provide people with relevant information from unstructured data. Smart rings process physiological information like heart rate, sleep patterns, and activity levels using machine-learning algorithms. By analyzing this data, smart rings can provide users with valuable insights into their health and well-being.

One of the key advantages of smart rings is the ability of machine-learning algorithms to find patterns and trends in the data. For example, over time, machine-learning algorithms may detect anomalies or unpredictable patterns in a user's heart rate data that may indicate potential health issues. This can be used by users to proactively address health problems and receive the attention they require.

Thanks to machine-learning algorithms, smart rings may also customize data processing and interpretation based on specific user characteristics. These variances in each user's physiological characteristics and behaviors can be accommodated by machine-learning techniques. By continuously learning from the user's data, the algorithms can enhance their models and provide more accurate and unique insights.

However, using machine-learning techniques in smart rings is not without its challenges. It takes a lot of computing power to manage and analyze enormous amounts of data in real time. The memory and processing power of smart rings are restricted, which limits the complexity of the algorithms that may be applied.

Another challenge is guaranteeing the algorithms' precision and dependability. Machine-learning algorithms are trained on massive datasets, and the quality and representativeness of the training data have a big influence on how well the algorithms work. It is essential to carefully curate and validate the training datasets in order to guarantee the algorithms' success and minimize any biases.

Last but not least, machine learning methods are necessary for smart rings to analyze and understand biometric data. Insightful data on users' health and wellness is provided by these algorithms, enabling proactive healthcare and supporting users in making informed lifestyle decisions. Current advances in machine-learning techniques continue to expand the capabilities of smart rings, paving the way for more individualized and effective user experiences despite computational resource constraints and algorithm accuracy being obstacles.

Context Awareness and Ambient Intelligence in Smart Rings

Due to their context awareness and ambient intelligence, smart rings can understand the user's surroundings, respond to them, and provide useful information and assistance. These features enhance the user experience by offering notifications, advice, and interventions that are timely and contextually appropriate. The ways in which context awareness and ambient intelligence are used in smart rings are discussed below.

Using sensors like accelerometers, gyroscopes, GPS, and environmental sensors, smart rings can gather contextual data about the user's surroundings. By analyzing this data in real time and combining it with other user-specific data, smart rings can provide recommendations and interventions that are aware of the context. A smart ring, for example, can automatically change the notification settings if it detects that the user is in a noisy environment to guarantee that important notifications are not missed.

Machine-learning algorithms are used by smart rings' ambient intelligence to understand user preferences, behavioral trends, and usage context. Smart rings may continuously analyze user data and learn from user interactions to provide individualized and proactive assistance. For instance, a smart ring may find out that a user frequently gets stressed out while engaging in certain activities or at particular hours of the day. Based on this data, the smart ring can provide the user with useful reminders or relaxation strategies to help them effectively manage their stress.

Challenges and Factors to Consider When Using AI in Smart Rings

When utilizing AI in smart rings, there are challenges and things to keep in mind. Here are some crucial issues to address:

- **Data quality and reliability:** Artificial intelligence algorithms heavily rely on trustworthy and high-quality data for accurate analysis and interpretation. Consistency must be maintained for smart rings to collect data that is accurate and trustworthy.

- **Power and processing limitations:** Smart rings are limited in both battery life and processing speed. It is essential to use AI algorithms that are computationally efficient and energy mindful to maintain peak performance and extend battery life.
- **Privacy vs. personalization:** To customize smart ring features, access to personal data is required. Finding the perfect balance between user privacy and personalization may be challenging. Manufacturers must develop strong privacy policies, data anonymization techniques, and secure data storage in order to protect user information.

AI-Powered Smart Rings' Ethical and Privacy Implications

Smart ring AI adoption raises serious privacy and ethical concerns. Here are some important factors to consider [10]:

- **Data security and privacy:** Smart rings gather private and delicate health information. It is crucial to ensure the security and privacy of this data in order to avoid unauthorized access, breaches, or exploitation of users' personal information. Strict access controls, reliable encryption, and safe data transport methods must all be used. Users must be made aware of the methods used to obtain personal data, the goals for which it will be used, and the benefits and drawbacks of utilizing AI-powered services. Open communication and obtaining users' informed consent are necessary for establishing trust and preserving moral standards.
- **Fairness and bias:** Biases present in the training data may make AI algorithms susceptible. Prejudices must be addressed and diminished in order to ensure that consumers from all demographic groups are treated equally and impartially.
- **Accountability and explainability:** The AI algorithms used in smart rings ought to be transparent and totally understandable. Users should be able to intervene if they believe the AI system is not performing as intended since the decision-making process of the system should be clear to them.

User Experience and Interface in Smart Rings

User interface (UI) and interaction methods must be taken into account while designing the user experience for smart rings. Smart rings combine a range of interface modalities to enable fluid and natural user interactions. In the context of smart rings, let's examine the numerous user experience and interface elements.

User interfaces and methods of interaction:

- **Touchscreen:** Smart rings frequently have a small touchscreen that enables users to scroll through menus and interact with settings and apps. Touchscreen interaction offers a simple and recognizable user interface like that of a smartphone or smartwatch.

- **Gestures:** Gyroscopes and accelerometers incorporated into smart rings are used to detect gestures. Users can create motions with their fingers by tapping, swiping, or twisting them to begin certain actions or browse menus. The use of gesture-based communication is convenient and gesture-based communication is both convenient and allows for hands-free interaction.
- **Voice control:** Users can control smart rings using voice commands and voice recognition technology. Users can initiate tasks, do searches, or interact with virtual assistants by voicing their commands. In particular, when employing touch-based interaction is difficult or when hands are busy, voice control increases accessibility and convenience.

Customizing options:

Users of smart rings can personalize their experiences in a number of ways to suit their preferences. Among these choices are:

- **Watch faces:** Users can choose from a selection of watch faces to customize the appearance of their smart ring's display. To suit your interests and needs, a range of patterns, colors, and levels of intricacy are available.
- **App settings:** Users can adjust settings on the loaded apps on their smart rings. Users have the ability to alter an app's preferences, notifications, and functionality.
- **Personalization:** Smart rings offer tailored functionality based on user preferences and behavior. Users can choose to define activity objectives, customize activity tracking features, or choose their preferred notification methods, for example. Personalization enhances the user experience by tailoring the device to the user's preferences.

Battery life and charging methods:

The battery life of smart rings has a big impact on their usefulness and usability. While the size of the battery is constrained by the compact form factor, smart ring manufacturers use several techniques to maximize battery life:

- Energy-efficient components and software algorithms are used to build smart rings, which use less electricity. By adopting power management techniques, such as reducing display brightness and background activities, battery life is extended.
- Wireless charging is a feature that many smart rings have, allowing users to simply set their gadgets down on a charging pad to recharge them. Because wireless charging does not require physical connectors or wires, it is practical and easy to use.
- Battery-saving features on smart rings frequently block specific features or reduce the frequency of data syncing. These settings aid in preserving battery life when users anticipate going without access to charging for prolonged periods of time.

Smart rings' user interface and experience are designed to allow for extensive customization and fluid interactions. Voice control, gestures, and touch screens enable intuitive and interactions. To provide a more distinctive touch and broaden their usefulness, smart rings can be personalized with watch faces, app settings, and personalization. Additionally, maximizing battery life and putting in place practical charging techniques results in increased usage time and user convenience.

CHALLENGES AND FUTURE TRENDS

Smart rings are convenient and small, but they also have significant flaws that must be fixed if they are to be widely adopted and offer a better user experience [11].

1. **Limited display and input options:** Smart rings have a small screen size and fewer input choices than smartphones or even smartwatches. It is challenging to use the device and display complex information due to this restriction. It is crucial to design user-friendly interfaces that are intuitive and take these limitations into account. Manufacturers must carefully consider the size and placement of buttons or touch-sensitive areas on the ring in order to increase usability and decrease errors.

2. **Battery life:** Because of their compact size, smart rings cannot accommodate large batteries. Therefore, battery life presents a significant issue. To enable smart rings to be worn all day without requiring frequent recharges, it's crucial to strike a balance between battery consumption and functionality. Manufacturers are constantly looking into energy-efficient components, improving software algorithms, and even taking into account alternative energy sources like solar or kinetic energy to tackle this issue.

3. **Sensor accuracy and reliability:** In order to gather information such as activity tracking, heart rate monitoring, sleep patterns, and gesture recognition, smart rings rely on sensors. Considerations like sensor precision and dependability are crucial. Ensuring the caliber, dependability, and consistency of sensor measurements is essential for providing precise insights and a seamless user experience. To improve the accuracy and dependability of the data collected, manufacturers engage in sensor calibration, sensor fusion techniques, and continual algorithm improvement.

4. **Connectivity and integration:** Smart rings must be simple to integrate with various systems and ecosystems in order to function properly. Problems could arise with smartphones, smartwatches, and other devices due to compatibility, interoperability, and synchronization concerns. Manufacturers must make sure that connectivity protocols (like Bluetooth) operate without a hitch and are compatible with well-liked operating systems in order to enable seamless data transmission, notification synchronization, and app integration.

Security and Privacy Issues

It is crucial to address privacy and security issues regarding the sensitive data that smart rings collect about users, including health information, activity patterns, and personal preferences, in order to boost user confidence and trust in these devices.

1. **Data protection:** Smart-ring producers must utilize robust data encryption methods and secure storage practices to protect user data from unauthorized access or data breaches. This includes both data at rest (stored on the device) and data in transit (transmitted to auxiliary programs or cloud services).
2. **User consent and control:** Users of smart rings should have control over the data they share and how it is gathered. Allowing users to make informed decisions about their data by providing them with granular control options and open permission procedures. It is the responsibility of manufacturers to ensure that their products have clear privacy policies, simple privacy settings, and mechanisms for users to change their data preferences.
3. **Anonymization and aggregation:** To lessen privacy issues, smart rings can employ techniques like data anonymization and aggregation. By removing personally identifiable information and privately integrating data, meaningful insights can still be obtained while maintaining individual identities. Manufacturers should follow applicable data protection legislation and privacy best practices in order to secure user privacy.
4. **Integration with third parties:** Data sharing rules should be transparent and in line with industry best practices when smart rings interact with third-party platforms or services. Manufacturers believe that in data-sharing agreements with third parties, the users' security and privacy should come first. Clear user consent and data-anonymization practices might help to minimize privacy issues when using external services.

PROSPECTS FOR SMART RING TECHNOLOGY AND EMERGING TRENDS

A number of recent innovations that are currently taking shape are influencing the direction of smart ring technology. These advancements might make smart rings more practical and easier to use.

1. Advanced health monitoring functions for smart rings are being added at a quick rate, including electrocardiogram (ECG) measurements, body temperature sensing, stress detection, and blood oxygen level monitoring. These advancements enable users to keep tabs on their health and provide information that can help them make better-educated decisions about their welfare.
2. Thanks to advancements in sensor technology and machine-learning algorithms, smart rings are now capable of reliably recognizing and comprehending the wearer's motions. This enables simple control of a wide range

of devices and applications, including smart TVs, smartphones, and home automation systems. For example, a simple hand gesture can be used to answer calls, control music playing, or change smart home settings.

3. The IoT ecosystem and the user can be connected by smart rings. To give contextual data and enable seamless automation, they can exchange data with adjacent connected devices and sensors. For instance, when a smart ring detects the user's presence, it can adjust the lighting, temperature, and other settings in a room accordingly.

4. Smart rings can be used for secure biometric authentication. Smart rings analyze distinguishing biometric qualities, such as fingerprints or vein patterns, to verify the wearer and give access to secure systems, apps, or physical regions. The lack of passwords and PINs enhances convenience and security.

POSSIBILITIES FOR DEVELOPMENT AND INTEGRATION WITH OTHER WEARABLE TECHNOLOGY

The potential for smart ring technology to evolve and be integrated with other wearables is exciting.

1. **Improvements in sensor technology:** By continuing to progress sensor technology, including miniaturization, increased precision, and new types of sensors, the potential of smart rings will be expanded. For example, they would be capable of incorporating cutting-edge sensors for tracking hydration, keeping an eye on blood pressure, or even non-invasively monitoring glucose levels.

2. **Smartwatch and smartphone integration:** Smart rings can be added to and integrated with current wearables like smartwatches and smartphones. A smooth user experience can be provided by the smart ring by enhancing the capabilities of the other devices. For example, a smart ring can track fitness data while the wearer is wearing a smartwatch, display notifications from a connected smartphone, and enable quick connection with smart home devices.

3. **Integration of ecosystems:** Larger technology ecosystems, such as platforms for smart homes, health monitoring, and virtual assistants, can communicate with smart rings. Users may benefit from a consistent user experience and smooth data sharing across a range of platforms and devices thanks to this connection.

4. **Integration of fashion and design:** As smart ring technology advances, manufacturers are looking into collaborations with fashion labels to create aesthetically beautiful designs that can accommodate varied personal styles. This blend of technology and fashion can increase the adoption of smart rings by luring more customers who value both functionality and style.

The connectivity, battery life, sensor accuracy, and display size of smart rings are constrained. Privacy and security concerns must be resolved if user data is to be protected and trust is to be fostered. Emerging technologies include advanced health monitoring, gesture recognition, IoT system integration, and biometric authentication. Future advancements in sensor technology, the integration of wearable technology, and collaborations with the fashion sector are all conceivable. These advancements will influence the design of future smart ring technology, enhancing its usability, security, and convenience of use.

CONCLUSION

Smart rings have evolved into potent and versatile wearable technologies with significant applications in a variety of fields, revolutionizing how we interact with technology and enhancing our daily lives. Their impact can be seen in a number of industries, including communication, access control, personalization, and health and fitness. Because of their exquisite aesthetics, cutting-edge technology, and distinctive functionality, smart rings have become essential items that satisfy a variety of needs and interests.

Smart rings have been crucial in enabling people to assume personal responsibility for their health and fitness. Through features like heart rate monitoring, sleep tracking, activity tracking, and stress management, smart rings offer users real-time insights into their physical well-being. Users can utilize these tools to make informed decisions about their lifestyle, workout routines, and general health. Additionally, smart rings can serve as early warning systems, spotting potential health issues and urging wearers to get help from a doctor. By bridging the gap between technology and healthcare, smart rings have the potential to enhance preventative healthcare procedures and contribute to better overall health outcomes.

Smart ring features enhance connectivity and communication. Users no longer need to regularly check their cell phones because they may receive call alerts, message notifications, and social media updates instantly on their rings. Users can stay connected while concentrating on other tasks thanks to this seamless connection, which enables discrete and practical communication. Additionally, contactless smart rings offer a safe and practical substitute for conventional payment methods. By tapping their rings, customers may make purchases without using physical wallets or credit cards. People's daily activities are made simpler and more effective by these elements.

Smart ring technology and design have evolved tremendously as well. Smart rings put a strong emphasis on aesthetics and user comfort while being small, light, and visually appealing. Customers can choose a design that best meets their personal taste from a number of styles, including minimalist bands, conventional ring forms, and programmable modules. To give durability and aesthetic appeal, materials like metals, ceramics, and composites are used. Heart rate monitors, accelerometers, NFCs, and microprocessors are just a few of the sensors and parts found in smart rings. Now that precise and dependable data can be transmitted, smart rings can operate and provide a positive user experience.

Smart rings offer a wide array of features and functionalities that cater to different user needs. Smart rings provide a wide range of technologies, from fitness and health tracking to notification and communication capabilities, to enhance the user experience. They provide consumers with heart rate monitoring, sleep monitoring, and activity monitoring so they can keep an eye on and improve their physical well-being. Users may stay connected without constantly being tethered to their devices by using their rings to receive call alerts, message notifications, and social media updates. Smart rings are an excellent tool for in-store purchases since their contactless payment technology facilitates quick and secure transactions. Furthermore, smart rings can include functions like sedentary reminders, silent alarms, and quick response options, which increase their utility and usability.

Smart rings' usage of AI broadens the range of potential applications even further. Machine learning algorithms may analyze and interpret the data from smart rings to provide users with tailored insights and suggestions. Thanks to AI-powered personalization and adaptive intelligence, smart rings may learn and adapt to specific tastes and behaviors, improving the user experience. Due to context awareness and ambient intelligence, smart rings can provide pertinent information based on the user's surroundings and wants. These AI capabilities enhance the functionality and usability of smart rings, making them simpler to use.

On the other hand, integrating AI into smart rings comes with its own set of challenges. Privacy and security issues are essential to avoid since AI algorithms rely on collecting and analyzing personal data. Manufacturers must develop efficient data protection measures in order to safeguard user information and maintain user confidence. For seamless integration and a positive user experience, compatibility and interoperability with other hardware and software are equally essential factors. To ensure the ethical development and application of AI-powered smart rings, ethical issues like bias in AI algorithms or possible data exploitation must be adequately addressed.

The future of smart rings seems promising in the long run. New developments in AI algorithms, interaction with other wearable technology, and expanded personalization options are some of the emerging themes. The development of AI algorithms will lead to better user experiences, more precise and knowledgeable data analysis, and personalized health advice. A whole ecosystem of networked devices will be created by integration with other wearables, such as smartwatches or fitness trackers, enabling efficient data sharing and general functioning. Additionally, users will be able to further personalize their bands to better reflect their own sense of style and preferences thanks to enhanced customization options for smart rings.

Last but not least, smart rings have proven to be crucial accessories that unite fashion and technology by offering a number of capabilities that enhance many facets of our lives. In areas including communication, access control, personalization, and health and fitness, they have found application and relevance. Thanks to their fashionable designs, cutting-edge technology, and AI capabilities, smart rings provide users with personalized insights, straightforward communication, safe transactions, and a pleasant user experience. The future of smart rings has enormous promise for advancements and integration into our increasingly interconnected world, despite challenges and ethical issues. Smart rings will transform how we interact with wearable electronics as technology improves and improves the quality of our daily lives.

REFERENCES

1. Marc, Kurz, Robert, Gstoettner and Erik, Sonnleitner, "Smart Rings vs. Smartwatches: Utilizing Motion Sensors for Gesture Recognition", *Applied Sciences*, 2021. https://doi.org/10.3390/app11052015
2. Bogdan-Florin, Gheran, Jean, Vanderdonckt and Radu-Daniel, Vatavu, "Gestures for Smart Rings: Empirical Results, Insights, and Design Implications", *Association for Computing Machinery*, 2018. https://doi.org/10.1145/3196709.3196741
3. Magno, M., Salvatore, G. A., Jokic, P. and Benini, L. "Self-Sustainable Smart Ring for Long-Term Monitoring of Blood Oxygenation", *IEEE Access*, 2019;7:115400–115408. https://doi.org/10.1109/ACCESS.2019.2928055
4. Asgari Mehrabadi, M., Azimi, I., Sarhaddi, F., Axelin, A., Niela-Vilén, H., Myllyntausta, S., Stenholm, S., Dutt, N., Liljeberg, P. and Rahmani, A. "Sleep Tracking of a Commercially Available Smart Ring and Smartwatch Against Medical-Grade Actigraphy in Everyday Settings: Instrument Validation Study", *JMIR Mhealth Uhealth*, 2020;8(11):e20465. https://doi.org/10.2196/20465. https://mhealth.jmir.org/2020/11/e20465
5. Feng, S., Mäntymäki, M. and Salmela, H. "Affordances of Sleep-Tracking: Insights from Smart Ring Users", in: Papagiannidis, S., Alamanos, E., Gupta, S., Dwivedi, Y. K., Mäntymäki, M. and Pappas, I. O. (eds.), *The Role of Digital Technologies in Shaping the Post-Pandemic World. I3E 2022. Lecture Notes in Computer Science* (vol. 13454). Springer, Cham, 2022. https://doi.org/10.1007/978-3-031-15342-6_27
6. Bilius, L. B., Vatavu, R. D. and Marquardt, N. "Exploring Application Opportunities for Smart Vehicles in the Continuous Interaction Space Inside and Outside the Vehicle", in Carmelo Ardito,Rosa Lanzilotti, Alessio Malizia, Helen Petrie, Antonio Piccinno, Giuseppe Desolda, Kori Inkpen (Eds.) *Human-Computer Interaction–INTERACT 2021: 18th IFIP TC 13 International Conference, Bari, Italy, August 30–September 3, 2021, Proceedings, Part II 18* (pp. 140–149). Bari, Italy: Springer International Publishing, 2021.
7. Takezawa, Shiho and Horie, Masatsugu. "Smart Rings Seen as New Frontier for Cashless Payments", *Bloomberg.com*. 25 May 2021. Retrieved November 11, 2021.
8. "Learn how Oura Ring Works | Go inside | Oura Ring", *Oura Ring*. Retrieved November 27, 2018.
9. Plummer, Libby. "Smart Rings: The Good, the Bad and the Ugly in Smart Jewellery", *Wearable.com*. 7 July 2016. Retrieved November 21, 2016.
10. HUFFPOST. "This Ring Lets You Feel Your Spouse's Heartbeat in Real Time", *HUFFPOST*. 9 August 2016. Retrieved November 11, 2021.
11. "This NFC Ring Puts Wireless Transfer Tech on Your Finger So You Can Fist-Bump Phones", *TechCrunch*. 23 July 2013. Retrieved November 12, 2021. https://en.wikipedia.org/wiki/Smart_ring

15 Smart Glasses with Sensors

Xavier Borah, A. Thangam, and Neelam Kumari

INTRODUCTION

In today's world, technology in every field is growing at a very fast pace. The way we use technology, makes us think in a certain way, making us behave in a certain way. There is the technology of smart glasses which is developing at a rapid rate. Smart glasses have the potential to change the way we perceive and interact with our surroundings. These technologically advanced wearable devices have the capability to merge the both digital and physical worlds. Because of this, they can create a world that we have never seen before, they can open whole new possibilities that we have never seen prior. Smart glasses are equipped with an array of sensors that are capable of detecting and interpreting the environment around us, they can provide us with real-time data of our surroundings, are capable of giving us contextual information, and basically, can give us an experience we have never experienced before.

In 2013, Google launched its first prototype of smart glasses. It achieved a very significant milestone in technology, especially in the history of smart glasses. The prototype featured an optical head-mounted display and built-in cameras, it also came with voice activation [1]. Users could do a plethora of activities like taking photos and videos of their surroundings, accessing the internet, seeing maps, and many other things. These Google glasses created a very significant interest in people's minds. But they faced a lot of privacy concerns and they had very little battery life. So, because of flaws and privacy concerns, Google withdrew them from the consumer market in 2015. Although, in 2017, Google took out an enterprise edition of these glasses, and also launched a second edition of them in 2019. Following Google, many companies and different startups started to develop their own smart glasses for different industries like healthcare, sports, and augmented reality (AR) technology.

Microsoft also put out their first set of smart glasses devices by the name HoloLens. This was an AR headset that was developed and manufactured completely by Microsoft. It used the word lens rather than the word glasses. It combined the real world with the virtual elements, allowing the users to see digital content and holograms overlaid onto their present physical surroundings. It came with multiple sensors such as cameras, depth sensors, and motion sensors. The motion sensors were there to track users' movements and by that, it gathered information about the physical world around them. "HoloLens" came with built-in speakers and also built-in microphones for both audio inputs as well as audio outputs. HoloLens deployed

DOI: 10.1201/9781032686714-15

with vast promising industry applications like gaming, and entertainment, but especially for medical training and defense sectors. In an effort to improve soldier readiness, Microsoft received a nearly $21.88 billion deal over 10 years to ship 120,000 HoloLens glasses for US military use [1].

Lenovo has also contributed to the sector of smart glasses as they brought in the idea of multitasking through smart glasses with their product Think Reality A3. Lenovo's augmented reality smart glasses (ARSGs) aimed at the current-day workforces, were specially designed to cater to the emerging trend of "distributed workforces and hybrid work models" [1]. They created a virtual 1080p monitor specifically for users who need a private workspace anywhere they want. Initially, they offered five virtual monitors, all arranged side by side.

Apple also recently entered the market with its new product called Vision Pro, which offers an impressive range of capabilities. Equipped with its own powerful chipset (M2), the Vision Pro is essentially a wearable Mac with augmented reality and virtual reality (VR) capabilities in the form of glasses with built-in speakers [2]. This innovative device allows users to enjoy a seamless AR experience that boasts a multitude of functionalities. With its sleek design and advanced technology, the Vision Pro opens up a world of possibilities for its users. There is only one drawback of the product and that is its battery capacity but, in the next two or three iterations, that issue also will be fixed by Apple [2].

Components and functionalities of smart glasses

Smart glasses come with a variety of components, which make them very versatile across different industries. These components normally include small-size displays, high-definition (HD) cameras (with or without optical image stabilizer (OIS) for capturing images and videos, motion sensors for gathering environmental data, wireless connectivity for seamless internet access, built-in speakers for audio output, voice communication, and commands. They come with a built-in microphone and a battery to power all the integrated features. Additionally, smart glasses are equipped with powerful processors to run applications, possess their own memory, and come with unique user interfaces developed by different companies. Thanks to these components, smart glasses offer a wide array of functionalities, including hands-free communication, AR and VR experiences, multimedia consumption, navigation assistance, activity tracking, and much more. They provide users with an unprecedented level of flexibility, enabling them to seamlessly interact with their surroundings while engaging in various tasks simultaneously [3]. Consequently, smart glasses prove to be highly beneficial for productivity. Their ability to seamlessly merge the physical and digital realms enhances user convenience and delivers a truly immersive experience.

TYPES OF SENSORS IN SMART GLASSES

OPTICAL SENSORS

Optical sensors have the most important role in smart glasses, mostly in capturing and displaying visual information or data in our eyes. Optical sensors are those that use light or electromagnetic radiation especially to detect, measure, and analyze all different properties of users' surroundings. Optical sensors are used in a variety of applications including smart glasses, industrial automation, imaging systems, environmental monitoring purposes, and also scientific research.

Overview and Applications of Optical Sensors

There are numerous applications of optical sensors. One such application is proximity sensors, which detect the presence or absence of objects within a specific range. Proximity sensors are utilized in various devices such as smartphones, automated doors, robotics, and touchless interfaces for object detection. Another type of optical sensor is the ambient light sensor, also known as photodetectors. These sensors typically measure the ambient light available in the environment [4]. They are commonly found in devices with display screens, such as phones or laptops, to adjust screen brightness automatically [4]. When discussing ambient light sensors, it's important to mention image sensors as well. These sensors capture real-time images and convert them into electrical signals. They are used in cameras, surveillance systems, and medical imaging devices for capturing and processing information. Another optical sensor worth mentioning is the fiber optic sensor, which transmits and detects light signals through optical fibers. Fiber optic sensors find applications in temperature sensing, strain measurement, and intrusion detection in security systems [4]. Biometric sensors are also commonly used optical sensors these days. They employ optical techniques to capture and analyze the behavioral characteristics of individuals. We frequently encounter them in fingerprint scanners, facial recognition technology, and iris recognition systems, primarily for authentication purposes, especially in mobile devices. There are also optical pressure sensors that measure pressure using the principles of light reflection or transmission. These sensors are employed in industries such as those of aerospace, automotive, and healthcare to monitor and control pressure. Motion and gesture sensors are other types of sensors that detect and interpret human gestures and movements [4]. They find applications in areas like virtual reality systems, smart home devices, and gaming consoles. Additionally, spectroscopy techniques utilize light-matter interaction to identify and characterize substances. Automotive sensors are also important, playing a role in collision detection, lane departure warning, adaptive lighting, and advanced driver assistance systems (ADAS) to enhance safety in automotive applications.

Examples of Optical Sensors in Smart Glasses

As we now know, smart glasses come with various optical sensors to enhance or enable advanced functionality and also for better interaction. Smart glasses come with built-in camera sensors to capture real-time data such as pictures and videos. These sensors are also engaged in AR experiences by overlaying digital content onto the real world. There is also an ambient light sensor that measures the intensity of light surroundings. According to the ambient lighting conditions, to give optimal visibility to the wearer it automatically adjusts the brightness of the display. There are proximity sensors in smart glasses to detect the real-time presence or absence of any objects in the range of proximity to the device. They are used to activate or deactivate specific features or to switch between different modes based on the wearer's interaction or when the glasses are put on or taken off. To enable hands-free control and interaction with the device, smart glasses use infrared sensors. By tracking the movement of the wearer's eyes, they allow users to navigate through menus, scroll, and perform actions using just their gaze. Also, to detect and interpret hand movements and gestures made by the wearer, smart glasses come with gesture recognition sensors, such as a gyroscope or an accelerometer. It allows the wearer to do touchless interaction with the device's interface or control external applications. Advanced smart glasses may include sensors to monitor the wearer's pupil dilation. This data can provide insights into the user's cognitive and emotional state, enabling applications like attention tracking, stress detection, or personalized feedback based on the wearer's responses.

ENVIRONMENTAL SENSORS

Environmental sensors are specially designed to detect various parameters and also measure various parameters or conditions in the surrounding environment of the wearer. These types of sensors play a crucial role in monitoring and understanding the state of the environment. They enable us to make informed decisions and take appropriate actions to mitigate environmental challenges [5].

Overview and Applications

These environmental sensors have the capability to measure the concentration of pollutants such as particulate matter (PM2.5, PM10), carbon monoxide (CO), nitrogen dioxide (NO2), ozone (O3), and other harmful gases. This type of information helps in assessing air quality, identifying pollution sources, and implementing strategies to improve air quality. These kinds of sensors are also used to measure different parameters like barometric pressure, temperature, humidity, wind speed, and wind direction. Weather monitoring stations utilize these sensors to collect data for weather forecasting, climate studies, and meteorological research. These sensors can also measure different kinds of parameters in water bodies, such as temperature, pH level of the water, dissolved oxygen, turbidity, conductivity, and levels of contaminants in the water such as nutrients and heavy metals. This data is essential for assessing water quality, detecting pollution sources, and managing water resources effectively. Also, sensors placed in the soil can measure the moisture from the soil, soil temperature, and soil nutrient levels. This type of information helps in optimizing irrigation practices and crop growth, and it also prevents water wastage. Similarly, there are various other environmental sensors for different cases like noise pollution monitoring or radiation monitoring, etc.

Examples of Environmental Sensors in Smart Glasses

Smart glasses come with environmental sensors because they have the capability to give users real-time environmental data and can enhance their understanding of the surrounding environment. While smart glasses are mainly known for their AR capabilities, the inclusion of environmental sensors can further increase their functionality [5]. Some smart glasses come with a UV sensor, which has the capability to measure the level of ultraviolet (UV) radiation in the environment. By monitoring UV exposure, smart glasses can provide real-time UV index readings and alert users when they need to apply sunscreen or take other protective measures to avoid overexposure. Also, there are temperature and humidity sensors, which can provide users with information about the ambient temperature and humidity levels [5]. These sensors can assist users in understanding the comfort level of their surroundings and adjusting their clothing or activities accordingly.

BIOMETRIC SENSORS

Biometric sensors are the kinds of sensors that come with a combination of mechanical and electrical technology that work to capture human biometric data. This can be palm, iris, or it can be face data. It captures the information digitally in a specific way and turn into a biometric template [6].

Overview and Applications

Biometric sensors are devices that analyze the unique physical characteristics of individuals as well as their behavioral characteristics. These sensors capture and analyze biological information of an individual to authenticate or identify them based on their distinct traits. Biometric sensors are commonly used for access control systems in various settings such as airports, offices, and secure facilities. They can utilize fingerprint, iris, face, or palm recognition to verify an individual's identity and grant or deny access accordingly. Nowadays, most smartphones, tablets, and laptops come with built-in biometric sensors, such as fingerprint scanners or facial recognition [6]. These biometrics provide secure and convenient authentication, replacing traditional password methods. Biometric sensors also find applications in verifying a person's identity for financial transactions, online services, and government programs. Furthermore, biometric sensors have the capability to monitor various health parameters, including heart rate, blood pressure, body temperature, and oxygen saturation. These sensors are often integrated into wearable devices to analyze or track our vital signs. Additionally, biometric sensors are frequently used for payment and transaction purposes [6]. They are used in companies to accurately track the attendance and working hours of their employees.

Examples of Biometric Sensors in Smart Glasses

Smart glasses come with a plethora of biometric sensors, especially in AR technology. Eye-tracking sensors track the constant movements of the user's eyes, which can be utilized for various purposes such as controlling the virtual display or analyzing user behavior. Smart glasses also include iris scanners, which capture the pattern of the user's eyes, aiding in secure authentication and identity verification. These sensors can also analyze the user's facial patterns or features, serving as a means of user authentication. Smart glasses are also equipped with accelerometers and gyroscopes to detect head movements and enable gesture controls.

MOTION SENSORS

Motion sensors are a type of sensor that recognizes and measures movement or changes in position. They are specifically designed to capture physical motion and translate the data into electrical signals, allowing them to be used for various purposes. Nowadays, motion sensors are commonly used in smart devices, enabling functionalities such as motion detection, gesture recognition, and activity tracking. They can be found in a variety of applications, including security systems used in homes, gaming consoles and controllers, new-age smartphones, and fitness trackers like smartwatches or smart bands. By analyzing user movements and gestures, motion sensors can provide precise data tracking of the user's activity and enhance the user experience by enabling intuitive control mechanisms.

Examples of Motion Sensors in Smart Glasses

Different kinds of motion sensors are found in smart glasses. Usually, these glasses come with an accelerometer that measures movement and acceleration in three dimensions (X, Y, and Z).It detects a range of movements such as tilting, shaking, and rotation. Smart glasses also incorporate a gyroscope, which aids in measuring angles, rotational velocity, and device orientation. This facilitates the detection of rotational movements. Additionally, smart glasses are equipped with a magnetometer to detect changes in the device's magnetic field. This feature helps determine the device's orientation and direction, making it useful for navigation and compass functionality. To provide comprehensive motion-tracking capabilities, smart glasses also incorporate an inertial measurement unit (IMU) that combines the accelerometer, gyroscope, and magnetometer. The IMU enables the measurement of linear acceleration, angular velocity, and magnetic field strength.

CHALLENGES AND LIMITATIONS

PRIVACY AND SECURITY CONCERNS

Privacy and security are very important factors when using smart glasses. Smart glasses are capable of collecting and storing personal data, including visual and audio information of other human beings. Because of that, there is the risk of unauthorized access to this data, which can potentially compromise user privacy. Smart glasses have the ability to record images and videos. This can raise concerns about privacy because there is a possibility that other individuals may be recorded. This can have the capability to infringe on the privacy of those people who are around the user. Like all other connected devices that we use, it can be very vulnerable to cyber-attacks. If smart glasses have support from third-party applications, they can possibly collect additional information or access sensitive information without obtaining the user's approval. It is very important to look closely whenever giving permissions to third-party apps.

POWER AND ENERGY EFFICIENCY

Power and energy efficiency have so many challenges and limitations in smart glasses. Smart glasses rely on internal batteries or they can connect to a power source and be used like a power bank. Internal batteries have limited capacity due to space issues. Due to so many functionalities in different components, like the display, all the sensors, and wireless connectivity, the battery drains quickly. This leads to shorter usage time and frequent recharging. Smart glasses use high-quality display technologies such as liquid crystal display (LCD) or organic light-emitting diodes (OLEDs), which are power hungry. Bright and high-resolution displays consume a lot of battery power, which impacts the overall battery life. Smart glasses require a lot of processing power to run all applications and to perform complex tasks. However, nowadays most of the chipset manufacturers make their chipsets at 4nm technology so it is getting more powerful day by day and also getting less power hungry. Intensive use of powerful processing power and high-resolution display technology generates heat, which impacts both the user's comfort and device performance. As mentioned above, smart glasses use various sensors, such as accelerometers, gyroscopes, magnetometers, and cameras. All these sensors consume power at a higher rate when the sensor is working. Smart glasses most of the time rely on wireless connectivity, which can be Wi-Fi or Bluetooth, for communication purposes or data transfer. These kinds of wireless technologies consume a lot of power, and continuous connectivity has a major impact on the overall device's energy efficiency.

USER ACCEPTANCE AND DESIGN CONSIDERATIONS

User Acceptance

Smart glasses have built-in sensors and cameras, which lead to concerns about the invasion of other people's privacy. Users can be hesitant to adopt smart glasses technology due to the potential for unauthorized recording or surveillance. Wearing a set of smart glasses in public places may attract attention and maybe it will be not socially acceptable in different situations. Users will feel self-conscious and uncomfortable wearing them, which can easily affect the acceptance of smart glasses. Smart glasses often introduce updated user interfaces and interaction paradigms and users have to invest time and effort in learning to use them. Users will have to show interest in the software so that they understand how to operate the smart glasses.

Design Considerations

The design of the smart glasses is a vital factor in their acceptance. Users may be hesitant to wear bulky or less attractive devices, making it essential to prioritize aesthetics and fashion appeal which will be a little difficult. Smart glasses should be lightweight, well-balanced, and comfortable to wear for extended periods. Poor ergonomics can lead to discomfort or even physical strain, reducing user satisfaction and acceptance. Smart glasses usually have a small display area, which makes it important to confirm high-quality visuals. The display should offer the needed amount of brightness, clarity, and resolution to provide a smooth AR experience.

Power management is a critical consideration. Users expect smart glasses to at least last a day without frequent recharging. A longer duration of battery life enhances the experience of the user and reduces the inconvenience of frequent charging. Smart glasses need easy-to-use interfaces. Gestures, voice commands, or other hands-free interaction methods should be implemented to ensure ease of use and to minimize distraction from the real-world environment. Smart glasses have to accommodate different visual impairments, hearing impairments, or motor limitations to ensure inclusivity and equal access.

INTEGRATION WITH ARTIFICIAL INTELLIGENCE AND MACHINE LEARNING

Integration with artificial intelligence (AI) and machine learning (ML) in smart glasses opens up a wide range of possibilities and applications that will enhance functionality and take the capabilities of these wearable devices to the next level. AI algorithms can be used to enable object and image recognition capabilities in smart glasses [7]. This will allow users to identify and gather information about the objects that are in sight of them, it can also detect landmarks, people, or text simply by looking at them. ML algorithms will continuously learn and will continuously improve the recognition accuracy over time [7]. AI and ML will also be employed to provide real-time, context-aware AR overlays on the smart glasses' display. Enhancing the visual experience of the glasses is achievable by analyzing the user's surroundings and utilizing machine learning models. This capability enables the glasses to add pertinent information or virtual objects to the user's field of view. Smart glasses with built-in AI can be able to utilize facial recognition technology to identify individuals in real time for better safety and security. This will be used particularly in various professional fields such as law enforcement, security, or customer service, where quick identification of people is important [7]. By integrating natural language processing (NLP) capabilities, it will be possible to understand and interpret voice commands, enabling hands-free operation. Users can interact with the glasses by giving verbal instructions, asking questions, or even having conversations with AI assistants [7]. With AI algorithms, they will be able to analyze user preferences, user behavior, and historical data to provide personalized recommendations directly on the smart glasses' display. This can be useful for suggesting relevant content to the user, as well as product recommendations, or services based on the user's interests and context. ML algorithms that are used in smart glasses will be able to monitor and analyze health-related data of the user such as heart rate, blood pressure, or even eye movements. This type of information can be used to give users real-time feedback, or personalized recommendations in terms of health, or it can alert the user of potential health issues. AI-powered smart glasses will have capabilities to offer real-time translation, which will help users understand different languages, and from this users will be able to communicate in different languages [7]. By leveraging speech recognition and translation algorithms, the glasses can provide instant translations of spoken or written text. AI and ML algorithms can be very useful in enabling gesture recognition and will be able to interpret hand gestures, which will allow users to interact with the device through gestures instead of traditional input methods. By this, the experience of the user is going to be far more convenient. The possibilities are vast, and because of ongoing advancements in AI and ML technologies will continue to expand the capabilities of smart glasses in the future.

CONCLUSION AND FUTURE PROSPECTS

In conclusion, with the help of AI and ML technologies in smart glasses, the capabilities and potential applications of smart glasses have been increased. With the combination of AI algorithms and wearable technology, many features have been enabled such as object and image recognition for the user to understand better, augmented reality overlays to help with multitasking and a variety of purposes, facial recognition technology to secure users' data, natural language processing, personalized recommendations, health monitoring, real-time translation, and gesture recognition. Because of these advancements, smart glasses are more versatile, intuitive, and useful in different domains, including healthcare, education, industrial settings, and daily life. It is very possible that in the future smart glasses will incorporate additional sensors to expand or increase their sensory capabilities. Also, advanced AI assistants put into smart glasses can become even more sophisticated, they will be able to provide proactive assistance and anticipate user needs based on contextual cues and user behavior patterns. One of the biggest challenges facing smart glasses is they have limited battery life and bulkiness. Future research will definitely have to focus on developing more efficient power management systems and designing sleeker, lightweight smart glasses for better user comfort and longer usage time. Smart glasses will be able to enable collaborative experiences by allowing wearers to share their AR views with others or collaborate on tasks in real time. Future research can explore collaborative applications and develop platforms that facilitate shared experiences. Research should also consider the ethical and social implications of the widespread adoption of smart glasses. It is important to address concerns related to privacy, surveillance, and the impact of AR on human interactions and behavior.

REFERENCES

1. S. Salloum, A. Aburayya and A. Alsharhan, "Technology Acceptance Drivers for AR Smart Glasses in the Middle East: A Quantitative Study," *International Journal of Data and Network Science* 6.1 (2022): . 16, 2021.
2. "Wikipedia - Apple Vision Pro," *Wikipedia*, [Online]. Available: https://en.wikipedia.org/wiki/Apple_Vision_Pro.
3. B. R. Amana, "A Comprehensive Review of Smart Glasses Technology- Future of Eyewear," *Turkish Journal of Computer and Mathematics Education*, vol. 12, p. 12, 2021.
4. K. A. E. H. S. Fidanboylu, "Fiber Optic Sensors and Their Applications," *ResearchGate* 3 (2009), p. 6.
5. "Environmental Sensors," *Sunbird*, [Online]. Available: https://www.sunbirddcim.com/glossary/environmental-sensors.

6. "Bioetric Sensor," *Iproov*, [Online]. Available: https://www.iproov.com/biometric-ency-clopedia/biometric-sensor#:~:text=Biometric%20sensors%20are%20a%20mechanical ,is%20a%20fingerprint%20pad%2Fsensor.

7. "AI Smart Glasses: Redefining Possibilities for the Visually Impaired," *Techopedia*, [Online]. Available: https://www.techopedia.com/ai-smart-glasses-redefining-possi-bilities-for-the-visually-impaired#:~:text=With%20AI%20and%20ML%20at,similar %20to%20a%20recommendation%20engine.

16 Smart Earphones

Emmanuel Lawenrence, Sourav Chakraborty, and Chinna Swamy Dudekula

INTRODUCTION

Smart earphones have completely revolutionized the way we interact with audio devices by introducing a wide range of advanced features. Thanks to technological progress these compact gadgets have surpassed their original purpose of music playback. Now equipped with wireless connectivity and smart assistants they have become versatile tools that greatly enhance our audio experience. This chapter aims to shed light on the evolution of smart earphones and speculate about their promising future.

They have evolved from traditional wired earphones to sophisticated wireless devices packed with advanced features. The introduction of technologies like Bluetooth, active noise cancellation (ANC), touch controls, and voice assistant integration has propelled the growth of smart earphones, enhancing their functionality, convenience, and user experience.

The impact of smart earphones extends beyond personal audio consumption. In the music industry, smart earphones have transformed music listening habits, contributing to the rise of streaming services and the decline of physical media sales. They have influenced music production techniques, pushing artists and engineers to consider optimizing the listening experience specifically for smart earphone users. Additionally, smart earphones have found applications in various fields such as the workplace, education, healthcare, transportation, and gaming, improving productivity, learning outcomes, patient care, safety, and entertainment experiences.

Looking ahead, the future of smart earphones holds immense potential. Emerging technologies and trends continue to shape their trajectory. Artificial intelligence (AI) integration enables personalized audio experiences and voice-controlled interactions. Augmented reality (AR) and virtual reality (VR) integration offer immersive audio-visual experiences. Health monitoring capabilities advance their role in healthcare. Edge computing and brain-computer interface integration push the boundaries of what smart earphones can achieve. As the technology progresses, smart earphones will likely become even more intuitive, intelligent, and seamlessly integrated into our daily lives.

THE EMERGENCE OF SMART PHONES

In examining the history of smart earphones, we discover a captivating tale of their rise in popularity and technological advancements. Step by step, we'll explore how

DOI: 10.1201/9781032686714-16

these innovative devices have evolved over time, from their humble beginnings to the cutting-edge designs we see today. The origin story of wearable audio devices takes us back to the early 20th century when the introduction of headphones marked a pivotal moment. These unwieldy and wired contraptions served as essential tools for professionals such as telephone operators and radio workers. Gradually, headphones became more compact and convenient, making way for the birth of earphones that catered to a wider audience.

The initial advancement of earphone technology was motivated by the need for portable audio solutions. The shift from headphones to earphones brought about enhanced convenience and mobility, allowing individuals to personalize and access audio experiences easily. A significant milestone in the history of smart earphones was marked by the incorporation of wireless technology. Bluetooth, specifically, played a crucial role in transforming earphones. Earphones equipped with Bluetooth granted users the liberty to move around without being confined to their audio source. This groundbreaking innovation popularized wireless earphones and set a strong foundation for future advancements in smart earphone technology. [1]

The introduction of wireless earphones brought about a significant shift in audio technology history. The advent of wireless connectivity not only offered convenience but also revolutionized music listening and audio consumption practices for individuals all around. What began as wirelessly connected earphones evolved progressively into smart earphones, incorporating advanced features that further enhanced their utility. [2]

By incorporating built-in microphones and controls integrated into the earpieces themselves, individuals can effortlessly answer phone calls, adjust volume levels, and manage their music playback. This fusion of features not only significantly enhances convenience levels but also improves overall user-friendliness for all users involved. The researchers emphasize the revolutionary development of smart earphones and their integration of advanced capabilities. These advancements have truly transformed ordinary earphones into versatile devices that offer an immersive audio experience while providing additional functionalities. [3]

Sound quality has perpetually been an exceedingly pivotal aspect that cannot be overlooked or undermined when it concerns the realm of earphones. The domain of sound technology has undergone significant progressions, rendering an unparalleled enhancement to the overall auditory encounter. An influential feature known as ANC technology has gracefully transformed into a fundamental and intrinsic component that one can find in numerous smart earphones. The consequential outcome of incorporating this cutting-edge technology is an unprecedented reduction in ambient background noise, culminating in the creation of an unequivocally immersive auditory environment that captivates the senses. Additionally, the realm of audio playback has been enriched with remarkable clarity and fidelity, all owing to the revolutionary development of high-quality audio codecs.

The domain of audio technology has witnessed revolutionary strides, which have invariably disrupted and transformed the smart earphone industry. The remarkable implementation of ANC technology, coupled with the advent of high-fidelity audio codecs, has indubitably revolutionized and reinvented the audial landscape. The end result is a truly transformative user experience, an encounter that is characterized by an unprecedented level of immersion and the seamless reproduction of audio content with utmost precision and excellence. [4]

In recent years, the landscape of smart earphones has not just confined itself to the solitary realm of audio capabilities. It has, in fact, transcended the traditional boundaries and has emerged as a multifaceted device that boasts an array of health and fitness tracking features. The meticulous integration of built-in sensors has enabled the monitoring of physiological aspects, such as heart rate, step tracking, and the measurement of various other biometric data. The ramifications of this seamless amalgamation have been truly profound and far-reaching. Smart earphones have, without a shadow of a doubt, emerged as the go-to choice for fitness enthusiasts, as they are bestowed with real-time feedback and a profound understanding of exercise regimens and endeavours, thereby enabling them to attain unprecedented insights and progress in their physical pursuits.

Certain investigations conducted have shown groundbreaking advancements in smart earphone technology. Their research delves into the integration of health monitoring capabilities within these innovative audio devices, elucidating the extraordinary prospects they offer in terms of fitness tracking and overall health supervision. [5]

The annals of smart earphones epitomize an incessant voyage of innovation and technological progress in the realm of auditory apparatuses. From the nascent era of wired headphones to the contemporary marvels of cutting-edge smart earphones, the evolution has been nothing short of remarkable. These awe-inspiring gadgets have become versatile and multifaceted accessories, thanks to the assimilation of wireless technology, intelligent functionalities, avant-garde acoustic enhancements, and health surveillance capabilities. As the relentless march of technology continues unabated, we eagerly await the forthcoming chapters of this extraordinary narrative, which promise to elevate our acoustic encounters and expand the horizons of smart earphone functionality.

THE PROFOUND IMPACT OF SMART EARPHONES ON THE MUSIC INDUSTRY

Smart earphones have engendered a profound transformation in the domain of music consumption, bestowing upon enthusiasts personalized and immersive auditory experiences. The forthcoming segment embarks upon an exploratory journey, unraveling the far-reaching consequences of these ingenious devices on the music industry. It delves into the myriad ways in which they have influenced patterns of music consumption, shaped the progression of audio technology, and revolutionized the revenue models of artists and music streaming services.

Smart earphones have heralded a paradigm shift in the patterns of music consumption. Empowered by the proliferation of streaming platforms and the portability afforded by smart earphones, music aficionados now luxuriate in unrestricted access to vast libraries of melodious wonders, irrespective of time or place. This phenomenon has spurred an upsurge in music streaming and witnessed a commensurate decline in the sales of physical albums. Listeners revel in the ability to curate personalized playlists, unearth fresh musical gems, and relish on-demand access to an expansive tapestry of artists and genres. [6]

The advent of smart earphones has precipitated remarkable advancements in audio technology, meticulously tailored to satisfy the discerning needs of music enthusiasts. Manufacturers have prioritized the enhancement of sound quality, employing state-of-the-art audio codecs and pioneering sophisticated audio processing algorithms. The result is an augmented audio fidelity, an expanded dynamic range, and an engrossing soundscape, ultimately immersing users in an auditory experience of unparalleled engagement. [7]

The soaring popularity of smart earphones has exerted a discernible influence on the production and mixing processes employed in the creation of music. Modern-day music producers and engineers demonstrate heightened cognizance regarding the intricacies of how music resonates across disparate listening environments, including smart earphones. Consequently, they meticulously optimize the mixing and mastering phases to ensure that the music exudes an irresistible allure, irrespective of whether it emanates from studio monitors, car speakers, or the hallowed confines of smart earphones. This relentless focus on optimizing the auditory experience across diverse devices has assumed the status of an indispensable cog in the music production workflow. [8]

Smart earphones have simultaneously presented artists and music streaming services with a wealth of challenges and opportunities. On one hand, artists now enjoy the ability to reach a global audience through streaming platforms, thereby affording their music greater discoverability and shareability. In turn, streaming services derive considerable benefits from amplified user engagement and the resultant surge in subscription revenue. On the other hand, artists grapple with the novel conundrum of fair compensation, as the remuneration they receive from streaming services typically hinges on the number of streams their creations accumulate, which may not always serve as an accurate reflection of their artistic merit. [9]

The transformative influence of smart earphones has compelled artists and the music industry to adapt their revenue models, necessitating the embrace of novel approaches. Artists now place a heightened emphasis on live performances, merchandise sales, brand collaborations, and licensing agreements to supplement the income generated by music streaming. They strive to create immersive and unparalleled live experiences, conceiving interactions with their fan bases that transcend the realm of recorded music, thereby forging unbreakable bonds and offering value that extends far beyond the auditory realm. [10]

Smart earphones have irrevocably revolutionized the music industry, reshaping the contours of music consumption, propelling advancements in audio technology, and fundamentally altering the revenue models of artists and music streaming services. Their omnipresence has conferred upon listeners an unprecedented degree of convenience and personalization, while simultaneously impelling artists and the industry at large to traverse uncharted territories in their quest for novel revenue streams. As the grand tapestry of the musical landscape unfurls before our very eyes, artists, streaming services, and technology companies alike must exhibit an unyielding resolve to navigate these winds of change, thereby ensuring the creation of a sustainable and flourishing ecosystem that caters to the needs of both creators and listeners.

EXPLORING THE MULTIFACETED WORLD OF SMART EARPHONES: FEATURES, BENEFITS, AND DRAWBACKS

Smart earphones have evolved into dynamic audio devices, offering a myriad of features that enhance the audio experience and provide unparalleled convenience to users. In this chapter, we embark on a journey to explore the key features of smart earphones, delve into their associated benefits, and address the inherent drawbacks. By gaining a comprehensive understanding of these aspects, users can make well-informed decisions when selecting their preferred audio devices, ensuring an optimal audio experience tailored to their needs and preferences.

FEATURES OF SMART EARPHONES

- **Active noise cancellation (ANC):** Active noise cancellation, a prominent feature in smart earphones, serves as a sonic shield against unwanted background noise and distractions. Through the utilization of built-in microphones, these earphones detect ambient sounds, leveraging advanced ANC technology to generate sound waves that effectively cancel out these interferences. The result is a heightened and more immersive listening experience, enriched by an oasis of focused soundscapes. [11]
- **Voice assistant integration:** Smart earphones embrace the power of voice assistants, seamlessly integrating the likes of Siri, Google Assistant, or Alexa into their functionality. This integration empowers users to effortlessly control their earphones and access a multitude of functions through simple voice commands. From making calls and sending messages to managing music playback and retrieving information, users can navigate their audio experience with remarkable ease, unburdened by the need for manual interactions. [12]
- **Touch controls:** The advent of touch-sensitive controls on smart earphones ushers in a new era of user convenience and intuitive interaction. With a mere tap or a well-placed gesture on the earpieces, users can effortlessly manage audio playback, adjust volume levels, handle incoming or outgoing calls, and even summon voice assistants to their aid. These touch controls revolutionize the way users interact with their earphones, bestowing upon them an unparalleled sense of control and a gateway to a seamless audio experience. [13]
- **Sensors:** The inclusion of various sensors within smart earphones elevates their functionality to new heights, imbuing them with an enhanced user experience. These sensors possess the remarkable ability to detect when the earphones are donned, automatically pausing or resuming audio playback

based on user preferences. In certain models, proximity sensors enable a seamless and effortless audio control, recognizing when the earphones are removed or placed back on, ensuring a continuous and uninterrupted audio journey. [14]

- **Battery life:** The longevity of battery life is an indispensable consideration when evaluating smart earphones. Users seek extended usage periods without the constant need for recharging. Manufacturers employ diverse strategies to optimize battery performance, including the integration of power-efficient components, advanced power management systems, and rapid-charging capabilities. It is important to note that the actual battery life of smart earphones may vary depending on individual usage patterns and the specific features engaged. [15]

BENEFITS OF USING SMART EARPHONES

- **Improved audio quality:** Smart earphones harness the power of advanced audio technologies, boasting high-quality drivers and audio codecs that elevate sound reproduction to new heights. Users are treated to an audio experience enriched by enhanced clarity, detailed sound imaging, and an expanded frequency range. These benefits culminate in a listening experience that is not only immersive but also highly enjoyable, transcending the boundaries of conventional audio devices. [16]
- **Hands-free control:** The integration of voice assistants and touch controls ushers in an era of hands-free control, empowering users with newfound freedom. Tasks such as managing earphone functions, controlling audio playback, adjusting settings, and accessing information can now be accomplished through simple voice commands or intuitive gestures. The elimination of manual interactions enhances convenience, allowing users to focus on their audio journey with effortless ease. [12, 13]
- **Convenience:** Smart earphones epitomize the pinnacle of convenience, offering unparalleled portability, lightweight design, and pocket-friendly dimensions. The absence of cumbersome cables eliminates the hassle of tangled wires, providing users with the freedom to revel in their audio content without the encumbrance of physical limitations. Whether indulging in music, podcasts, or other audio delights, smart earphones deliver convenience at every turn. [17]
- **Safety:** Smart earphones prioritize user safety, incorporating features designed to enhance situational awareness. Certain models include ambient sound modes that enable external sounds to permeate the listening experience, enabling wearers to remain attuned to their surroundings. This feature assumes critical significance in outdoor or potentially hazardous environments, ensuring users remain vigilant and secure amidst their audio indulgences. [18]

Drawbacks of Using Smart Earphones

- **Cost:** Smart earphones boasting advanced features and cutting-edge technologies often carry a higher price tag when compared to their traditional counterparts. The cost of smart earphones can vary significantly based on factors such as brand, model, and the specific functionalities they offer. Users are encouraged to carefully consider their budgetary constraints while prioritizing the features that align with their individual needs and preferences. [19]
- **Battery life:** Despite remarkable strides in battery technology, the battery life of smart earphones may still present a potential limitation. Certain features, such as active noise cancellation or continuous streaming, can drain the battery at an accelerated rate. Users are advised to be mindful of their usage patterns and ensure access to charging solutions whenever necessary. [15]
- **Connectivity issues:** Occasionally, users may encounter connectivity issues when using smart earphones. Factors such as signal interference, range limitations, or compatibility problems with specific devices can affect the stability of the wireless connection. While such issues are generally infrequent, they can cause temporary inconvenience and necessitate troubleshooting to restore optimal functionality. [20]
- **Privacy Concerns:** The integration of microphones and sensors in smart earphones raises legitimate privacy concerns. These devices collect and process user data, prompting the need for caution regarding unauthorized access or potential data breaches. It is prudent for users to choose reputable brands and manufacturers that prioritize robust data security measures. Additionally, users should remain cognizant of the privacy implications associated with the use of smart earphones and exercise due diligence in protecting their personal information. [11]

Smart earphones represent an unparalleled fusion of features and benefits that enhance the audio experience while offering convenience and safety. However, it is essential to acknowledge the potential drawbacks, including cost considerations, battery life limitations, occasional connectivity issues, and privacy concerns. Armed with this comprehensive understanding of the features, benefits, and drawbacks, users can make informed decisions when selecting and utilizing smart earphones, ensuring an optimal and personalized audio journey.

EXPLORING THE ENVIRONMENTAL IMPACT OF SMART EARPHONES

In today's world, where environmental sustainability takes center stage, it is imperative to scrutinize the environmental impact of our technological devices. This chapter delves into the environmental implications of smart earphones, evaluating their life cycle, material composition, energy in consumption, and potential e-waste concerns.

By comprehending these factors, we can make informed decisions and pave the way for a more sustainable future.

- **Life cycle assessment:** A life cycle assessment (LCA) serves as a valuable tool in evaluating the environmental impact of a product throughout its entire life cycle, encompassing stages such as raw material extraction, manufacturing, distribution, use, and disposal. While specific LCAs for smart earphones may be limited, we can draw insights from similar electronic devices to understand their environmental footprint.
- **Material composition:** Smart earphones comprise various components, including plastic, metal, electronic circuitry, batteries, and cables. The extraction and processing of raw materials for these components can have environmental consequences, such as habitat destruction, resource depletion, and energy consumption. Moreover, the manufacturing processes involved, such as assembly, and soldering, can contribute to greenhouse gas emissions and waste generation. [21]
- **Energy consumption:** The energy footprint of smart earphones encompasses both their usage and charging requirements. These devices rely on battery power, necessitating regular charging. The source of energy used for charging, such as electricity from the grid, can have varying environmental impacts depending on the energy mix and generation methods in a particular region. Additionally, the energy efficiency of the charging process and the standby power consumption of the earphones themselves influence their overall energy consumption. [22]
- **E-waste concerns:** The proliferation of electronic waste, or e-waste, poses a significant environmental challenge. Smart earphones, like any electronic device, contribute to the growing e-waste stream when they reach the end of their usable life. Improper disposal or inadequate recycling practices can lead to hazardous substances leaching into the environment, soil contamination, and the release of greenhouse gases during incineration. Proper e-waste management, including recycling and responsible disposal, is crucial in mitigating these environmental impacts. [23]
- **Sustainable practices:** While smart earphones have environmental implications, several sustainable practices and solutions can help mitigate their impact.
- **Material Selection:** Manufacturers can prioritize the use of sustainable materials, such as recycled plastics and bio-based alternatives, reducing reliance on virgin resources and minimizing environmental impact during the production phase.
- **Energy Efficiency:** Enhancing the energy efficiency of smart earphones and their charging systems can reduce energy consumption and associated greenhouse gas emissions. This can be achieved through optimized circuitry, power management features, and encouraging the use of renewable energy sources for charging.

- **Product lifespan:** Encouraging product durability, repairability, and upgradability can extend the lifespan of smart earphones, reducing the need for frequent replacements and minimizing e-waste generation.

RESPONSIBLE DISPOSAL AND RECYCLING

Establishing effective e-waste collection and recycling programs ensures that end-of-life smart earphones are properly managed. This involves implementing safe recycling practices to recover valuable materials and prevent environmental contamination.

By embracing sustainable design principles, enhancing energy efficiency, prolonging product lifespan, and implementing responsible disposal and recycling practices, we can minimize the environmental impact of smart earphones. Together, we can work towards a more sustainable future, where technology and environmental consciousness coexist harmoniously.

EXPLORING THE SECURE UTILIZATION OF SMART EARPHONES: SAFEGUARDING AUDITORY HEALTH AND CULTIVATING ENVIRONMENTAL AWARENESS

Smart earphones offer a captivating auditory expedition, but it is of paramount importance to place safety as our topmost priority during their utilization. This segment concentrates on delineating guidelines encompassing the secure utilization of smart earphones. It encompasses sagacious pointers aimed at averting hearing impairment, emphasizing the significance of intermittent pauses, and fostering attentiveness toward our surroundings. By adhering to these practices, we can revel in the benefits bestowed by smart earphones while safeguarding our overall well-being.

1. **Exercising restraint with elevated volume levels:** It is crucial to abstain from employing smart earphones at exorbitant volume levels as such practices may inflict harm upon our auditory faculties. Safeguarding our ears necessitates maintaining a safe volume threshold. The World Health Organization (WHO) endorses a threshold of 85 decibels (dB) or lower, not exceeding 8 hours per diem. Enabling the volume limit feature, an attribute available in select smart earphones or smartphone settings can assist in preserving optimal sound levels. [21]
2. **Embracing regular respite:** Extended and unbroken utilization of smart earphones can strain our auditory apparatus, amplifying the likelihood of encountering hearing predicaments. To forestall fatigue and circumvent potential harm, it is advisable to integrate intermittent respites into our intelligent earpiece usage routine. Experts advocate following the 60/60 rule, wherein listening sessions should be limited to a maximum of 60 minutes at 60% of the device's utmost volume, followed by a 5 to 10-minute

hiatus. This intermission facilitates a period of reprieve and rejuvenation for our ears. [22]

3. **Attentive mindfulness of surroundings:** While engrossed in the auditory wonders rendered by smart earphones, it is of utmost significance to remain cognizant of our environment. Maintaining an acute awareness of our surroundings, particularly during driving, cycling, or engaging in activities necessitating situational attentiveness, is of paramount importance. Avoiding the utilization of smart earphones in potentially perilous situations, such as when traversing thoroughfares or operating machinery, is highly recommended. By remaining vigilant, we secure our personal well-being as well as the well-being of others.

4. **Prudent utilization of noise-cancellation features:** Smart earphones often possess the capacity to nullify ambient noise, augmenting the auditory experience. However, judicious employment of this feature is essential. Complete annulment of external noise can engender isolation from our surroundings, potentially impeding our ability to perceive crucial sounds like approaching vehicles or emergency sirens. It is advisable to exercise noise-cancellation functionality in secure and controlled settings, such as during air travel or within tranquil chambers.

Promoting the safe utilization of smart earphones is crucial for the preservation of auditory health and the sustenance of environmental consciousness. By circumventing excessive volumes, integrating periodic respites, and nurturing mindfulness of our surroundings, we can partake in immersive auditory encounters while upholding our overall well-being. Remember, responsible and secure usage of smart earphones engenders a protracted and gratifying auditory experience.

THE UPCOMING ERA OF SMART EARPHONES: ANTICIPATED TRENDS, EMERGING TECHNOLOGIES, AND POTENTIAL APPLICATIONS

Smart earphones have seamlessly integrated into our daily lives, offering an array of remarkable features and capabilities. As we look towards the future, a world of exhilarating possibilities unfolds for smart earphones, presenting emerging technologies, noteworthy trends, and potential applications spanning diverse domains. In this chapter, we embark on an exploration of the future of smart earphones, delving into their profound implications within the realms of workplaces, education, healthcare, transportation, and gaming.

ANTICIPATED TRENDS

1. **Integration of AI:** The future of intelligent earpieces hinges upon harnessing the power of AI to elevate user experiences to unprecedented heights. AI algorithms possess the prowess to analyze user preferences, adapt sound

profiles accordingly, and furnish personalized recommendations, thereby crafting an immersive and tailored auditory milieu. Moreover, the integration of AI can bolster voice recognition accuracy, facilitating more efficient and seamless voice-controlled interactions. [24]

2. **Fusion of AR and VR:** The convergence of AR and VR technologies with smart earphones heralds a new era of immersive audio encounters. Users can revel in audio content that seamlessly blends with virtual environments, imparting an added layer of richness and interactivity. Contextual information can be superimposed onto the real world, transforming audiovisual experiences into captivating and engaging journeys. [24]

3. **Health monitoring and biometric analysis:** The future of intelligent earpieces encompasses an expansion of their role in health monitoring and biometric analysis. By incorporating advanced sensors, these earpieces can track vital signs, monitor activity levels, and provide real-time analysis of biometric data. Such information can prove instrumental in health monitoring, fitness tracking, and even early detection of certain medical conditions. [25]

EMERGING TECHNOLOGIES

1. **Leveraging edge computing:** The future of smart earphones lies in harnessing the capabilities of edge computing, enabling the processing of data locally on the earpieces themselves. This paradigm shift ensures swifter response times, reduced latency, and heightened privacy by minimizing the transmission of sensitive data to external servers. Edge computing empowers advanced audio processing, real-time translation, and personalized audio experiences. [24]

2. **Integration of brain-computer interface (BCI):** BCI technology boasts immense potential for the future of smart earphones. By establishing a direct interface with the user's brain activity, these earpieces can decipher neural signals, delivering personalized audio experiences based on cognitive and emotional states. The integration of BCI can revolutionize audio content interaction, unlocking entirely new modes of communication. [25]

POTENTIAL APPLICATIONS

1. **Workplaces embracing smart earphones:** Intelligent earpieces possess the capacity to redefine productivity in workplaces. Through the employment of voice assistants and AI integration, employees can utilize their earpieces for hands-free communication, scheduling, task management, and information retrieval. Noise-cancellation features foster an environment conducive to concentration, facilitating enhanced focus and productivity. [26]

2. **Smart earphones transforming education:** The domain of education stands to benefit immensely from the transformative potential of intelligent earpieces. With AI-powered learning platforms and customized audio

content, students gain access to tailored educational materials and interactive lessons. Language translation capabilities foster multilingual learning, while the integration of augmented reality ensures immersive educational experiences. [27]

3. **Smart earphones revolutionizing healthcare:** The impact of intelligent earpieces on the healthcare landscape is nothing short of revolutionary. By monitoring vital signs, tracking patient health, and delivering personalized audio feedback, these earpieces enhance patient care. Remote patient monitoring, early detection of health conditions, and telemedicine consultations are all made possible, thereby enhancing healthcare accessibility and efficiency. [28]

4. **Smart earphones enabling advanced transportation:** Intelligent earpieces assume a pivotal role in transportation, furnishing audio-based navigation, real-time traffic updates, and safety alerts. Augmented reality integration overlays navigation instructions onto the user's visual field, heightening situational awareness and engendering intuitive navigation experiences. [26]

5. **Gaming enhanced by smart earphones:** Gaming experiences can be elevated to new heights with the aid of intelligent earpieces. Immersive audio technologies, including 3D audio and spatial sound, bestow realistic audio cues and enrich the gaming ambiance. Biometric sensors integrated into intelligent earpieces introduce biofeedback gaming experiences, where in-game actions respond to the user's physiological responses. [28]

The future of intelligent earpieces brims with vast potential, characterized by emerging technologies, noteworthy trends, and a myriad of applications. AI integration, AR/VR capabilities, health monitoring features, edge computing, and BCI integration collectively steer the evolution of intelligent earpieces. In workplaces, education, healthcare, transportation, and gaming, intelligent earpieces serve as catalysts, revolutionizing the realms of communication, learning, well-being, navigation, and audio content engagement. As these technologies continue to advance, we eagerly anticipate an era of redefined audio experiences.

CONCLUSION

The epoch-making advent of smart earphones has unequivocally revolutionized the audio-sphere, thereby endowing us with an extensive repertoire of features and advantages that impeccably augment our quotidian existence. Throughout the course of this ex, we have meticulously traversed the annals of history, delved into the multifarious facets of smart earphones, scrutinized the future trends that lie in store, and unfurled their manifold applications across an array of domains. Evidently, the adoption of smart earphones has engendered an indelible impact on the realms of music, workplace efficacy, education, healthcare, transportation, and gaming.

In the music industry, the advent of smart earphones has wrought a paradigm shift in the patterns of music consumption, bequeathing us with the unparalleled ability to save an inexhaustible reservoir of harmonies and melodies at our beck and call, while simultaneously basking in the personalized allure of bespoke playlists. The convenience and portability that smart earphones offer have irreversibly transformed our interaction with music, precipitating a discernible dip in the realm of physical album sales, while concurrently engendering an augmented reliance on the manifold benefits proffered by music streaming services. Furthermore, the nascent influence of smart earphones on music production and sound engineering is unequivocally palpable, as artists and producers now meticulously tailor the auditory experience to cater to the proclivities and predilections of smart earphone users.

In other spheres of human enterprise, smart earphones have found themselves veritably embedded in an array of applications, each of which possesses the latent potential to heighten various facets of our existence. In the realm of the workplace, these ingenious devices serve as an invaluable catalyst, exponentially augmenting productivity through their seamless facilitation of hands-free communication, task management, and the boon of noise cancellation technology. In the realm of education, smart earphones emerge as an indispensable tool, paving the way for hyper-personalized learning experiences, the facilitation of linguistic translations, and the seamless integration of augmented reality, thereby bequeathing students with a holistic and immersive pedagogical landscape. In the realm of healthcare, these ingenious marvels of engineering present unparalleled opportunities for remote patient monitoring, the nuanced analysis of biometric data, and the prodigious potential of telemedicine, thus effectively democratizing access to quality healthcare. In the realm of transportation, smart earphones gracefully step into the fray, providing us with an auditory tapestry of audio-based navigation, judicious safety alerts, and the wondrous overlays of augmented reality, thereby effectuating a transformative experience in the realm of mobility. In the realm of gaming, the all-encompassing influence of smart earphones is indisputable, as they inexorably contribute to the realm of immersive experiences through the seamless integration of spatial audio and the judicious amalgamation of biometric feedback.

While the manifold benefits of smart earphones are undeniable, it is prudent for us to judiciously acknowledge and evaluate the accompanying drawbacks that may arise from their extensive usage. Pertinent factors such as the matter of cost, the efficacious management of battery life, the vexing specter of connectivity conundrums, and the perennial concern over privacy must be meticulously weighed and factored into our deliberations. In the inexorable march of technological evolution, it is incumbent upon the manufacturers of smart earphones to confront and surmount these challenges, as they zealously strive towards an unremitting pursuit of progress, championing causes such as affordability, battery efficiency, the seamless facilitation of connectivity, and the inviolable sanctity of data security.

Ultimately, it is an irrefutable verity that smart earphones have emphatically transmogrified the very fabric of our interaction with audio content and the indomitable terrain of technology itself. Their pervasive impact transcends the arbitrary boundaries of industries, permeating various facets of our lives with unwavering determination and unrelenting efficacy. Through the conscientious act of keeping ourselves apprised of the ever-evolving trends, the nascent technologies, and the

need for responsible usage, we as individuals can effectively maximize the manifold benefits tendered by smart earphones, while concurrently mitigating any potential drawbacks that may arise. As the inexorable march of time unfurls before us,
smart earphones shall persist in their unwavering resolve to shape our audio-spheric
experiences, catapulting us into an era characterized by incessant innovation and an
unrelenting advance towards an increasingly immersive and meticulously tailored
auditory tapestry.

REFERENCES

1. Y. L. X. & J. Y. Liu, "Wearable device management with body sensor network for smart
   earphone." *IEEE Access*, 8 (2020), pp. 120084–120094.
2. M. & C. H. Chen, "Wireless earphone system with a human–machine interface based
   on electrooculography." *Journal of Electronic Science and Technology*, 17(4) (2019),
   pp. 354–362.
3. S. & G. M. A. Al-Fayyadh, "Development of a smart earphone with health monitoring capability." 7th International Conference on Control, Automation and Robotics
   (ICCAR), Singapore, 2021, pp. 154–159.
4. A. & S. R. Urick, "The future of headphone technology: A review." Audio Engineering
   Society Conference: 2021 AES Audio Engineering Month, San Francisco, CA, USA,
   2021.
5. H. L. H. & L. X. Du, "Wearable intelligent earphone for monitoring health status."
   IEEE Access, 8 (2020), pp. 15529–15538.
6. M. Mulligan, "Consumer electronics and smart speakers—user survey analysis."
   Journal of Research in Interactive Marketing 12. 4 (2018).
7. K. & T. M. Collins, "Audio branding in the era of smart speakers and voice assistants."
   Journal of Audio Engineering Society, 67(1/2) (2019), pp. 3–13.
8. A. J. & M. M. Müller, "Audio production in the age of mobile streaming: Mixing
   practices and aesthetic implications." *Journal of the Audio Engineering Society*, 67(6)
   (2019), pp. 398–409.
9. S. Dredge, "Streaming platforms and artists: How to achieve a fair deal." *WIPO
   Magazine* (2020).
10. H. Forbes, "The rise of smart earbuds: How the future of audio tech is changing the way
    we listen to music." *MusicTech* (2020).
11. D. H. & Y. C. H. Ko, "Analysis of hazardous noise caused by active noise cancel-
    lation earphones." *Journal of the Acoustical Society of America*, 146(4) (2019), pp.
    EL330–EL336.
12. H. Q. & B. A. Dinh, "The rise of voice assistants in the internet of things." *IEEE
    Pervasive Computing*, 17(3) (2018), pp. 9–12.
13. W. & L. S. Zhang, "Design and realization of touch control intelligent earphones based
    on C3I technology." *Journal of Physics: Conference Series*, 1808(1) (2021), p. 12116.
14. K. C. P. & W. L. Chen, "Wearing detection and recognition for intelligent wireless
    earphone." *Multimedia Tools and Applications*, 80(23) (2021), pp. 35011–35026.
15. A. & S. R. Urick, "The future of headphone technology: A review." Audio Engineering
    Society Conference: 2021 AES Audio Engineering Month, San Francisco, CA, USA ,
    2021.
16. Zhong, Xiao-li & Xie, Bo-sun, "Head-Related Transfer Functions and Virtual Auditory
    Display." *Soundscape Semiotics - Localization and Categorization* 12.4 (2021).
17. "Life cycle assessment of consumer electronics: A case study of laptop computers."
    Environmental Science & Technology, 45(13) (2011), pp. 5560–5567.

18. S. M. E. & S. S. Neugebauer, "Power consumption and energy efficiency of Bluetooth low energy." Proceedings of the 5th International Conference on the Internet of Things (IoT 2015), Seoul, Korea (South), 2015, pp. 55–62.

19. C. P. F. V. G. V. K. R. & S. P. Baldé, *The Global E-waste Monitor 2017: Quantities, Flows, and Resources*. United Nations University (UNU)/United Nations Institute for Training and Research (UNITAR), 2017.

20. World Health Organization, "Make listening safe." (2019) https://www.who.int/.

21. Dehankar, Shruti S. & Gaurkar, Sagar S. "Impact on Hearing Due to Prolonged Use of Audio Devices: A Literature Review". *Soundscape Semiotics - Localization and Categorization* 14.11 (2022: e31425).

22. X. & L. Y. Chen, "Intelligent sound recognition method for smart earphones based on convolutional neural networks." *Electronics*, 10(11) (2021), p. 1312.

23. S. & V. G. Vassallo, "A wearable system for augmented hearing in virtual reality applications." *Future Internet*, 13(2) (2021), p. 39.

24. H. L. H. & L. X. Du, "Wearable intelligent earphone for monitoring health status." *IEEE Access*, 8 (2020), pp. 15529–15538.

25. W. & J. M. Zhang, "Edge computing-based wearable device for real-time emotion recognition in smart earphones." *Future Internet*, 12(11) (2020), p. 187.

26. M. K. K. M. & A. I. Azeem, "Brain–computer interface for auditory systems. Smart earphones: A comprehensive review." *International Journal of Information Management*, 50 (2020), pp. 345–356.

27. J. & J. A. Smith, "The future of smart earphones: Trends and predictions." *Journal of Technology and Society*, 15(2) (2018), pp. 78–92.

28. C. & C. S. Wong, "Augmented reality audio for smart earphones." Proceedings of the International Conference on Human-Computer Interaction, Donostia Gipuzkoa Spain, 2019, pp. 47–55.

17 AI-Enabled Glasses

Aluri Anand Sai, Pokala Pranay Kumar,
and Ravali Gunda

INTRODUCTION

In recent years, the field of artificial intelligence (AI) has made remarkable progress, enabling technologies that were once considered science fiction to become a reality. One such innovation that has gained significant attention is AI-enabled glasses. These futuristic eyewear devices have the potential to revolutionize the way we perceive and interact with the world around us [1].

The year is 2023. AI-enabled glasses are the latest trend in wearable technology. These glasses are equipped with a variety of sensors and AI algorithms that allow them to do everything from translating languages to providing real-time information about the world around you. Background AI-enabled glasses, also known as smart glasses or augmented reality glasses, are a form of wearable technology that combines traditional eyewear with advanced AI capabilities. These glasses incorporate various sensors, cameras, and display systems to provide users with real-time information and augmented experiences [2]. AI algorithms and technologies embedded in these glasses enable functionalities such as object recognition, natural language processing, and augmented reality (AR) interactions.

In this chapter, we will delve into the capabilities, applications, and implications of AI-enabled glasses, exploring their transformative impact on various aspects of our lives. We will explore the potential of AI-enabled glasses and discuss some of the ways that they are being used today. We will also look at some of the challenges that need to be addressed before AI-enabled glasses can become truly mainstream.

OBJECTIVES

The primary objective of this chapter is to provide a comprehensive overview of AI-enabled glasses, including their history, technological components, and applications. AI-enabled glasses help to enhance our natural vision. Through computer vision algorithms, these glasses can analyze the visual input from the environment and provide valuable insights and contextual information to the wearer. For example, they can recognize objects, faces, and text, and display relevant details in real-time, effectively providing an augmented reality experience. Additionally, this chapter aims to analyze the challenges and ethical implications of AI-enabled glasses. The chapter concludes with a discussion of future trends and advancements in the field [3].

DOI: 10.1201/9781032686714-17

SCOPE

The scope of this chapter encompasses the development of AI-enabled glasses from early concepts to current advancements. It covers the key components of these glasses, including optical display systems, sensors, connectivity features, and AI processing units. The chapter explores the application of various AI algorithms and technologies in the context of AI-enabled glasses, focusing on computer vision, natural language processing, gesture recognition, and augmented reality integration. It also discusses the potential applications of AI-enabled glasses in sectors such as healthcare and retail. Furthermore, the chapter addresses the challenges and ethical considerations associated with AI-enabled glasses and presents future trends in this field.

OVERVIEW

WHAT ARE AI-ENABLED GLASSES?

AI-enabled glasses, also known as smart glasses or AR glasses, are wearable devices that incorporate AI capabilities to enhance the user's visual experience. These glasses typically consist of a display unit, sensors, a processor, and connectivity options. They combine the functionality of traditional glasses with advanced technologies to provide users with augmented reality, contextual information, and interactive features. AI-enabled glasses are a type of wearable technology that is equipped with a variety of sensors and AI algorithms. These sensors allow the glasses to track your gaze, your facial expressions, and your surroundings [4]. The AI algorithms then use this information to provide you with a variety of services, such as:

- **Translation:** AI-enabled glasses can translate text from one language to another in real time. This is a great way to communicate with people who speak different languages.
- **Information:** AI-enabled glasses can provide you with real-time information about the world around you. This information can include things like the weather, traffic, and nearby businesses.
- **Navigation:** AI-enabled glasses can help you navigate your way around unfamiliar places. They can do this by providing you with turn-by-turn directions and by showing you a map of your surroundings.
- **Productivity:** AI-enabled glasses can help you be more productive. They can do this by allowing you to take notes, access your calendar, and control your devices hands-free.

AI-enabled glasses are wearable devices that combine advanced optics, computing power, and artificial-intelligence algorithms to enhance and augment our visual experience. Equipped with sensors, cameras, and micro displays, these glasses have the ability to process visual data in real time and overlay digital information onto the wearer's field of view.

How Are AI-Enabled Glasses Being Used Today?

AI-enabled glasses are still in the early stages of development, but they are already being used in a variety of ways. Some of the ways that AI-enabled glasses are being used today include:

- **Business:** AI-enabled glasses are being used by businesses to improve customer service, provide training, and collect data.
- **Education:** AI-enabled glasses are being used by schools to provide students with access to information, help them learn new languages, and track their progress.
- **Healthcare:** AI-enabled glasses are being used by healthcare professionals to provide remote care, diagnose diseases, and monitor patients.
- **Entertainment:** AI-enabled glasses are being used to provide immersive entertainment experiences, for activities such as watching movies or playing games.

History and Evolution of AI-Enabled Glasses

The concept of AI-enabled glasses and wearable displays has been explored and developed over several decades [5]. Here's a brief history and evolution of AI-enabled glasses:

- **1965:** The first mention of a wearable computer-like device can be traced back to Ivan Sutherland's pioneering work on "The Ultimate Display" concept. He envisioned a head-mounted display that would overlay virtual information onto the user's field of view.
- **1980s:** Research and development of head-mounted displays (HMDs) continued, primarily for military and industrial applications. However, these early versions were bulky, heavy, and lacked advanced AI capabilities.
- **1990s:** The term "smart glasses" started to gain traction. Steve Mann, often regarded as the "father of wearable computing," developed a series of head-mounted wearable computers, including the EyeTap, which incorporated computer vision and display technologies.
- **Early 2000s:** Companies like MicroOptical Corporation and Microvision began introducing commercial head-mounted displays, targeting niche markets such as the military, medical, and industrial sectors. These devices provided basic display capabilities but lacked advanced AI functionalities.
- **2012:** Google unveiled Google Glass, a high-profile project that aimed to bring AI-enabled glasses to the mainstream. Google Glass featured a head-up display (HUD), a camera, voice recognition, and the ability to display contextual information. However, due to privacy concerns and limited consumer adoption, Google Glass was later repositioned for enterprise applications.

- **2015:** Microsoft introduced HoloLens, an augmented reality headset that allowed users to overlay digital content onto their real-world view. HoloLens utilized spatial mapping, gesture recognition, and voice commands to provide an immersive augmented reality experience.
- **2016:** Snapchat launched Spectacles, a pair of camera-equipped glasses that allowed users to capture photos and videos from their perspective. Although not AI-enabled in the traditional sense, Spectacles sparked interest in wearable camera technology.
- **2019:** Several companies, including Vuzix, North (acquired by Google), and Focals by North, introduced smart glasses aimed at consumers. These glasses featured more discreet designs and integrated AI capabilities, such as voice assistants, real-time information displays, and basic augmented reality functionalities.
- **2021 and beyond:** AI-enabled glasses continue to evolve, with advancements in computer vision, natural language processing, and machine learning technologies. Companies like Apple, Facebook (now Meta), and various startups are actively working on developing next-generation smart glasses with advanced AI capabilities, aiming to deliver more immersive augmented reality experiences and practical applications.

KEY MILESTONES IN DEVELOPMENT

This section highlights significant milestones in the development of AI-enabled glasses, including the introduction of Google Glass and the subsequent advancements in the field [6], [7]. It discusses how these milestones have shaped the evolution of AI-enabled glasses and paved the way for their integration with AI technologies.

COMPONENTS OF AI-ENABLED GLASSES

- **Display unit:** AI-enabled glasses have a small display positioned in front of the user's eyes, often near the eyepiece. This display can be transparent, allowing the user to see the real world while overlaying digital information or graphics onto their field of view. It provides a visual interface for presenting augmented reality content.
- **Sensors:** These glasses incorporate various sensors to gather information about the user's surroundings and enable interaction. Common sensors include cameras for capturing images and videos, depth sensors for spatial mapping, gyroscopes and accelerometers for motion detection, and microphones for audio input.
- **Artificial intelligence algorithms:** AI-enabled glasses employ AI algorithms to process data from the sensors, interpret the user's environment, and generate real-time responses. These algorithms often involve computer vision, natural language processing, and machine learning techniques to analyze visual and auditory data, recognize objects or gestures, and provide intelligent insights.

- **Connectivity:** AI-enabled glasses can connect to the internet or other devices through wireless technologies like Wi-Fi or Bluetooth. This connectivity enables data exchange, remote communication, and access to cloud-based services, enhancing the glasses' capabilities and allowing for seamless integration with other systems.
- **Interaction methods:** AI-enabled glasses provide various methods for user interaction. This can include touch-sensitive surfaces, voice commands, gesture recognition, or even eye-tracking. These interaction methods allow users to control the glasses, navigate through menus, and interact with digital content in an intuitive manner.
- **Augmented reality and overlays:** One of the main features of AI-enabled glasses is the ability to overlay digital information onto the user's real-world view. This can include displaying text, images, videos, or 3D objects that are relevant to the user's context. These overlays can provide real-time data, contextual information, or interactive elements that enhance the user's perception and understanding of their surroundings.
- **Personalized content and services:** AI-enabled glasses can offer personalized content and services based on the user's preferences, location, or previous interactions. This can include customized recommendations, personalized notifications, or tailored information, creating a more personalized and relevant experience for the user.
- **Optical display systems:** Optical display systems are essential components of AI-enabled glasses. This section examines the various display technologies used in these glasses, such as head-up displays, organic light-emitting diode (OLED) displays, and waveguide displays. It discusses the advantages and limitations of each technology and their impact on the user experience.
- **Sensors and cameras:** AI-enabled glasses incorporate a range of sensors and cameras to capture and process real-world information. This section explores the types of sensors and cameras commonly found in AI-enabled glasses, including depth sensors, RGB (red, green, and blue wavelength) cameras, and eye-tracking sensors. It discusses how these sensors contribute to functionalities such as object recognition, gaze tracking, and environment mapping.
- **Connectivity features:** Connectivity features allow AI-enabled glasses to connect to external devices and networks, facilitating data transfer and communication. This section examines the connectivity options available in AI-enabled glasses, including Wi-Fi, Bluetooth, and cellular connectivity. It discusses the implications of these connectivity features on data transmission, security, and user interactions.
- **Processing units and AI chips:** AI-enabled glasses rely on powerful processing units and AI chips to perform complex computations and run AI algorithms. This section explores the types of processors and AI chips commonly used in AI-enabled glasses, such as system-on-chip (SoC) designs and specialized AI chips. It discusses the significance of these components in enabling real-time AI capabilities in the glasses.

Overall, AI-enabled glasses combine AI capabilities with wearable technology to provide an augmented reality experience, real-time information, and interactive features, making them a versatile tool with applications in areas such as healthcare, manufacturing, education, entertainment, and more [8], [9].

AI ALGORITHMS AND TECHNOLOGIES FOR GLASSES

- **Computer vision and object recognition:** Computer vision algorithms empower AI-enabled glasses to analyze and interpret visual information from the surrounding environment. This section discusses the role of computer vision in object recognition, scene understanding, and facial recognition in the context of AI-enabled glasses. It also explores the challenges and advancements in computer vision algorithms for glasses.

  Computer vision algorithms enable AI-enabled glasses to perceive and understand the user's visual environment. These algorithms analyze images or video streams captured by the glasses' cameras to recognize objects, detect features, track motion, and perform tasks like image segmentation, object recognition, or facial recognition. Computer vision is fundamental for overlaying augmented reality content onto the real-world view.

- **Natural language processing (NLP):** NLP algorithms allow AI-enabled glasses to understand and process human language inputs. This section examines how NLP algorithms are utilized in AI-enabled glasses to enable voice commands, speech-to-text conversion, and language translation functionalities. It discusses the advancements and limitations of NLP in the context of glasses.

  NLP algorithms can analyze spoken or written language to extract meaning, identify entities, and perform tasks. NLP capabilities allow users to interact with the glasses using voice commands or receive spoken instructions or information.

- **Gesture and emotion recognition:** Gesture and emotion recognition technologies enhance user interactions with AI-enabled glasses. This section explores the use of sensors and computer vision algorithms to detect and interpret gestures and emotions. It discusses the applications of gesture and emotion recognition in AI-enabled glasses, such as hands-free control and personalized user experiences.

  Gesture recognition technologies allow AI-enabled glasses to interpret hand or body movements as commands or interactions. These algorithms analyze the visual data captured by the cameras or the motion data from gyroscopes and accelerometers to recognize specific gestures. Gesture recognition allows users to control the glasses or interact with augmented reality content without physical manipulation.

- **Augmented reality (AR) and virtual reality (VR) integration:** AR and VR technologies have become integral to AI-enabled glasses, enhancing the user experience and enabling immersive interactions. This section examines the integration of AR and VR in AI-enabled glasses, including the challenges of overlaying virtual content on the real world and the potential applications of AR and VR in different sectors.

- **Machine learning and deep learning:** Machine learning and deep learning algorithms form the foundation of AI-enabled glasses' intelligence and adaptive capabilities. This section explores how these algorithms are applied in AI-enabled glasses for tasks such as personalized recommendations, user behavior analysis, and predictive analytics. It also discusses the training and optimization processes for AI models in glasses.

  Machine learning algorithms play a crucial role in AI-enabled glasses by enabling them to learn and adapt from data. These algorithms can train models using large datasets to recognize patterns, make predictions, and improve performance over time [10]. Machine learning is used in various aspects of AI-enabled glasses, such as gesture recognition, object detection, personalization, or even eye-tracking algorithms [11].

- **Sensor integration:** AI-enabled glasses incorporate a variety of sensors to gather data about the user's environment. This includes cameras, depth sensors, gyroscopes, accelerometers, microphones, and more. The integration of sensor data allows AI algorithms to perceive the user's context, track movements, capture visual or auditory information, and provide relevant augmented reality content or intelligent insights.

- **Connectivity and cloud computing:** AI-enabled glasses often leverage connectivity options such as Wi-Fi or Bluetooth to connect to the internet or other devices. This connectivity enables the glasses to access cloud-based services, offload computation tasks, and exchange data in real-time. Cloud computing provides additional processing power and storage capacity, allowing AI algorithms to leverage more extensive resources and access sophisticated AI models or services.

- **Eye tracking:** Eye-tracking technologies track the movement and gaze of the user's eyes. By using cameras or specialized sensors, AI-enabled glasses can determine where the user is looking, how their eyes move, and even estimate their attention or emotional state. Eye tracking is utilized for gaze-based interactions, user-attention analysis, or enhancing the display of information based on the user's focus.

APPLICATIONS OF AI-ENABLED GLASSES

AI-enabled glasses have a wide range of applications across various industries and domains. Here are some notable applications [12], [13]:

- **Industrial and manufacturing:** AI-enabled glasses can provide real-time information and instructions to workers in manufacturing or industrial settings. They can overlay step-by-step instructions, safety guidelines, or equipment maintenance procedures, enhancing productivity and reducing errors. AI-enabled glasses can also facilitate remote assistance, where experts can guide on-site workers through complex tasks or troubleshoot issues.

- **Healthcare and medicine:** AI-enabled glasses have potential applications in healthcare, including telemedicine, surgery, and training. They can enable remote consultations by allowing healthcare professionals to see the patient's perspective through the glasses' camera. In surgical procedures, AI-enabled glasses can overlay vital signs, patient data, or surgical instructions, improving precision and reducing reliance on external displays. They can also aid in training medical professionals by providing real-time guidance and feedback.

- **Education and training:** AI-enabled glasses can transform the learning experience by providing interactive and immersive educational content. They can overlay relevant information, 3D models, or simulations onto the user's field of view, enhancing understanding and engagement. AI-enabled glasses can also assist in vocational training, such as guiding individuals in learning new skills or providing real-time feedback during practice sessions.

- **Field service and maintenance:** AI-enabled glasses can support field service technicians by providing them with real-time access to manuals, troubleshooting guides, or equipment specifications. They can overlay relevant information onto the technician's view, enabling hands-free access to instructions and reducing downtime. AI-enabled glasses can also leverage computer vision to recognize and identify components or equipment, assisting in maintenance and repair tasks.

- **Navigation and logistics:** AI-enabled glasses can enhance navigation and logistics operations by overlaying directions, route information, or location-based data onto the user's view. This can be useful for delivery personnel, warehouse workers, or travelers who need real-time guidance and information about their surroundings. AI-enabled glasses can also provide data on inventory management, item picking, or barcode scanning, streamlining logistics processes.

- **Gaming and entertainment:** AI-enabled glasses can create immersive gaming experiences by overlaying virtual objects, characters, or game-related information onto the real-world view. They can enable augmented reality games that interact with the user's environment or provide a heads-up display for gaming statistics and notifications. AI-enabled glasses can also enhance live events by overlaying additional information or providing personalized content to attendees.

- **Accessibility and assistive technology:** AI-enables glasses can assist individuals with disabilities by providing real-time audio descriptions of the user's surroundings, reading text aloud, or offering navigation support. They can help visually impaired individuals navigate public spaces, recognize faces, or read signs. AI-enabled glasses can also facilitate communication by transcribing and translating spoken language for users with hearing impairments.

- **Retail and customer service:** AI-enabled glasses can enhance the retail experience by providing personalized recommendations, product information, or virtual try-on experiences. They can assist sales associates in identifying products, accessing inventory information, or processing payments. In customer service, AI-enabled glasses can provide real-time access to customer data, support ticketing systems, or deliver contextual information to improve interactions.

These are just a few examples of the applications of AI-enabled glasses. As the technology continues to advance, new applications and use cases will emerge, further expanding the potential of AI-enabled glasses across various industries and domains.

ADVANTAGES OF AI-ENABLED GLASSES:

AI-enabled glasses offer numerous advantages across various domains. Here are some key advantages:

- **Enhanced AR experience:** AI-enabled glasses can provide an immersive AR experience by overlaying digital information on the real world. This opens up possibilities for improved navigation, real-time data visualization, and interactive digital content.
- **Hands-free operation:** Unlike traditional devices such as smartphones or tablets, AI-enabled glasses allow users to interact with digital content without the need for physical manipulation. This enables hands-free operation, enhancing convenience and multitasking capabilities.
- **Real-time information:** AI-enabled glasses can leverage AI algorithms to process and display real-time information directly in the user's field of view. This can be particularly useful in scenarios where accessing information quickly is crucial, such as in industrial settings or emergency response situations.
- **Contextual awareness:** AI algorithms integrated into glasses can analyze the user's environment and provide relevant contextual information. For example, AI-enabled glasses can identify objects, recognize people's faces, or translate foreign languages, providing users with valuable insights and facilitating seamless interactions.
- **Personalized assistance:** AI-enabled glasses can serve as personal assistants, offering personalized recommendations, reminders, and notifications based on the user's preferences, location, and activities. This helps users stay organized, productive, and informed. From retrieving information from the internet to managing calendars and reminders, the glasses become a hands-free interface for accessing and interacting with digital information.
- **Accessibility features:** AI-enabled glasses can assist individuals with disabilities, such as visual impairments. By leveraging computer vision and natural language processing, these glasses can describe the surroundings, read text aloud, and provide audio-based assistance, thus enhancing accessibility and inclusivity.

- **Training and skill development:** In fields such as medicine or manufacturing, AI-enabled glasses can provide real-time guidance and training to professionals. For instance, during surgical procedures, these glasses can overlay instructions, display vital signs, or offer remote expert assistance, ultimately improving training outcomes and reducing errors.
- **Data collection and analysis:** AI-enabled glasses equipped with sensors can collect and analyze data in real-time. This capability has applications in various domains, including sports performance analysis, quality control in manufacturing, and environmental monitoring, providing valuable insights and supporting informed decision-making.
- **Seamless connectivity:** AI-enabled glasses can connect to the internet or other devices wirelessly, enabling seamless integration with existing systems or Internet of Things (IoT) infrastructure. This connectivity allows for efficient data sharing, collaboration, and remote communication, expanding the glasses' capabilities.
- **Potential for innovation:** AI-enabled glasses serve as a platform for developers and innovators to create new applications and services. As the technology advances, more developers can leverage the AI capabilities of these glasses to build innovative solutions, leading to continuous improvements and expanding possibilities.
- **Accessibility and inclusivity:** One of the most significant advantages of AI-enabled glasses is their potential to enhance accessibility for individuals with visual impairments. By employing advanced object recognition and text-to-speech technologies, these glasses can audibly describe the wearer's surroundings, helping them navigate their environment more effectively. Additionally, they can provide real-time translations of text, making foreign languages more accessible.
- **Professional applications:** AI-enabled glasses hold immense potential in various professional domains. In healthcare, they can assist surgeons during complex procedures by overlaying medical images or vital signs directly onto their field of view. In manufacturing, these glasses can provide real-time instructions to workers, improving efficiency and reducing errors. Furthermore, they can aid in remote collaboration by allowing experts to see what the wearer sees, facilitating teleconferencing and remote support.

It's worth noting that the advantages of AI-enabled glasses are continually evolving as the technology progresses, opening up new opportunities and applications in various industries.

CHALLENGES AND ETHICAL CONSIDERATIONS

Despite the potential of AI-enabled glasses, there are still a number of challenges and ethical considerations that need to be addressed before they can become truly mainstream.

CHALLENGES

- **Cost:** AI-enabled glasses are still relatively expensive. This is a barrier to adoption for many people.
- **Battery life:** The battery life of AI-enabled glasses is still limited. This means that they cannot be used for extended periods of time.
- **Privacy and data security:** The adoption of AI-enabled glasses raises concerns about privacy and data security. This section examines the potential risks and challenges associated with the collection and processing of personal data through AI-enabled glasses. It explores privacy protection measures and the importance of data security in the context of AI-enabled glasses.

ETHICAL IMPLICATIONS

The use of AI algorithms and technologies in glasses introduces ethical considerations that need to be addressed. This section discusses issues such as bias in AI algorithms, transparency and explainability, and the impact of AI-enabled glasses on social interactions. It explores the ethical frameworks and guidelines that can govern the development and use of AI-enabled glasses. While AI-enabled glasses offer numerous benefits, their adoption raises ethical concerns that must be addressed. Privacy and data security are paramount, as these glasses constantly collect and process visual and auditory information. Safeguarding user data and ensuring transparent data-usage practices are essential. Additionally, issues surrounding addiction, distraction, and the potential for misuse of AI capabilities need to be carefully considered and regulated.

FUTURE TRENDS AND ADVANCEMENTS

This section presents future trends and advancements in AI-enabled glasses, including advancements in AI algorithms, miniaturization of components, and integration with other wearable devices. It discusses the potential applications of AI-enabled glasses in fields such as education, gaming, and industrial sectors. It also explores emerging technologies that may shape the future of AI-enabled glasses, such as brain–computer interfaces and edge computing. As AI technologies continue to evolve, the potential of AI-enabled glasses will only expand. Advancements in AI algorithms, sensor miniaturization, and battery technology will result in smaller, more powerful devices with extended battery life. Moreover, the integration of AI-enabled glasses with other emerging technologies such as 5G connectivity, edge computing, and IoT will unlock new possibilities, enabling seamless integration into smart environments.

CONCLUSION

This chapter provides a comprehensive overview of AI-enabled glasses, exploring their history, components, applications, challenges, and ethical considerations. It highlights the significant advancements in AI algorithms and technologies that have enabled the integration of AI capabilities into glasses. The chapter concludes with insights into future trends and advancements, emphasizing the transformative potential of AI-enabled glasses in various sectors.

AI-enabled glasses represent a significant milestone in the development of wearable technology and artificial intelligence. By merging the physical and digital realms, these glasses empower users with augmented vision, personalized assistance, and enhanced accessibility. While ethical considerations remain, the potential applications and transformative impact of AI-enabled glasses make them a promising avenue for future exploration. With further advancements, these glasses could reshape how we perceive and interact with the world, offering a glimpse into a future where the boundaries between the real and virtual are blurred.

REFERENCES

1. https://www.sciencedirect.com/science/article/pii/S1878875020314406
2. https://ceramics.onlinelibrary.wiley.com/doi/abs/10.1111/ijag.15881
3. https://dl.acm.org/doi/abs/10.1145/3491102.3501925
4. https://link.springer.com/article/10.1007/s40940-020-00132-8
5. https://www.sciencedirect.com/science/article/pii/S0921510722002495
6. https://www.sciencedirect.com/science/article/pii/S1474034621001816
7. https://www.nature.com/articles/s41591-020-0931-3
8. https://www.sciencedirect.com/science/article/pii/S2351978920310556
9. https://www.mdpi.com/2076-3417/13/3/1394
10. https://ieeexplore.ieee.org/abstract/document/9720526/
11. https://ieeexplore.ieee.org/abstract/document/9873400/
12. https://www.tandfonline.com/doi/abs/10.1080/09506608.2020.1815394
13. https://ieeexplore.ieee.org/abstract/document/9055973/

18 AI in Smart Fitness Trackers

Pisipati Sai Anurag, Pankaj Kumar Singh, Marepalli Radha, and Karthik Ramesh

INTRODUCTION

In recent years, smart fitness trackers have gained significant popularity, becoming an integral part of people's lives, helping them monitor their physical activities and accomplish their fitness goals. These wearable devices have evolved significantly with the integration of artificial intelligence (AI). This chapter explores the role of artificial intelligence in smart fitness trackers, discussing how AI enhances the accuracy and functionality of these devices. We delve into the various techniques employed in smart fitness trackers such as machine learning algorithms, data analytics, and personalized recommendations. Additionally, we highlight the challenges and future directions of AI-enabled smart fitness trackers [1].

OVERVIEW OF SMART FITNESS TRACKERS

Smart fitness trackers are wearable devices designed to track and monitor various aspects of an individual's physical activity and health. They have become a popular companion for fitness enthusiasts, health-conscious individuals, and those looking to lead a more active lifestyle. These trackers typically include sensors such as accelerometers, heart rate monitors, gyroscopes, and global positioning system (GPS) modules to gather data on user's movements, heart rate, sleep patterns, and location [2].

Apart from real-time monitoring, fitness trackers often offer features such as step counting, calorie expenditure, distance traveled, and activity recognition based on the type of activity performed. Many of these devices are also water resistant, allowing users to also track their swimming activities. The collected data is usually synced to a mobile app or a cloud-based platform, enabling users to analyze their progress and set fitness goals.

THE EVOLUTION OF SMART FITNESS TRACKERS

The integration of AI in smart fitness trackers marks a significant milestone in the development of these wearable devices. Early versions of fitness trackers primarily relied on rule-based algorithms and heuristics to interpret the collected data.

However, with advancements in AI, fitness trackers have transitioned to more intelligent and data-driven systems [3].

Machine learning algorithms have been instrumental in the evolution of fitness-tracking devices. Initially, supervised learning algorithms were employed to recognize basic activities like walking, running, and cycling. As datasets grew and AI models improved, fitness trackers became better at accurately identifying a wider set of activities, including yoga, weightlifting, and even dancing.

Unsupervised learning techniques have also been applied to allow the device to identify irregular patterns in data, such as sudden change in heart rate or sleep disturbances. This capability helps users become more aware of potential health issues and prompts them to seek appropriate medical attention whenever needed.

Significance of AI in Improving Fitness Tracking

AI has brought several key advantages to smart fitness trackers, enhancing their effectiveness and usability [4].

1. **Enhanced accuracy:** By leveraging AI algorithms, fitness trackers can provide more accurate and reliable data. Machine learning models continuously learn from user inputs and sensor data, making predictions and interpretations more precise over time.
2. **Personalized insights:** AI enables smart fitness trackers to generate personalized insights and recommendations based on individual user data. These insights go beyond generic fitness advice, helping users set realistic goals, make healthier lifestyle choices, and tailor their workouts to achieve optimal results.
3. **Real-time feedback:** With AI, fitness trackers can provide real-time feedback during workouts, helping users maintain proper form, pace, and intensity. This immediate feedback fosters a sense of accountability and motivation, which can lead to better workout performance and adherence.
4. **Seamless integration:** AI allows smart fitness trackers to integrate with other health-related devices and applications, creating a comprehensive ecosystem for health and wellness management. This integration enables users to track their progress across multiple platforms and gain a holistic view of their health data.

AI TECHNIQUES IN SMART FITNESS TRACKERS

Artificial intelligence techniques play a crucial role in enhancing the functionality and accuracy of smart fitness trackers. These techniques enable the devices to analyze and interpret data collected from various sensors, providing valuable insights and personalized recommendations to users. In this section, we explore the specific AI techniques employed in smart fitness trackers, including machine learning algorithms, data analytics, and personalized recommendations.

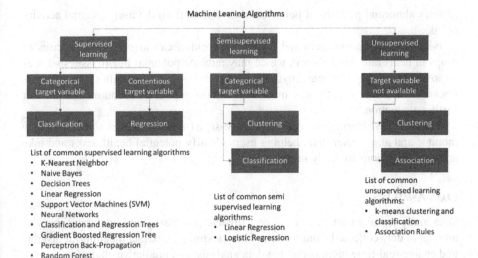

FIGURE 18.1 Machine learning algorithms in smart fitness trackers

MACHINE LEARNING ALGORITHMS

Machine learning algorithms are at the core of AI-enabled smart fitness trackers. They enable the devices to learn from collected data and make predictions or classifications based on patterns and relationships in the data [5], [6]. Figure 18.1 depicts three types of machine learning algorithms: supervised learning, semi-supervised learning, and unsupervised learning

Supervised Learning for Activity Recognition

Supervised learning algorithms are commonly used in smart fitness trackers to recognize and classify different types of physical activities performed by the user. These algorithms are trained on labeled data sets, where each activity is associated with specific features extracted from the sensor data.

For example, a supervised learning algorithm can be trained on a dataset that includes accelerometer and gyroscope readings collected during activities like walking, running, cycling, or yoga. By analyzing patterns and features in data, the algorithm can accurately classify the ongoing activity in real-time.

Supervised learning algorithms continually improve their accuracy as they receive feedback from users and collect more labeled data. This iterative learning process enables the smart fitness trackers to better recognize and differentiate activities, providing users with more precise activity tracking.

Unsupervised Learning for Anomaly Detection

Unsupervised learning algorithms are employed in smart fitness trackers for anomaly detection. These algorithms analyze the data collected from various sensors and identify abnormal patterns of behaviors that deviate from the user's normal activity profile.

For instance, an unsupervised learning algorithm can identify sudden rises or drops in heart rate during sleep, which may indicate potential health issues such as sleep apnea or irregular heart rhythms. By detecting these anomalies, smart fitness trackers can alert users to seek medical attention to provide recommendations for further evaluation.

Unsupervised learning algorithms enable smart fitness trackers to continuously monitor and assess user data, helping users identify potential health risks and take appropriate actions for early intervention.

DATA ANALYTICS

Data analytics in smart fitness trackers involve processing and analyzing the vast amount of data collected from sensors. These techniques extract meaningful insights and enable real-time monitoring, big data analysis, and predictive analytics.

Real-Time Data Processing and Analysis

Real-time data processing and analysis are essential components of AI-enabled smart fitness trackers. By processing sensor data in real time, these devices can provide immediate feedback and guidance to users during workouts or activities.

For example, during a running session, the smart fitness tracker can analyze the user's pace, stride length, and heart rate in real time. Based on this analysis, the device can provide audio or visual cues to help the user maintain an optimal pace or adjust their running technique to prevent injuries.

Real-time data analysis also allows smart fitness trackers to monitor users' heart rate zones during exercise, ensuring they stay within the desired intensity range for maximum effectiveness and safety.

Smart fitness trackers generate a significant amount of data over time creating a valuable resource for big-data analysis and predictive analytics. By aggregating and analyzing this data, AI algorithms can identify long-term trends, patterns, and correlations that help users gain deeper insights into their fitness and health.

Predictive analytics algorithms can leverage this historical data to forecast future outcomes and provide personalized recommendations. For example, based on an individual's workout history. Sleep patterns, and nutrition data, a smart fitness tracker can predict the optimal workout duration and recovery period for achieving specific fitness goals.

PERSONALIZED RECOMMENDATIONS

AI enables smart fitness trackers to provide personalized recommendations tailored to individual users. These recommendations cover various aspects, including fitness plans, nutrition, and diet suggestions [7].

Individualized Fitness Plans

Based on user data, including activity levels, fitness goals, and historical workout performance, AI algorithms can generate individualized fitness plans. These plans consider factors such as the user's current fitness level, preferences, and time constraints and recommend suitable workouts and training regimens.

The smart fitness tracker can provide guidance on exercise types, duration, frequency, and intensity to optimize progress toward specific goals, such as weight loss, muscle gain, or cardiovascular fitness. These recommendations adapt and evolve as the user progresses, ensuring ongoing motivation and effectiveness.

Tailored Nutrition and Diet Suggestions

AI-enabled smart fitness trackers can analyze user data related to nutrition and provide personalized dietary recommendations. By integrating nutritional databases, the devices can track users' calorie intake, macronutrient distribution, and even micronutrient deficiencies.

Based on this data, the smart fitness tracker can offer tailored suggestions for meal planning, portion control, and healthier food choices. It can also provide reminders to drink water, consume post-workout recovery meals, or avoid certain foods based on specific dietary restrictions or allergies.

These personalized recommendations help users make informed choices about their nutrition, promoting a healthy and balanced diet that aligns with their fitness goals.

IMPROVED ACCURACY AND PERFORMANCE

AI plays a pivotal role in enhancing the accuracy and performance of smart fitness trackers. By leveraging AI techniques, these devices can provide more precise and detailed information regarding activity tracking, heart rate monitoring, sleep analysis, and advanced sports performance metrics. In this section, we explore how AI improves the accuracy and performance of smart fitness trackers in these areas [8].

ENHANCED ACTIVITY TRACKING

Smart fitness trackers with AI capabilities offers enhanced activity tracking, accurately recognizing and categorizing various physical activities. With the integration of machine learning algorithms, these devices can differentiate between activities like walking, running, cycling, swimming, weightlifting, and yoga.

AI algorithms continuously learn from user data and patterns, improving the accuracy of activity recognition over time. This ensures that users receive more accurate data on the duration, intensity, and type of physical activities they engage in, allowing them to monitor their progress and set achievable fitness goals.

Additionally, AI-powered smart fitness trackers can distinguish between different variations of activities. For example, they can identify whether a user is doing a slow jog or a sprint, or if they are performing high-intensity interval training (HIIT) workouts. This level of specificity enables users to track their efforts more precisely and make informed decisions about their training routines.

ACCURATE HEART RATE MONITORING

Heart rate monitoring is a vital feature of smart fitness trackers, providing valuable insights into cardiovascular health and workout intensity. AI techniques significantly improve the accuracy of heart rate monitoring, allowing users to obtain more reliable and precise measurements.

Using machine learning algorithms, smart fitness trackers can refine heart rate monitoring by accounting for factors such as individual variations, motion artifacts, and environmental conditions. These algorithms learn from the data collected by the heart rate sensor and can filter out noise, resulting in more accurate readings.

SLEEP TRACKING ANALYSIS

Sleep tracking is another area where AI improves the accuracy and analysis capabilities of smart fitness trackers. By incorporating AI algorithms, these devices can capture and analyze sleep patterns with greater precision, offering insights into sleep quality, duration, and sleep stages.

Machine learning algorithms can detect sleep-related events such as sleep onset, wakeups, and sleep stages (e.g., light sleep, deep sleep, REM (rapid eye movement) sleep). These algorithms learn from a combination of accelerometer and heart rate data, identifying patterns that correspond to different sleep stages

The AI algorithms also enable sleep trackers to provide sleep recommendations and personalized insights. For example, the devices may suggest adjustments to sleep duration, bedtime routines, or environmental factors to improve sleep quality based on individual sleep patterns and preferences.

ADVANCED SPORTS PERFORMANCE METRICS

AI enhances smart fitness trackers by providing advanced sports performance metrics, helping users analyze and optimize their athletic performance. These metrics go beyond basic activity tracking and provide more in-depth insights into specific sports and exercises [9].

With AI algorithms, smart fitness trackers can track and analyze metrics such as power output, cycling efficiency, and swimming-stroke analysis. This information enables users to fine-tune their technique, improve efficiency, and monitor progress over time.

AI algorithms can also provide personalized recommendations for training and recovery based on sports-specific data. For instance, a smart fitness tracker may suggest interval training sessions, provide recovery advice, or recommend specific exercises to address weaknesses in a user's performance.

By incorporating AI algorithms, smart fitness trackers become comprehensive tools for athletes and fitness enthusiasts, offering detailed insights and personalized guidance to enhance sports performance and achieve fitness goals.

AI-ENABLED FEATURES AND FUNCTIONALITIES

AI brings a range of features and functionalities to smart fitness trackers, enhancing the user experience and providing valuable tools for achieving fitness goals. In this section, we explore some of the AI-enabled features and functionalities found in smart fitness trackers. Figure 18.2 gives a clear understanding of the different features and functionalities of smart fitness trackers.

VIRTUAL COACHES AND TRAINERS

AI-powered smart fitness trackers can act as virtual coaches and trainers, providing personalized guidance and support during workouts. These virtual coaches leverage machine learning algorithms to analyze user data, monitor performance, and offer real-time feedback and motivation.

For example, during a workout session, a smart fitness tracker with a virtual coach feature can provide audio cues, coaching tips, and encouragement based on users' performance and goals. The virtual coach can analyze metrics such as heart rate, pace, and form to guide users toward optimal performance and technique.

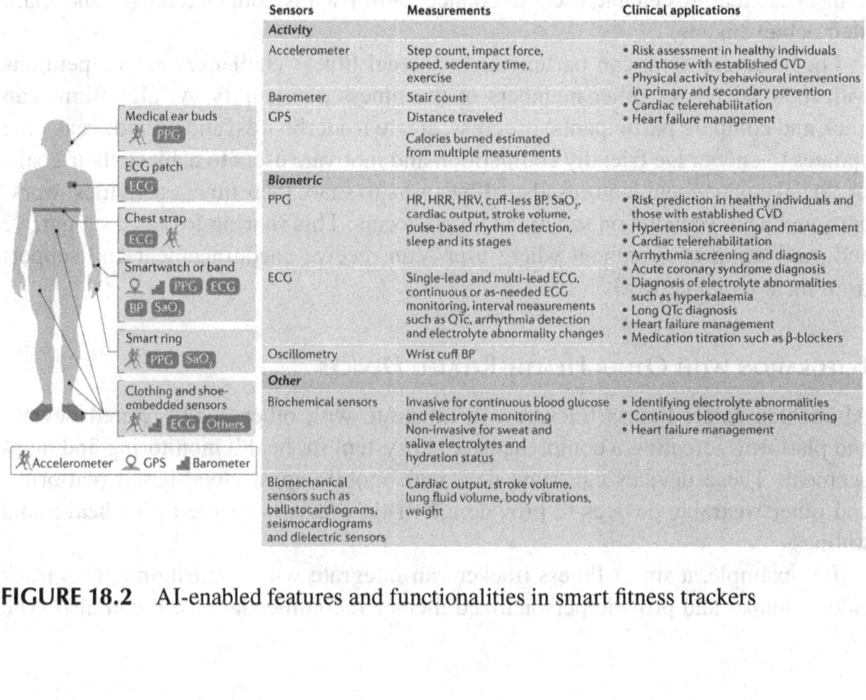

Sensors	Measurements	Clinical applications
Activity		
Accelerometer	Step count, impact force, speed, sedentary time, exercise	• Risk assessment in healthy individuals and those with established CVD • Physical activity behavioural interventions in primary and secondary prevention • Cardiac telerehabilitation • Heart failure management
Barometer	Stair count	
GPS	Distance traveled	
	Calories burned estimated from multiple measurements	
Biometric		
PPG	HR, HRR, HRV, cuff-less BP, SaO₂, cardiac output, stroke volume, pulse-based rhythm detection, sleep and its stages	• Risk prediction in healthy individuals and those with established CVD • Hypertension screening and management • Cardiac telerehabilitation • Arrhythmia screening and diagnosis • Acute coronary syndrome diagnosis • Diagnosis of electrolyte abnormalities such as hyperkalaemia • Long QTc diagnosis • Heart failure management • Medication titration such as β-blockers
ECG	Single-lead and multi-lead ECG, continuous or as-needed ECG monitoring, interval measurements such as QTc, arrhythmia detection and electrolyte abnormality changes	
Oscillometry	Wrist cuff BP	
Other		
Biochemical sensors	Invasive for continuous blood glucose and electrolyte monitoring Non-invasive for sweat and saliva electrolytes and hydration status	• Identifying electrolyte abnormalities • Continuous blood glucose monitoring • Heart failure management
Biomechanical sensors such as ballistocardiograms, seismocardiograms and dielectric sensors	Cardiac output, stroke volume, lung fluid volume, body vibrations, weight	

Body diagram labels: Medical ear buds (PPG); ECG patch (ECG); Chest strap (ECG); Smartwatch or band (PPG, ECG, BP, SaO₂); Smart ring (PPG, SaO₂); Clothing and shoe-embedded sensors (ECG, Others). Legend: Accelerometer, GPS, Barometer

FIGURE 18.2 AI-enabled features and functionalities in smart fitness trackers

Virtual coaches can also offer customized workout plans and routines tailored to the user's fitness level, preferences, and goals. They can adapt and adjust the training programs based on user feedback and progress, providing a personalized training experience comparable to having a human coach.

SMART GOAL SETTING AND TRACKING

AI-enabled smart fitness trackers facilitate smart goal setting and tracking, empowering users to set realistic and achievable fitness goals. These devices utilize machine learning algorithms to analyze user data, track progress, and provide recommendations for goal setting.

Smart fitness trackers can consider various factors such as current fitness level. Activity history and individual preferences to recommend appropriate goals. AI algorithms can assess the user's capabilities and suggest incremental milestones to ensure gradual progress and minimize the risk of injury.

The trackers continuously track the user's progress toward their goals, providing real-time feedback and insights. Users can view their achievements, set new targets, and receive personalized recommendations for adjusting their workouts or activities to stay on track.

SOCIAL AND COMMUNITY ENGAGEMENT

AI-powered smart fitness trackers often include social and community engagement features, leveraging AI algorithms to create a sense of community and foster motivation. These features enable users to connect with friends, join challenges, and share their achievements.

For instance, users can participate in virtual fitness challenges or competitions with their friends or other members of the fitness community. AI algorithms can track and compare participants' progress, create leaderboards, and provide real-time updates to encourage friendly competition and motivate users to achieve their goals.

Social engagement features also allow users to share their fitness activities, workouts, and achievements on social media platforms. This sharing fosters a supportive and motivating environment where users can receive encouragement and support from their social network.

INTEGRATION WITH OTHER HEALTH-RELATED DEVICES

AI-enabled smart fitness trackers can integrate with other health-related devices and platforms, creating a comprehensive ecosystem for health monitoring and management. These devices can sync data with mobile apps, cloud-based platforms, and other wearable devices to provide a holistic overview of the user's health and wellness.

For example, a smart fitness tracker can integrate with a nutrition app to track calorie intake and provide personalized dietary recommendations. It can also sync with a sleep monitor to analyze sleep patterns and provide insights into the user's overall health and well-being.

Furthermore, integration with medical devices such as heart rate monitors, blood pressure cuffs, or glucose monitors enables smart fitness trackers to collect and analyze additional health data. AI algorithms can process and correlate this data to provide more comprehensive health insights and recommendations.

The integration of various health-related devices and platforms enhances the user experience by consolidating the data, offering more accurate analysis, and providing actionable recommendations for improved health and fitness.

CHALLENGES IN AI-ENABLED SMART FITNESS TRACKING

While AI brings numerous benefits to smart fitness trackers, there are several challenges that need to be addressed, in this section, we discuss some of the key challenges associated with AI-enabled smart fitness trackers. Table 18.1 gives a brief understanding of the challenges faced in AI-enabled smart devices, the impact of those challenges, and their potential solutions.

DATA PRIVACY AND SECURITY CONCERNS

AI-enabled smart fitness trackers collect and process a significant amount of personal data, including health-related information. This data can be sensitive and private, raising concerns about data privacy and security.

Users must trust their data is stored safely and securely, and that appropriate measures are taken to protect it from unauthorized access or misuse. Fitness tracker manufacturers need to implement robust data encryption, secure storage practices, and strong access controls to safeguard user data.

TABLE 18.1

Challenges in AI-enabled smart fitness trackers

Challenge	Impact	Potential Solution
Data privacy and security concerns	Risk of unauthorized access	Implement robust data encryption, user-content mechanisms, and secure data storage practices.
Ethical considerations	Potential misuse of user data	Develop and follow ethical guidelines for data collection, usage, and sharing.
Interpretability and transparency	Difficulty in understanding of AI algorithms.	Use explainable AI techniques, provide transparency in algorithms and decision-making process.
User adoption and acceptance	Resistance to adopting new technologies.	Educate users about the benefits, address concerns, and provide user-friendly and intuitive designs.

Additionally, transparency regarding data collection, storage, and usage is crucial. Users should be informed about how their data is being utilized and should be able to control and manage their privacy settings.

ETHICAL CONSIDERATIONS AND USER TRUST

As fitness trackers become more advanced with AI capabilities, ethical considerations come into play. For example, AI algorithms may make recommendations that affect users' health and well-being. Ensuring the accuracy, reliability, and fairness of these recommendations is crucial.

There is also a need to address potential biases in AI algorithms, as biased recommendations can have negative consequences for users. Bias can arise from imbalances in the training data or algorithmic design, leading to disparities in the way recommendations are provided to different user groups.

Transparency and explainability of AI algorithms can help build user trust. Users need to understand how the algorithms make decisions and provide recommendations. Clear communication about the limitations and uncertainties associated with AI-enabled features can help manage user expectations and foster trust in the system.

INTERPRETABILITY AND TRANSPARENCY IN AI ALGORITHMS

AI algorithms used in fitness trackers can be complex and difficult to interpret. This lack of interpretability can pose challenges in understanding how the algorithms arrive at their recommendations or classifications. It becomes important to ensure that AI algorithms are transparent and explainable to users.

The ability to explain how AI algorithms make predictions or provide recommendations is crucial for user understanding and acceptance. Researchers and developers are exploring methods for making AI algorithms more interpretable, such as using techniques like rule-based explanations or generating human-readable summaries of the underlying decision-making process.

By enhancing the interpretability and transparency of AI algorithms, users can better comprehend, and trust, recommendations provided by smart fitness trackers.

SMART ADOPTION AND ACCEPTANCE

The adoption and acceptance of AI-enabled smart fitness trackers pose challenges as well. While some users embrace the technology, others may have concerns about privacy and accuracy of, or the reliability on, technology for health-related issues.

Ensuring the smart fitness trackers are user-friendly, intuitive, and provide tangible benefits is crucial for user adoption. Clear communication about the benefits and limitations of AI-enabled features can help users understand the value proposition and make informed decisions about using these devices.

User education and engagement play a pivotal role in promoting acceptance. Educating users about the capabilities and functionalities of smart fitness trackers, as well as addressing their concerns and providing support, can help build trust and encourage users to integrate these devices into their fitness routines.

FUTURE OPPORTUNITIES AND ADVANCEMENTS

Despite the challenges, the field of AI-enabled fitness trackers holds immense potential for future advancements. Some opportunities for improvements and innovation include:

- Continued research and development in AI algorithms to enhance accuracy, interpretability, and personalization of recommendations
- Integration with emerging technologies such as augmented reality (AR) and virtual reality (VR) to provide immersive and interactive fitness experiences
- Collaboration with healthcare professionals and researchers to leverage AI capabilities in preventive healthcare and chronic disease management
- Enhancing the user experience through innovative user interfaces, voice assistants, or natural language processing for seamless interaction with smart fitness trackers
- Leveraging AI to analyze aggregated and anonymized data from multiple users to identify trends, patterns, and insights that can inform public health initiatives or contribute to scientific research

By addressing these challenges and embracing future opportunities, AI-enabled smart fitness trackers have the potential to revolutionize the way individuals monitor and improve their fitness levels, leading to healthier lifestyles and improved overall well-being.

FUTURE DIRECTIONS AND OPPORTUNITIES

The field of AI-enabled smart fitness trackers holds immense potential for further advancements and innovations. In this section, we explore some of the key areas for future development and opportunities.

CONTINUOUS IMPROVEMENT OF AI ALGORITHMS

One of the primary areas of future advancement is the continuous improvement of AI algorithms used in smart fitness trackers. Ongoing research and development can focus on refining existing algorithms and developing new ones to enhance accuracy, reliability, and personalization of recommendations.

Researchers can explore advanced machine learning techniques such as deep learning and reinforcement learning, to improve the capabilities of smart fitness trackers. These techniques can enable devices to learn from larger datasets, adapt to individual user preferences more efficiently, and provide more sophisticated and context-aware recommendations.

Additionally, efforts can be directed toward addressing biases in AI algorithms to ensure fair and equitable recommendations for all users, regardless of their demographic or background.

MULTI-MODAL SENSOR INTEGRATION

Smart fitness trackers can benefit from the integration of multiple sensors to gather a more comprehensive range of health and fitness data. This multi-modal sensor integration can enable the trackers to capture data beyond the traditional accelerometer and heart rate sensors.

Integration with sensors like gyroscopes, barometers, skin temperature sensors, or electromyography (EMG) sensors can provide additional insights into factors such as body movement, environmental conditions, and muscle activity. By combining data from different sensors, AI algorithms can generate a more accurate and detailed analysis of user activities, exercise form, and overall fitness level.

AUGMENTED REALITY AND VIRTUAL REALITY APPLICATIONS

Integration of AI-enabled smart fitness trackers with healthcare systems presents significant opportunities for preventive healthcare and disease management. By connecting with electronic health records (EHRs) and healthcare providers' systems, smart fitness trackers can greatly contribute to a more holistic approach to health monitoring and management.

The integration can enable healthcare professionals to access real-time health and fitness data, allowing them to make more informed decisions, monitor patient progress remotely, and provide personalized recommendations. This integration can be particularly beneficial for individuals with chronic conditions or those undergoing post-rehabilitation programs.

Furthermore, the aggregated and anonymized data from smart fitness trackers can contribute to population health analysis, enabling healthcare systems to identify patterns, trends, and potential health risks at a larger scale.

CONCLUSION

In conclusion, AI has had a transformative impact on smart fitness trackers, enhancing their capabilities and providing users with personalized insights and advanced tracking and health monitoring features. The integration of AI algorithms has improved accuracy, enabled data-driven recommendations, and empowered individuals to make informed decisions about their fitness and well-being.

As the field continues to evolve, it is crucial to address challenges related to data privacy, ethical considerations, interpretability, and user acceptance. By doing so, smart fitness trackers can build trust, foster user adoption, and maximize the potential of AI in improving health outcomes.

In the future, we can expect AI to play a more pivotal role in smart fitness trackers, with continuous advancements in algorithms, sensor integration, immersive technologies, and integration with healthcare systems. These advancements will further empower individuals to monitor and optimize their fitness levels, prevent health risks, and lead a healthier lifestyle.

It is important for manufacturers, researchers, and healthcare professionals to collaboratively innovate in this space, ensuring that AI smart fitness trackers are user-friendly, secure, and provide meaningful insights. With responsible development, these devices can become invaluable tools in promoting individual wellness and contributing to broader public health initiatives.

In conclusion, AI has the potential to revolutionize the field of smart trackers, providing users with personalized insights, advanced tracking capabilities, and health monitoring features. By addressing the challenges, embracing future opportunities, and prioritizing user-centric design, AI-enabled smart fitness trackers can empower individuals to lead healthier lifestyles, improve their overall well-being, and achieve their fitness goals.

REFERENCES

1. Chen, L., Zhang, G., & Zhou, G. (2019). Artificial intelligence in wearable health devices: Algorithms, challenges, and perspectives. *Journal of Healthcare Engineering*, 2019, 1–16.
2. Kerner, Y., & Halevy, E. (2020). Artificial intelligence in sports and fitness: An interdisciplinary systematic review. *Artificial Intelligence in Medicine*, 105, 101844.
3. Mueller, F. F., & Agnisarman, S. (2019). Artificial intelligence applications in sports and health. *International Journal of Computer Science in Sport*, 18(2), 79–95.
4. Orwat, C., & Graefe, A. (2020). Artificial intelligence and wearables in sports and health: Where are we now and where are we heading? *Journal of Medical Internet Research*, 22(4), e16429.
5. Ravi, D., Wong, C., Deligianni, F., Berthelot, M., Andreu-Perez, J., & Lo, B. (2017). Deep learning for health informatics. *IEEE Journal of Biomedical and Health Informatics*, 21(1), 4–21.
6. Salminen, J., & Ravaja, N. (2020). Machine learning for predicting, tracking, and improving users' sports performance: A systematic literature review. *ACM Computing Surveys*, 53(6), 1–41.
7. Shoaib, M., Bosch, S., Incel, O. D., Scholten, H., & Havinga, P. J. (2015). Fusion of smartphone motion sensors for physical activity recognition. *Sensors*, 15(9), 19653–19676.
8. Sushma, P. G., & Manikandan, M. S. (2018). Artificial intelligence and internet of things based smart health monitoring systems for elderly people. In *2018 IEEE 4th International Conference on Electrical Energy Systems (ICEES)* (pp. 1–6). IEEE.
9. Theodoridis, T., Xefteris, S., & Bardis, N. (2019). Machine learning algorithms for activity recognition in smartwatches and fitness trackers. In *2019 IEEE 11th International Conference on Intelligent Data Acquisition and Advanced Computing Systems: Technology and Applications (IDAACS)* (pp. 1057–1061). IEEE.

19 Smart Helmets

Vanshhita Jaju and Saurav Sulam

INTRODUCTION

DEFINITION AND MEANING

Presently many bicycle accidents that occur on the road result in death. People can get injured or even killed while cycling and one of the reasons is not wearing a helmet. Smart helmets have emerged as a revolutionary advancement in the realm of personal protective equipment (PPE), integrating state-of-the-art technologies to enhance safety, provide real-time data, and foster connectivity. These helmets are equipped with various sensors, wireless communication capabilities, and advanced features that go beyond the traditional purpose of protecting the head [1, 2]. By leveraging the power of technology, smart helmets have the potential to significantly improve safety in a wide range of industries and activities, including construction, sports, and transportation [3].

The evolution of smart helmets has been driven by the need to address safety concerns and improve overall user experience. Traditional helmets have served as a passive barrier against head injuries, but they often lack the ability to provide real-time information or detect potential hazards [3, 4]. With advancements in sensor technology, wireless communication, and data analysis, smart helmets have emerged as a proactive solution that can actively monitor and mitigate risks [2].

The primary purpose of smart helmets is to enhance safety in a variety of industries and activities [3]. By actively monitoring the wearer's surroundings and vital signs, these helmets can provide early warnings, prevent accidents, and mitigate risks [4]. They also offer real-time feedback and performance metrics, helping individuals optimize their performance and make informed decisions. Moreover, smart helmets facilitate communication between users and external parties, enabling coordination, collaboration, and emergency response [3].

HISTORICAL BACKGROUND

- **Early development:** In the late 1990s and early 2000s, initial attempts were made to incorporate technology into helmets for specific applications. For instance, the military began experimenting with head-mounted displays and communication systems integrated into helmets for soldiers.

 DOI: 10.1201/9781032686714-19

Sports enthusiasts and professionals also started exploring the integration of sensors and communication devices into helmets to enhance safety and performance [5].

- **Sensor integration:** As sensor technology advanced, smart helmets started incorporating various sensors to monitor vital signs, detect impacts, and measure environmental conditions. These sensors could provide real-time feedback and alerts to the wearer.

Accelerometers, gyroscopes, and other motion sensors were integrated into smart helmets to detect sudden movements, impacts, or changes in posture [6].

- **Communication and Connectivity:** With the rise of wireless communication technologies, smart helmets began incorporating communication systems, such as Bluetooth, to enable hands-free communication between riders, team members, or emergency services.

This allowed for seamless communication without the need for external devices or the removal of the helmet [6, 7].

- **Augmented reality (AR) integration:** Recent advancements in AR technology have led to the integration of heads-up displays (HUDs) and AR capabilities into smart helmets.

AR-enabled smart helmets overlay digital information onto the wearer's field of view, providing real-time data, navigation instructions, and visualizations without the need to divert attention from the environment [8].

- **Advanced safety features:** Smart helmets have continued to evolve in terms of safety features. They now include features like LED (light emitting diode) lights, turn signals, and brake lights to enhance visibility and promote rider safety on the road.

Some smart helmets also incorporate emergency alert systems that automatically notify designated contacts in case of an accident or fall [5].

- **Data collection and analysis:** Modern smart helmets often include data collection capabilities, allowing users to track and analyze their performance metrics, such as speed, distance traveled, heart rate, and more.

This data can be synchronized with smartphone apps or cloud platforms to provide users with insights and help them monitor their progress [5, 6].

EVOLUTION OF SMART HELMETS

The evolution of smart helmets has been driven by the continuous advancements in technology and the increasing demand for improved safety, connectivity, and functionality. Let's explore the key stages in the evolution of smart helmets:

- **Early sensor integration:** The initial phase of smart helmet development involved the integration of basic sensors, such as accelerometers and gyroscopes, to detect impacts and measure head movements. These sensors provided the foundation for early safety features, enabling impact detection and basic data collection.
- **Wireless communication and connectivity:** As wireless communication technologies progressed, smart helmets started incorporating Bluetooth and Wi-Fi connectivity. This allowed for seamless communication between the helmet and external devices like smartphones, wearable's, or other smart devices. Wearers could receive real-time alerts, share data, or communicate hands-free.
- **Advanced safety features:** Smart helmets began incorporating more advanced safety features to enhance protection. These features included proximity sensors to detect obstacles, collision detection systems to analyze impacts, and advanced warning systems to provide timely alerts to wearers. Some helmets also introduced rear-facing cameras for improved situational awareness.
- **Integration of global positioning system (GPS) and navigation:** To enhance navigation and route guidance, smart helmets started integrating GPS technology. This enabled wearers to receive turn-by-turn directions, track their routes, and locate points of interest. GPS integration also facilitated emergency response systems, allowing precise location tracking in case of accidents or emergencies.
- **Advanced display technology:** Advancements in display technology led to the integration of HUDs and AR features in smart helmets. These displays project relevant information, such as speed, navigation instructions, or hazard alerts, directly onto the wearer's field of view. AR overlays digital content onto the real world, enhancing situational awareness and providing contextual information.
- **Health monitoring and performance metrics:** Smart helmets expanded their capabilities beyond safety by incorporating health monitoring and performance metrics. Heart rate monitors, temperature sensors, and other biometric sensors allowed wearers to track their physical condition and performance. This integration enabled athletes, workers, and individuals in various fields to optimize their performance, prevent fatigue, and assess overall well-being.
- **Integration with artificial intelligence (AI) and machine learning (ML):** Recent advancements in AI and ML have started to impact smart helmet development. AI algorithms can analyze sensor data, identify patterns, and make informed decisions. ML techniques can enable predictive analytics,

detect anomalies, or customize safety features based on individual wearers' preferences and behavior.

- **Integration with emerging technologies:** The evolution of smart helmets is an ongoing process, with advancements in technology, user feedback, and market demands driving further innovation. As safety, connectivity, and usability continue to improve, smart helmets are becoming an increasingly integral part of various industries, activities, and personal safety practices [1, 7].

FEATURES OF SMART HELMETS

SAFETY FEATURES

Smart helmets incorporate a range of advanced safety features to protect the wearer and mitigate risks. Some notable safety features include:

- **Impact detection and analysis:** Smart helmets are equipped with sensors, such as accelerometers and gyroscopes that can detect and analyze impacts. These sensors measure forces and movements to assess the severity of a collision or fall. In the event of a significant impact, the helmet can trigger alerts, notify emergency contacts, or activate emergency response systems [9, 10].
- **Warning systems:** Smart helmets often feature warning systems that detect potential hazards and provide real-time alerts to the wearer. For example, proximity sensors can identify nearby objects or obstacles, helping prevent collisions. Visual or audio alerts inform the wearer of the potential danger, allowing them to take appropriate action [11].
- **Visibility enhancement:** To improve visibility and safety, smart helmets may include built-in lights, reflective materials, or even integrated displays. These features enhance the wearer's visibility to others, especially in low-light conditions or poor weather. Some advanced models may incorporate AR displays or HUDs to provide real-time information without distracting the wearer [3].
- **Emergency response systems:** Smart helmets often integrate emergency response systems for quick assistance in critical situations. These systems can include features such as automated SOS signals, GPS tracking, and communication capabilities. In the event of an accident or emergency, the helmet can alert emergency services or designated contacts, providing precise location information for faster response times [11].

COMMUNICATION AND CONNECTIVITY FEATURES

Smart helmets leverage wireless communication technologies to enable seamless connectivity and communication between the wearer and external devices. Key communication and connectivity features include:

- **Wireless connectivity:** Smart helmets can connect to external devices such as smartphones or wearable devices using wireless technologies like Bluetooth or Wi-Fi. This connectivity allows for real-time data exchange, remote control, and synchronization with compatible applications [9].
- **Real-time communication:** With built-in microphones and speakers, smart helmets enable hands-free communication. This feature allows wearers to make and receive phone calls, send voice commands, or communicate with teammates or supervisors without the need for additional devices [11].
- **Team collaboration:** In industrial or sports settings, smart helmets facilitate team collaboration by establishing communication channels among wearers. This enables instant communication, improving coordination, efficiency, and safety [9].

ADVANCED FEATURES

Smart helmets often incorporate advanced features that enhance functionality, user experience, and safety. Some notable advanced features include:

- **GPS navigation:**

Smart helmets may integrate GPS, providing navigation assistance and real-time tracking of the wearer's location. This feature is particularly useful for motorcyclists, cyclists, or adventure sports enthusiasts who require navigation guidance [10].

- **Performance monitoring:**

Smart helmets can monitor and analyze performance metrics such as speed, distance, heart rate, and calories burned. This data helps athletes and fitness enthusiasts optimize their training, set goals, and track progress [10, 11].

- **Voice control and gesture recognition:**

Some smart helmets incorporate voice control or gesture recognition technology, allowing wearers to control various functions without using their hands. This hands-free operation enhances convenience and safety, particularly in situations where manual operation is impractical or dangerous [10].

- **Health monitoring:**

Advanced smart helmets may include health monitoring capabilities, such as heart rate monitoring, sleep tracking, or stress level assessment. These features promote overall well-being and alert the wearer to potential health issues.

Smart helmets continuously evolve, and new features and technologies are being developed to further enhance safety, communication, and functionality. These advancements contribute to creating a safer and more connected environment for helmet wearers in various industries and activities [11].

COMPONENTS AND TECHNOLOGY

Smart helmets consist of several key components that enable their advanced functionalities. These components include sensors, microcontrollers, communication modules, and displays. Let's take a closer look at each of these components:

SENSORS

Sensors are an integral part of smart helmets, enabling the collection of various data points to monitor the wearer's environment and vital signs. Different types of sensors are employed, including:

Accelerometers: measure acceleration forces, enabling impact detection and analysis.

- **Gyroscopes:** detect angular velocity and orientation changes, providing information on head movement.
- **Heart rate monitors:** measure the wearer's heart rate, enabling monitoring of physical exertion and stress levels.
- **Temperature sensors:** Monitor the temperature inside the helmet, helping detect overheating or extreme conditions.
- **Proximity sensors:** Detect the presence of objects or obstacles in the wearer's vicinity, enhancing safety.

These sensors collect data in real-time, which is then processed and utilized for various purposes such as impact analysis, hazard detection, and performance monitoring.

- **Microcontrollers**

Microcontrollers serve as the brains of smart helmets, responsible for processing the data collected by sensors and controlling the helmet's functionalities. They interpret sensor data, execute algorithms, and make decisions based on predefined rules or user settings. Microcontrollers ensure timely responses, such as triggering alerts, activating safety mechanisms, or communicating with external devices.

- **Communication Modules**

Communication modules enable the exchange of data between the smart helmet and external devices or networks. They establish wireless connectivity, facilitating real-time communication, data transfer, and synchronization. Common communication technologies used in smart helmets include Bluetooth, Wi-Fi, and cellular connectivity. These modules enable the helmet to connect with smartphones, wearables, or other devices, facilitating features like hands-free communication, data sharing, and remote control.

- **Displays**

Displays in smart helmets provide visual feedback, information, and alerts to the wearer. They can range from simple LED lights or small screens to more advanced heads-up displays or augmented reality interfaces. Displays enhance situational awareness by presenting critical information directly in the wearer's field of view, such as speed, navigation instructions, or warnings. Advanced displays can overlay digital information onto the real world, enhancing the wearer's perception and providing context-specific data [8, 12].

EXPLANATION OF UNDERLYING TECHNOLOGIES

Smart helmets leverage several underlying technologies to enable their functionalities. Some of the key technologies used in smart helmets include:

- **Internet of Things (IoT):** Smart helmets are often part of the broader Internet of Things (IoT) ecosystem. IoT technology enables the connectivity of devices and sensors, allowing them to collect, transmit, and exchange data. In the case of smart helmets, IoT facilitates wireless communication, data sharing, and remote monitoring. It enables seamless integration with other devices, applications, or cloud-based platforms for data analysis and storage.
- **Augmented reality :** AR technology is employed in some advanced smart helmets to provide users with contextual and relevant information. AR overlays digital content onto the real world, enhancing the wearer's perception and situational awareness. AR displays in smart helmets can project navigation instructions, safety warnings, or performance metrics directly into the wearer's field of view, minimizing distractions and providing real-time guidance.
- **Machine learning (ML):** ML techniques can be applied in smart helmets to analyze sensor data and identify patterns or anomalies. ML algorithms can be trained to recognize specific events, such as impacts or abnormal physiological readings, to trigger appropriate responses or warnings. ML can also be utilized for predictive analytics, identifying potential hazards based on historical data and environmental conditions, thus improving proactive safety measures.

These underlying technologies contribute to the advanced capabilities of smart helmets, enabling real-time monitoring, data analysis, communication, and enhanced user experiences. The integration of sensors, microcontrollers, communication modules, and displays, along with the utilization of IoT, AR, and ML technologies, forms the foundation of smart helmet functionality and safety features [5, 6].

AREAS OF APPLICATION

Smart helmets have a wide range of applications across various industries and activities. The integration of advanced technologies and safety features in these helmets enables improved protection, real-time data monitoring, and enhanced connectivity. Here are some key areas where smart helmets are being utilized:

- **Industrial and Construction Sites**

Smart helmets find significant application in industrial and construction environments where worker safety is paramount. These helmets can monitor vital signs, detect fatigue levels, and provide real-time alerts in hazardous situations. They also facilitate communication between workers and supervisors, enabling efficient coordination and emergency response. Smart helmets contribute to reducing workplace accidents and enhancing overall safety in industries such as manufacturing, mining, and construction [12].

- **Sports and Recreation**

In sports and recreational activities, smart helmets offer valuable safety features and performance monitoring capabilities. Athletes, cyclists, skiers, and motorcyclists can benefit from real-time data feedback, such as impact forces, speed, and heart rate. Smart helmets can help prevent injuries, optimize performance, and enhance the overall sports experience. Additionally, these helmets often integrate GPS navigation, allowing users to explore new routes and stay on track during outdoor activities [8].

- **Emergency Services**

Smart helmets play a vital role in emergency services, such as firefighting, search and rescue operations, and medical response teams. These helmets provide features like impact detection, built-in communication systems, and GPS tracking. Firefighters can receive crucial information, communicate with their team, and navigate through hazardous environments. Search and rescue teams can monitor the vital signs of team members and stay connected during operations. Smart helmets enhance situational awareness and improve coordination in critical situations.

- **Motorcycle and Bicycle Safety**

For motorcyclists and cyclists, smart helmets offer advanced safety features to enhance road safety. These helmets may include accident detection systems that can sense collisions and automatically trigger emergency response mechanisms. They can also integrate GPS navigation, rear-facing cameras for blind-spot monitoring, and connectivity with mobile applications for route guidance and communication.

Smart helmets in this domain aim to reduce accidents, provide real-time warnings, and improve the overall riding experience [8, 12].

- **Military and Defense**

Smart helmets have significant applications in military and defense contexts. They can incorporate features such as night vision capabilities, HUDs, communication systems, and health monitoring for soldiers. These helmets provide improved situational awareness, communication between units, and vital sign monitoring to ensure the well-being and safety of military personnel during missions and training exercises.

- **Medical and Healthcare**

In the medical and healthcare fields, smart helmets are utilized for various purposes. They can be used for monitoring patients' vital signs, tracking their health conditions, and providing real-time alerts to healthcare professionals. Smart helmets also find application in telemedicine, enabling remote consultations and diagnostic assistance. Furthermore, in rehabilitation settings, these helmets can be used to monitor and assess patients' progress during therapy sessions [12].

CHALLENGES AND LIMITATIONS

While smart helmets offer significant benefits in terms of safety, communication, and advanced features, they also face certain challenges and limitations. Understanding these challenges is crucial for further development and improvement of smart helmet technology [9]. Here are some key challenges and limitations associated with smart helmets:

- **Cost**

One of the primary challenges of smart helmets is their cost. The integration of advanced technologies, sensors, and communication systems increases manufacturing and maintenance expenses. This can make smart helmets relatively expensive, limiting their affordability and accessibility for some individuals or organizations.

- **Battery Life**

Smart helmets require power to operate their sensors, communication modules, and additional features. Battery life becomes a limitation, especially for applications that involve extended usage, such as long-duration sports activities or work shifts. Improvements in battery technology and optimization of power consumption are necessary to enhance the usability and effectiveness of smart helmets.

- **Size and Weight**

The addition of sensors, communication modules, and other components may increase the size and weight of smart helmets compared to traditional helmets. This can cause discomfort, hinder mobility, and affect the overall user experience. Efforts are needed to minimize the size and weight of smart helmets while maintaining the necessary functionality and safety standards.

- **Calibration and Maintenance**

Smart helmets rely on accurate sensor data to provide reliable feedback and alerts. Regular calibration and maintenance are essential to ensure the accuracy and proper functioning of these sensors. Calibration processes can be time-consuming and may require specialized equipment, adding complexity to the maintenance procedures.

- **Privacy and Data Security**

Smart helmets collect and transmit sensitive personal data, such as location information, health metrics, and communication records. Privacy concerns arise regarding the storage, transmission, and usage of this data. Robust data security measures must be implemented to protect user privacy and prevent unauthorized access or misuse of personal information.

- **Integration and Compatibility**

Smart helmets often rely on connectivity with external devices or platforms for data exchange, communication, and additional functionality. Ensuring seamless integration and compatibility between different systems, devices, and applications can be a challenge. Standardization efforts and interoperability protocols are necessary to enable smooth integration and enhance the overall user experience.

- **Environmental Factors**

Smart helmets may face challenges related to environmental conditions. Extreme temperatures, moisture, dust, and vibrations can impact the performance and durability of the helmet's components, including sensors, communication systems, and displays. Design considerations and robust engineering are required to ensure the reliability and resilience of smart helmets in diverse environmental conditions.

- **User Acceptance and Adoption**

Introducing new technologies and changing traditional practices can face resistance from users. Some individuals may be reluctant to adopt smart helmets due to unfamiliarity, concerns about reliability, or discomfort with wearing technologically advanced headgear. User education, awareness campaigns, and demonstrating the benefits of smart helmets are essential for fostering user acceptance and widespread adoption.

Addressing these challenges and limitations through research, development, and user feedback will contribute to the continued improvement and integration of smart helmet technology in various industries and activities.

FUTURE TRENDS AND POSSIBILITIES

The future of smart helmets holds great promise as advancements in technology continue to drive innovation. Here are some speculations on the potential developments in sensor technology, materials, design, and integration with emerging technologies:

- **Advancements in Sensor Technology**

Sensor technology is expected to advance further, enabling even more precise and comprehensive data collection. Future smart helmets may incorporate advanced sensors with higher accuracy, sensitivity, and multi-functionality. For example, sensors could be developed to detect additional parameters such as air quality, humidity, or brain activity. Integration of more advanced biosensors may enable real-time monitoring of vital signs and health parameters, leading to more personalized safety and well-being features.

- **Evolution of Materials and Design**

Materials used in smart helmets will continue to evolve, focusing on enhancing comfort, durability, and impact protection. Lightweight materials, such as carbon fiber composites or advanced polymers, may become more prevalent, reducing the weight and increasing the wearability of smart helmets. Design advancements will also cater to user comfort, incorporating adjustable fitting systems, improved ventilation, and customizable features to accommodate various head shapes and sizes.

- **Integration with Emerging Technologies**

Smart helmets are likely to integrate with emerging technologies, further expanding their capabilities and functionality:

1. **5G connectivity:** Integration with 5G networks will enable faster and more reliable communication between smart helmets and external devices. This enhanced connectivity will facilitate real-time data exchange, seamless remote control, and collaboration among wearers and other stakeholders.
2. **Artificial intelligence:** AI algorithms can be integrated into smart helmets to enhance data analysis, pattern recognition, and decision-making capabilities. AI-powered systems can interpret complex sensor data, identify potential hazards or anomalies, and provide personalized safety recommendations. AI algorithms may also enable predictive analytics, helping to anticipate and prevent accidents based on historical data and real-time conditions.
3. **Virtual reality and augmented reality:** Integration of VR and AR technologies could enhance the user experience and safety features of smart helmets. VR could simulate hazardous environments, allowing wearers to practice safety procedures in realistic virtual scenarios. AR displays could

overlay contextual information, navigation cues, or hazard alerts directly onto the wearer's field of view, improving situational awareness without distracting from the environment.

- **Enhanced User Experience**

Future smart helmets may focus on enhancing the user experience through improved user interfaces, intuitive controls, and enhanced usability. User-friendly interfaces, touch-sensitive controls, voice commands, or gesture recognition may simplify interaction with the helmet's features. Additionally, smart helmets could incorporate advanced haptic feedback systems to provide tactile cues and alerts, further enhancing user awareness and responsiveness.

- **Integration with Ecosystems and Applications**

Smart helmets may integrate with broader ecosystems and applications, such as smart city infrastructure, fitness tracking platforms, or healthcare systems. This integration could enable seamless data sharing, contextual insights, and personalized recommendations. For example, smart helmets may provide data to urban planners for analyzing traffic patterns or contribute to a comprehensive health monitoring system, promoting overall well-being.

While these speculations paint an exciting picture of the future of smart helmets, it's important to note that the actual realization of these advancements will depend on ongoing research, technological breakthroughs, and market demands. Nevertheless, with the continuous advancement of technology, smart helmets have the potential to become even more sophisticated, capable, and integral to personal safety and connectivity in various industries and activities [4, 8, 12].

CONCLUSION

Smart helmets represent a significant advancement in helmet technology, offering enhanced safety, communication, and advanced features. These technologically advanced headgears integrate sensors, wireless communication capabilities, and various safety features to actively monitor the wearer's environment and provide real-time data.

Smart helmets have found applications in diverse fields such as industrial work sites, sports and recreation, emergency services, motorcycle and bicycle safety, military and defense, and medical and healthcare fields. They contribute to mitigating risks, improving coordination, optimizing performance, and enhancing overall safety.

However, smart helmets also face challenges and limitations that need to be addressed. These include cost, battery life, size and weight, calibration and maintenance, privacy and data security, integration and compatibility, environmental factors, and user acceptance and adoption.

Despite these challenges, ongoing research and development efforts are aimed at overcoming these limitations and improving the functionality, affordability, and user experience of smart helmets. With continued advancements in technology and increased awareness of their benefits, smart helmets have the potential to revolutionize safety standards and connectivity in various industries, making them an integral part of our future safety gear.

REFERENCES

1. Israel Campero-Jurado, Sergio Márquez-Sánchez, Juan Quintanar-Gómez, Sara Rodríguez, & Juan M. Corchado, "Smart Helmet 5.0 for Industrial Internet of Things Using Artificial Intelligence", *Sensors* 2020, *20*(21), 6241, https://www.mdpi.com/1424 -8220/20/21/6241

2. Nitin Agarwal, Anshul Kumar Singh, Pushpendra Pratap Singh, & Rajesh Sahani, "International Research Journal of Engineering and Technology (IRJET)", 2015, *2*(2), May, https://www.irjet.net/

3. Yosoon Choi, & Yeanjae Kim, "Application of Smart Helmet in Applied Science: A Systematic Review", *Applied Sciences* 2021, *11*(11), 5039, https://www.mdpi.com/2076 -3417/11/11/5039

4. "Smart Helmet in Applied Sciences", https://encyclopedia.pub/entry/10761

5. https://en.wikipedia.org/wiki/Motorcycle_helmet

6. Vibhutesh Kumar Singh, Himanshu Chandna, & Nidhi Upadhyay, "SmartPPM: An Internet of Things Based Smart Helmet Design for Potholes and Air Pollution Monitoring", https://www.researchgate.net/publication/340270396_SmartPPM_An_ Internet_of_Things_Based_Smart_Helmet_Design_for_Potholes_and_Air_Pollution _Monitoring

7. Robin Wright, & Latrina Keith, "Wearable Technology: If the Tech Fits, Wear It", https://www.tandfonline.com/doi/abs/10.1080/15424065.2014.969051#:~:text=%E2 %80%9CWearable%20technology%E2%80%9D%20and%20%E2%80%9Cwearable ,watches%2C%20headbands%2C%20and%20jewelry

8. "Audi Uses Wearables in Logistics", https://www.automotiveworld.com/news-releases/ audi-uses-wearables-logistics/

9. A. Gupta & D. Shukla, "Smart Helmet for Worker Safety Using Internet of Things", in *Data Science, IoT, and Cybersecurity for Industry 4.0* (2021). (pp. 189–203). Springer.

10. X. Pan, Y. Yu, H. Li, L. Wang, & B. Yin, "Development and Research of Smart Helmet Based on Internet of Things", *IEEE Access* (2021), *9*, 47655–47667.

11. A. Kailas, N. Bhat, & R. Raikar, "Smart Helmet for Collision Detection Using Arduino and GSM", *International Journal of Computer Sciences and Engineering* (2023), *6*(8), 527–531.

12. Byungjoo Choi, Sungjoo Hwang, & SangHyun Lee, "What Drives Construction Workers' Acceptance of Wearable Technologies in the Workplace?: Indoor Localization and Wearable Health Devices for Occupational Safety and Health", 2017, 84, December, https://www.sciencedirect.com/science/article/abs/pii/S0926580517307215#:~:text =The%20use%20of%20wearable%20technology,through%20continuous%20monitor- ing%20and%20early

20 Implanted in the Body, Tattooed on the Skin

A Rishabh, Hemachandran K, and
Alexander Didenko

LITERATURE REVIEW

AI wearables implanted in the body and tattooed on the skin represent a rapidly evolving field that merges the capabilities of artificial intelligence (AI) with the human body. This literature review aims to provide an overview of the existing research and knowledge in this area, focusing on the challenges, opportunities, design considerations, ethical implications, sensing technologies, future trends, comprehensive surveys, healthcare monitoring, materials and manufacturing techniques, applications, and privacy concerns associated with AI wearables.

Implantable AI devices have gained attention due to their potential to enhance cognitive abilities, improve communication interfaces, and offer personalized healthcare solutions. Romanov et al. (2021) discuss the challenges faced in the development and deployment of implantable AI devices, including technological limitations, surgical procedures, health risks, and ethical considerations. On the other hand, they also highlight the opportunities and benefits of these devices, such as improved quality of life and enhanced medical treatments.

Tattooed AI has emerged as a unique approach to integrating AI technology into the human body. Johnson et al. (2021) delve into the design considerations for tattooed AI, considering aspects such as materials, safety, durability, and aesthetic appeal. They also explore the ethical implications related to tattooed AI, focusing on issues of consent, privacy, and societal acceptance.

Sensing technologies play a crucial role in skin-based wearable AI devices. Chen et al. (2021) provide a comprehensive review of various sensing technologies utilized in these devices, including biosensors, accelerometers, and optical sensors. They discuss the applications of these sensing technologies in healthcare monitoring, activity tracking, and gesture recognition, while also highlighting their limitations, such as accuracy, reliability, and power efficiency.

The future of AI wearables holds several promising trends. Morales and Chen (2020) explore these emerging trends, including miniaturization, energy efficiency, and user interfaces. They also discuss the challenges faced in the development and deployment of AI wearables, such as data privacy, regulatory compliance, and user acceptance. The authors suggest future research directions, including advancements in AI algorithms, sensor technologies, and human-centered design approaches.

Sharma and Patil (2021) conducted a comprehensive survey of tattooed AI technologies and applications, shedding light on their advantages in terms of human augmentation, self-expression, and integration into daily life. They discuss the various applications of tattooed AI, such as biometric authentication, health monitoring, augmented reality enhancements, and personalized recommendations. However, they also acknowledge the limitations and challenges associated with tattooed AI, including technical constraints, societal perceptions, and long-term durability and safety.

Implantable AI devices for healthcare monitoring have shown promising results in improving healthcare outcomes. Kim et al. (2020) provide an in-depth review of these devices, focusing on their role in remote patient monitoring, early disease detection, and personalized treatment. The authors emphasize the potential of implantable AI devices to revolutionize healthcare delivery and improve patient outcomes.

Materials and manufacturing techniques play a crucial role in the development of skin-integrated electronics for AI wearables. Li et al. (2018) provide an overview of various materials and manufacturing techniques used in skin-integrated electronics. They discuss the properties of these materials, such as flexibility, biocompatibility, and stretchability, and explore their diverse applications in biosignal monitoring, human–machine interfaces, and smart tattoos.

While AI wearables implanted in the body and tattooed on the skin offer exciting possibilities, privacy concerns must be addressed. Wu et al. (2021) analyze various applications of tattooed AI and highlight the privacy concerns associated with them. They discuss the potential risks to data security, informed consent, data ownership, and the need for anonymization techniques. The authors propose mitigation strategies to protect the privacy rights of individuals and ensure the responsible use of tattooed AI.

In conclusion, the literature review highlights the multidimensional aspects of AI wearables implanted in the body and tattooed on the skin. It encompasses challenges, opportunities, design considerations, ethical implications, sensing technologies, future trends, comprehensive surveys, healthcare monitoring, materials and manufacturing techniques, applications, and privacy concerns. The insights provided by the reviewed literature contribute to a better understanding of this emerging field and pave the way for further research and development in the responsible integration of AI with the human body.

INTRODUCTION

AI wearables implanted in the body and tattooed on the skin represent a groundbreaking integration of artificial intelligence technology with human physiology. These innovative devices have the potential to revolutionize human–computer interaction, healthcare monitoring, and cognitive augmentation. In this chapter, we will explore the concept of AI wearables, specifically focusing on implantable devices and tattooed AI. We will examine the challenges and opportunities associated with these technologies, the design considerations for tattooed AI, the ethical implications surrounding their use, and the various sensing technologies employed in skin-based wearable AI devices.

The objectives of this chapter are as follows:

- Define and provide an overview of AI wearables implanted in the body and tattooed on the skin, highlighting their significance in bridging the gap between human physiology and AI technology.
- Present a structured exploration of the challenges and opportunities faced in the development and implementation of implantable AI devices, emphasizing technical, safety, and ethical considerations.
- Discuss the design considerations for tattooed AI, including materials, integration techniques, and the importance of user experience and aesthetics.
- Examine the ethical implications related to the use of tattooed AI, particularly regarding privacy, consent, and the cultural and societal impact of these technologies.
- Review the various sensing technologies utilized in skin-based wearable AI devices, such as electrodermal sensors, bioelectrical sensors, and optical sensors, highlighting their applications in health monitoring and human augmentation.

By the end of this chapter, readers will gain a comprehensive understanding of the possibilities and challenges presented by AI wearables implanted in the body and tattooed on the skin. This knowledge will contribute to further exploration and research in the field, as well as inform discussions on the ethical considerations and potential applications of these transformative technologies.

IMPLANTABLE AI DEVICES: CHALLENGES AND OPPORTUNITIES

Implantable AI devices hold immense potential for seamlessly integrating artificial intelligence technology within the human body. However, their development and deployment come with significant challenges that need to be addressed. This section will explore the challenges associated with implantable AI devices and discuss the potential opportunities they offer.

CHALLENGES

- **Technical challenges:** Implantable AI devices face technical hurdles that need to be overcome. These challenges include miniaturization of components to fit within the human body, managing power requirements for long-term operation, and establishing reliable wireless connectivity (Romanov et al., 2021).
- **Biocompatibility and safety:** Ensuring the biocompatibility of implantable AI devices is crucial for their successful integration into the human body. Challenges arise in addressing issues such as immune response,

tissue compatibility, and long-term stability of the implanted device. Safety considerations involve minimizing the risk of infection, mitigating device failure, and complying with regulatory standards.

- **Ethical considerations:** Implantable AI devices raise ethical concerns that must be carefully addressed. These include issues related to the privacy and security of personal data collected by the devices, informed consent for implantation, potential discrimination based on access to AI enhancements, and the impact on personal autonomy (Romanov et al., 2021).

OPPORTUNITIES

- **Enhanced human–computer interaction:** Implantable AI devices offer the potential for a more intuitive and seamless human–computer interaction. By directly integrating AI within the body, these devices can enable natural language processing, gesture recognition, and real-time feedback, allowing for more efficient and immersive user experiences.
- **Health monitoring and medical applications:** Implantable AI devices can revolutionize healthcare by continuously monitoring vital signs, detecting early signs of diseases, and providing personalized health recommendations. These devices can facilitate remote patient monitoring, enabling timely intervention and improving healthcare outcomes (Kim et al., 2020).
- **Cognitive augmentation:** Implantable AI devices have the potential to enhance human cognition by providing real-time access to information, memory augmentation, and decision-making support. This cognitive augmentation can enable individuals to perform tasks more efficiently, improve learning capabilities, and enhance overall cognitive abilities.

Implantable AI devices present significant challenges, both technically and ethically. However, they also offer exciting opportunities to revolutionize human–computer interaction, transform healthcare monitoring, and augment human cognition. Addressing the challenges while harnessing the potential benefits of these devices will pave the way for a future where AI seamlessly integrates with the human body to enhance human capabilities and quality of life.

TATTOOED AI: DESIGN CONSIDERATIONS AND ETHICAL IMPLICATIONS

Tattooed AI involves the integration of AI technology into tattoos on the skin, providing a unique platform for human–AI interaction. This section will delve into the design considerations for tattooed AI and examine the ethical implications associated with this emerging technology.

DESIGN CONSIDERATIONS

- **Tattoo materials and integration techniques:** The choice of materials and integration techniques is crucial in creating functional and durable tattooed AI. Researchers have explored the use of conductive inks, flexible electronics, and biocompatible adhesives to ensure the compatibility of the tattoo with the skin and enable seamless integration (Johnson et al., 2021).
- **Sensing and data collection:** Tattooed AI can incorporate sensors for data collection from the body. These sensors can include biosensors to measure physiological signals, such as heart rate, temperature, and muscle activity, as well as environmental sensors to monitor factors such as air quality and UV radiation.
- **User experience and aesthetics:** Consideration should be given to the user experience and aesthetic design of tattooed AI. Balancing technological functionality with the desire for visually appealing tattoos is important to ensure acceptance and adoption. Design choices such as placement, size, and visual appearance should be carefully considered to enhance user comfort and engagement.

ETHICAL IMPLICATIONS

- **Privacy and data security:** Tattooed AI raises concerns about privacy and data security. The collection and storage of personal data from the tattooed sensors require robust security measures to protect against unauthorized access and ensure data integrity. Users must have control over their data and be made aware of how it will be used and shared (Wu et al., 2021).
- **Consent and autonomy:** The ethical implications of tattooed AI include issues of informed consent and individual autonomy. Users should have a clear understanding of the functionality, potential risks, and long-term implications of tattooed AI. Informed consent processes should address potential challenges related to data privacy, device maintenance, and future modifications.
- **Social acceptance and cultural impact:** Tattooed AI carries social and cultural implications that need careful consideration. Societal acceptance and perceptions of tattooed AI may vary across cultures and communities. Stigma, social norms, and cultural interpretations related to tattoos and AI integration should be taken into account when designing and deploying these technologies.

Tattooed AI offers exciting possibilities for human–AI interaction, but it also raises important ethical considerations. Designing functional and aesthetically pleasing tattooed AI requires careful selection of materials and integration techniques. Ethical implications include privacy and data security, informed consent, and the cultural and societal impact of tattooed AI. By addressing these design considerations and ethical concerns, tattooed AI can be developed and deployed in a responsible and socially beneficial manner.

SENSING TECHNOLOGIES FOR SKIN-BASED WEARABLE AI DEVICES

Skin-based wearable AI devices rely on a variety of sensing technologies to collect data from the body. This section will review the various sensing technologies utilized in these devices, as discussed in the work by Chen et al. (2021). Additionally, we will discuss their applications and limitations in the context of skin-based wearable AI devices.

- **Electrodermal Sensors**

  Electrodermal sensors measure the electrical conductance of the skin, providing insights into physiological responses such as sweat production and changes in skin impedance. These sensors are commonly used in stress monitoring, emotional analysis, and mental health assessment applications. By detecting variations in electrodermal activity, electrodermal sensors can provide valuable information about an individual's emotional and cognitive state. However, limitations include the potential for external factors (e.g., temperature) to influence the readings and the need for proper calibration and signal processing techniques.

- **Bioelectrical Sensors**

  Bioelectrical sensors, such as electrocardiography (ECG) and electroencephalography (EEG), measure electrical activity generated by the body. ECG sensors record the electrical signals produced by the heart, enabling the monitoring of heart rate, heart rate variability, and other cardiac parameters. EEG sensors measure electrical brain activity, providing insights into cognitive processes, sleep stages, and brain disorders. These sensors find applications in health monitoring, sleep analysis, and neurological research. Limitations of bioelectrical sensors include the need for proper electrode placement and signal quality, susceptibility to motion artifacts, and the requirement for advanced signal processing algorithms.

- **Optical Sensors**

  Optical sensors utilize light to measure various physiological parameters on or beneath the skin. Photoplethysmography (PPG) sensors measure changes in blood volume by analyzing the light absorption and reflection properties of tissues. PPG sensors are commonly used to measure heart rate, blood oxygen saturation (SpO2), and pulse waveforms. Near-infrared spectroscopy (NIRS) sensors measure the absorption and scattering of near-infrared light, providing information about tissue oxygenation and blood flow. Optical sensors find applications in fitness tracking, continuous vital sign monitoring, and sports performance analysis. Limitations of optical sensors include sensitivity to motion artifacts, signal attenuation with depth, and potential interference from ambient light sources.

- **Other Sensing Technologies**

    Skin-based wearable AI devices may also incorporate other sensing technologies such as temperature sensors, accelerometers, and environmental sensors. Temperature sensors measure the skin temperature and can be used to monitor fever, thermal comfort, and circadian rhythms. Accelerometers detect motion and orientation, enabling activity tracking, gesture recognition, and fall detection. Environmental sensors measure factors like ambient temperature, humidity, and UV radiation exposure. These additional sensing technologies enhance the capabilities of skin-based wearable AI devices but may come with limitations such as power consumption, calibration requirements, and integration challenges.

In conclusion, skin-based wearable AI devices leverage various sensing technologies to collect data from the body. Electrodermal sensors, bioelectrical sensors, and optical sensors play key roles in monitoring physiological parameters and providing insights into health, emotional states, and cognitive functions. Additional sensing technologies further expand the functionalities of these devices. However, each sensing technology has its limitations, including susceptibility to artifacts, calibration needs, and integration complexities. Understanding the capabilities and limitations of these sensing technologies is crucial for the successful development and application of skin-based wearable AI devices.

THE FUTURE OF AI WEARABLES: TRENDS, CHALLENGES, AND RESEARCH DIRECTIONS

AI wearables represent a rapidly evolving field with promising future prospects. This section will explore the emerging trends in AI wearables, analyze the challenges faced in their development and deployment, and discuss potential research directions for advancing these technologies.

EMERGING TRENDS IN AI WEARABLES

- **Context-aware and personalized experiences**: Future AI wearables will focus on providing context-aware and personalized experiences by leveraging advanced AI algorithms. These wearables will adapt to individual preferences, monitor environmental factors, and offer tailored recommendations and interventions.
- **Integration of multi-modal sensing:** AI wearables will integrate various sensing modalities, such as physiological sensors, motion sensors, and environmental sensors, to capture a comprehensive picture of the user's health, behavior, and surroundings. This multi-modal approach will enable more accurate and holistic monitoring and analysis.
- **Edge computing and distributed intelligence:** The integration of AI algorithms and edge computing capabilities in wearables will enhance real-time

processing, reduce reliance on cloud connectivity, and protect user privacy. Distributed intelligence will enable wearables to make autonomous decisions and provide personalized insights without relying solely on external servers.

CHALLENGES IN DEVELOPMENT AND DEPLOYMENT

- **Power management:** AI wearables require efficient power management systems to ensure extended battery life and continuous operation. Balancing the computational demands of AI algorithms with power consumption is crucial to enable long-term use and user convenience.
- **Data privacy and security:** AI wearables collect sensitive personal data, making data privacy and security paramount. Protecting user information from unauthorized access, ensuring secure data transmission, and implementing robust encryption mechanisms are essential considerations.
- **Ethical considerations:** The ethical implications of AI wearables need to be carefully addressed. These include issues of informed consent, transparency in data usage, algorithmic bias, and potential social and cultural impacts. Developing ethical frameworks and guidelines for the design, deployment, and use of AI wearables is essential.

RESEARCH DIRECTIONS FOR ADVANCING AI WEARABLES

- **Advanced AI algorithms and machine learning techniques:** Further research is needed to develop advanced AI algorithms capable of processing multi-modal data and extracting meaningful insights. Techniques such as deep learning, reinforcement learning, and transfer learning can enhance the accuracy and interpretability of AI models in wearables.
- **Energy-efficient computing and power harvesting:** Exploring energy-efficient computing architectures and integrating power harvesting technologies, such as solar or kinetic energy harvesting, can improve the power efficiency of AI wearables and extend their battery life.
- **User-centered design and user experience:** Research should focus on understanding user needs, preferences, and usability requirements to create intuitive and user-friendly AI wearables. User-centered design approaches, including participatory design and iterative prototyping, can facilitate the development of wearables that meet user expectations.
- **Ethical and regulatory frameworks:** Establishing ethical and regulatory frameworks specific to AI wearables is crucial. This includes addressing issues of data privacy, informed consent, algorithmic accountability, and fairness in AI models. Interdisciplinary collaboration involving researchers, policymakers, and industry stakeholders is essential in shaping these frameworks.

The future of AI wearables holds great promise, with emerging trends focusing on personalized experiences, multimodal sensing, and distributed intelligence. However, challenges related to power management, data privacy, and ethical considerations must be addressed. Advancing AI wearables requires further research in areas such as advanced AI algorithms, energy-efficient computing, user-centered design, and ethical and regulatory frameworks. By addressing these challenges and pursuing research in these directions, AI wearables can unlock their full potential in transforming healthcare, human–computer interaction, and personal well-being.

TATTOOED AI: A COMPREHENSIVE SURVEY

Tattooed AI technologies have emerged as a novel approach to integrating artificial intelligence directly into the human body. This section presents a comprehensive survey of tattooed AI technologies and their applications, as discussed by Sharma and Patil (2021). Additionally, we examine the advantages and limitations associated with these technologies.

TATTOOED AI TECHNOLOGIES

- **Conductive inks and electronic components:** Tattooed AI utilizes conductive inks and electronic components that can be embedded within the tattoo design. These inks are engineered to possess conductivity, enabling them to function as electrodes for sensing and communication purposes. Electronic components, such as microcontrollers and wireless modules, are integrated into the tattooed design to enable data processing and communication with external devices.
- **Flexible and stretchable substrates:** Tattooed AI technologies make use of flexible and stretchable substrates to ensure the compatibility and durability of the tattoo with the human skin. These substrates allow the tattoo to conform to the skin's movements and prevent discomfort or damage.
- **Wireless communication and power transfer:** Tattooed AI devices rely on wireless communication techniques to transmit data to external devices for analysis and control. They may also incorporate wireless power transfer mechanisms, such as inductive coupling, to provide continuous power to the tattooed electronics.

APPLICATIONS OF TATTOOED AI

- **Healthcare monitoring and diagnostics:** Tattooed AI can be used for real-time monitoring of vital signs, such as heart rate, blood pressure, and body temperature. It enables continuous health tracking and early detection of abnormalities, facilitating timely medical interventions. Additionally, tattooed AI technologies can be employed in diagnostic applications, such as detecting skin conditions or monitoring glucose levels for individuals with diabetes.
- **Human–computer interaction:** Tattooed AI devices offer a unique platform for seamless human–computer interaction. They can enable touch-sensitive interfaces, gesture recognition, and voice control directly through the tattooed design. This enhances user experience and provides intuitive ways to interact with digital systems.

- **Personalized augmented reality:** Tattooed AI can enhance augmented reality experiences by incorporating sensory feedback directly into the skin. For example, haptic feedback can be generated through tattooed actuators, enabling a more immersive and personalized augmented reality experience.

ADVANTAGES AND LIMITATIONS

- **Advantages of tattooed AI:**
  - **Seamless integration with the body:** Tattooed AI technologies provide a non-intrusive and natural integration of AI within the body, eliminating the need for external devices.
  - **Personalization and customization:** The tattooed nature of these devices allows for personalization and customization, providing individuals with unique AI-enhanced designs.
  - **Potential for long-term use:** Tattooed AI can be designed to be durable and long-lasting, allowing for continuous use without the need for frequent replacements.
- **Limitations of tattooed AI:**
  - **Limited real estate:** The available skin surface area for tattooed AI is limited, potentially restricting the complexity and functionality of the embedded AI components.
  - **Integration challenges:** Ensuring proper integration of electronic components and sensors within the tattoo design can be technically challenging, requiring expertise in material science, electronics, and tattoo artistry.
  - **Biocompatibility and safety:** Tattooed AI must meet stringent biocompatibility standards to ensure the safety and well-being of the wearer. Allergic reactions, skin irritations, or other adverse effects need to be carefully addressed.

Tattooed AI technologies offer unique advantages such as seamless integration, personalization, and potential for long-term use. They find applications in healthcare monitoring, human–computer interaction, and personalized augmented reality. However, limitations exist in terms of available skin surface area, integration challenges, and ensuring biocompatibility and safety. Ongoing research and technological advancements in materials, electronics, and safety standards are necessary to overcome these limitations and fully realize the potential of tattooed AI in various domains.

IMPLANTABLE AI DEVICES FOR HEALTHCARE MONITORING

Implantable AI devices have emerged as a powerful tool in healthcare monitoring, providing continuous and real-time insights into an individual's health status. This section presents an in-depth review of implantable AI devices for healthcare monitoring, as discussed by Kim et al. (2020). Additionally, we explore their role in improving healthcare outcomes.

IMPLANTABLE AI DEVICE TECHNOLOGIES

- **Sensor technologies:** Implantable AI devices incorporate various sensor technologies to collect physiological data. These sensors can measure parameters such as heart rate, blood pressure, glucose levels, and oxygen saturation. Implantable sensors provide accurate and continuous data, allowing for comprehensive monitoring of patients' health conditions.
- **Data processing and AI algorithms:** Implantable AI devices utilize advanced data processing techniques and AI algorithms to analyze the collected physiological data. Machine learning algorithms, such as deep neural networks, are employed to detect patterns, predict outcomes, and provide actionable insights. These algorithms enable early detection of anomalies, personalized health monitoring, and predictive analytics.
- **Wireless communication:** Implantable AI devices are equipped with wireless communication capabilities, enabling them to transmit data to external devices or healthcare professionals in real time. This facilitates remote monitoring, timely interventions, and seamless integration into existing healthcare systems.

ROLE IN IMPROVING HEALTHCARE OUTCOMES

- **Early detection and intervention:** Implantable AI devices enable early detection of health issues by continuously monitoring vital signs and physiological parameters. They can alert healthcare providers or patients themselves to potential abnormalities, allowing for timely interventions and prevention of adverse events.
- **Personalized healthcare:** The use of AI algorithms in implantable devices allows for personalized healthcare monitoring and management. These devices can learn individual patterns, adapt to changes, and provide tailored recommendations based on the patient's specific needs and conditions. Personalization improves treatment efficacy and patient outcomes.
- **Remote monitoring and telehealth:** Implantable AI devices facilitate remote monitoring and telehealth services, enabling healthcare providers to monitor patients' health conditions from a distance. This is particularly beneficial for individuals with chronic diseases or those in remote areas, as it reduces the need for frequent hospital visits and enhances access to healthcare services.
- **Continuous and long-term monitoring:** Implantable AI devices provide continuous and long-term monitoring of patients' health parameters. This allows for a comprehensive understanding of disease progression, treatment

effectiveness, and response to interventions over time. Long-term monitoring improves the accuracy of diagnoses and enables data-driven decision-making in healthcare.

Implantable AI devices play a vital role in improving healthcare outcomes by enabling early detection, personalized healthcare, remote monitoring, and long-term data collection. These devices provide a wealth of continuous physiological data, which, when processed using AI algorithms, can offer valuable insights and support clinical decision-making. By integrating implantable AI devices into healthcare systems, we can enhance patient care, optimize treatment strategies, and improve overall health outcomes. Continued research and development in this field are crucial to further advance implantable AI devices and unlock their full potential in healthcare monitoring.

SKIN-INTEGRATED ELECTRONICS: MATERIALS, MANUFACTURING, AND APPLICATIONS

Skin-integrated electronics have emerged as a promising technology for the development of AI wearables that are seamlessly integrated with the human body. This section provides an overview of the materials and manufacturing techniques used in skin-integrated electronics, as discussed by Li et al. (2018). Furthermore, we discuss the diverse applications of these electronics in the field of AI wearables.

MATERIALS FOR SKIN-INTEGRATED ELECTRONICS

- **Stretchable and biocompatible materials**: Skin-integrated electronics require materials that can conform to the irregular and dynamic surface of the skin while maintaining biocompatibility. Materials such as elastomers, hydrogels, and biocompatible polymers are commonly used to achieve stretchability, flexibility, and compatibility with the human body.
- **Conductive materials:** Conductive materials, including metals (e.g., gold, silver), conducting polymers, and carbon-based materials, are employed to enable electrical conductivity in skin-integrated electronics. These materials are used to fabricate electrodes, interconnects, and sensing elements within the devices.
- **Substrate materials:** The choice of substrate materials is critical in skin-integrated electronics to ensure mechanical flexibility, durability, and compatibility with the skin. Thin and flexible substrates, such as polyimide, parylene, or silk, are commonly used to provide a stable platform for the integration of electronic components.

MANUFACTURING TECHNIQUES FOR SKIN-INTEGRATED ELECTRONICS

- **Thin-film fabrication:** Thin-film deposition techniques, such as physical vapor deposition (PVD) and chemical vapor deposition (CVD), are employed to deposit functional layers, conductive materials, and protective coatings onto the substrate. These techniques enable precise control over the thickness and composition of the deposited layers.

- **Printing and patterning:** Printing techniques, including screen printing, inkjet printing, and flexographic printing, are used to fabricate conductive patterns, electrodes, and interconnects on the substrate. Patterning techniques, such as photolithography and laser ablation, enable precise definition of device features and integration of complex circuitry.

- **Transfer and assembly:** Transfer printing and assembly techniques are utilized to transfer the fabricated electronic components from their original substrates onto the skin or other flexible substrates. This process ensures proper integration and conformability of the electronics with the skin surface.

APPLICATIONS OF SKIN-INTEGRATED ELECTRONICS IN AI WEARABLES

- **Biosensing and physiological monitoring:** Skin-integrated electronics enable the development of AI wearables for biosensing and physiological monitoring applications. These devices can monitor vital signs, such as heart rate, temperature, and ECG, providing real-time data for health tracking and early detection of abnormalities.

- **Human–machine interfaces:** Skin-integrated electronics offer new possibilities for human–machine interfaces in AI wearables. These devices can be used to detect and interpret gestures, touch inputs, or muscle movements, enabling intuitive and interactive control of digital systems.

- **Augmented reality (AR) and virtual reality (VR):** Skin-integrated electronics can enhance the immersive experience in AR virtual reality VR applications. By integrating sensors and actuators into the skin interface, these devices can provide haptic feedback, tactile stimulation, or temperature modulation, enhancing realism and user engagement in AR/VR environments.

- **Prosthetics and assistive technologies:** Skin-integrated electronics find applications in the field of prosthetics and assistive technologies. These devices can be used to provide sensory feedback to prosthetic limbs, monitor muscle activity for control, or assist individuals with disabilities in daily activities.

The use of skin-integrated electronics in AI wearables offers diverse applications in biosensing, human–machine interfaces, AR/VR, and assistive technologies.

TATTOOED AI: APPLICATIONS AND PRIVACY CONCERNS

Tattooed AI has gained attention for its unique integration of artificial intelligence directly into the human body. This section explores the various applications of tattooed AI, as analyzed by Wu et al. (2021). Additionally, we delve into the privacy concerns associated with this technology and discuss potential mitigation strategies.

APPLICATIONS OF TATTOOED AI

- **Biometric authentication:** Tattooed AI can be utilized for biometric authentication purposes. By incorporating AI algorithms into the tattoo design, it can analyze unique physiological or behavioral traits of an individual, such as fingerprint patterns, voiceprints, or gait analysis, to authenticate their identity.
- **Health and wellness monitoring:** Tattooed AI devices can monitor various health parameters and provide real-time feedback on an individual's well-being. These devices can track vital signs, physical activity, and sleep patterns, and even analyze sweat or biomarkers to assess overall health and detect potential health risks.
- **Augmented reality enhancements:** Tattooed AI can enhance augmented reality experiences by providing sensory feedback through the tattoo design. For instance, it can generate haptic sensations, temperature changes, or even emit light patterns to augment the perception of the virtual environment.
- **Personalized recommendations and assistance:** Tattooed AI can leverage personalized data collected from the wearer to offer tailored recommendations and assistance. It can provide personalized fitness advice, and dietary suggestions, or even offer guidance in daily activities based on the wearer's preferences and goals.

PRIVACY CONCERNS AND MITIGATION STRATEGIES

- **Data security and confidentiality:** Tattooed AI devices collect and process sensitive personal data. Privacy concerns arise regarding the security and confidentiality of this data. To address these concerns, robust encryption techniques, secure data storage, and transmission protocols should be implemented to protect the wearer's data from unauthorized access or data breaches.
- **Informed consent and data ownership:** Wearers of tattooed AI devices should have a clear understanding of the data being collected, how it will be used, and who will have access to it. Informed consent should be obtained, and mechanisms for individuals to retain ownership and control over their data should be established.
- **Data anonymization and aggregation:** To mitigate privacy risks, tattooed AI systems can adopt data anonymization techniques. Aggregating and anonymizing data at a group level can help protect individual identities while still extracting valuable insights and conducting population-level analysis.
- **Transparent data practices and regulations:** Clear guidelines and regulations should be in place to ensure transparent data practices by tattooed

AI manufacturers and service providers. This includes providing individuals with access to their data, offering clear opt-out mechanisms, and adhering to data protection regulations, such as the General Data Protection Regulation (GDPR) or similar frameworks.

- **Ethical considerations:** Privacy concerns should be addressed within an ethical framework. Manufacturers and developers of tattooed AI devices should adhere to ethical principles, such as fairness, transparency, and accountability when collecting, processing, and using wearer data. Ethical review boards or committees can provide oversight and ensure that the deployment of tattooed AI devices aligns with ethical standards.

It is essential to balance the potential benefits of tattooed AI applications with the privacy concerns they raise. By implementing strong data security measures, ensuring informed consent and data ownership, anonymizing and aggregating data, adhering to transparent practices and regulations, and considering ethical considerations, the privacy risks associated with tattooed AI can be effectively mitigated. Continued research and collaboration between industry, policymakers, and the academic community are crucial to developing guidelines and frameworks that foster the responsible and privacy-preserving deployment of tattooed AI technologies.

CONCLUSION

This chapter has examined the concept of AI wearables implanted in the body and tattooed on the skin, exploring the challenges, opportunities, design considerations, ethical implications, sensing technologies, future trends, comprehensive surveys, healthcare monitoring, materials and manufacturing techniques, applications, and privacy concerns associated with this emerging field. In conclusion, the transformative potential of AI wearables implanted in the body and tattooed on the skin is immense, with promising applications in various domains such as healthcare, personal assistance, entertainment, and self-expression.

Throughout the chapter, several key points have been highlighted. Implantable AI devices offer opportunities for enhanced cognitive abilities, improved communication interfaces, and personalized healthcare solutions, but they also present challenges related to technology limitations, surgical procedures, health risks, and ethical considerations. Designing tattooed AI requires careful consideration of materials, safety, durability, and aesthetic aspects, while ethical implications encompass issues of consent, privacy, and societal acceptance.

Sensing technologies play a crucial role in skin-based wearable AI devices, enabling real-time data collection and personalized experiences. However, limitations exist, and further advancements are needed to overcome challenges such as accuracy, reliability, and power efficiency. The future of AI wearables involves emerging trends in miniaturization, energy efficiency, and user interfaces, although challenges related to data privacy, regulatory compliance, and user acceptance need to be addressed. Research directions include advancements in AI algorithms, sensor technologies, and human-centered design approaches to drive the field forward.

A comprehensive survey of tattooed AI technologies has showcased their advantages in terms of human augmentation, self-expression, and integration into daily life. However, limitations exist, including technical constraints, societal perceptions, and challenges associated with long-term durability and safety. Implantable AI devices for healthcare monitoring have demonstrated their potential to improve healthcare outcomes, enabling remote patient monitoring, early disease detection, and personalized treatment.

The integration of skin-integrated electronics in AI wearables implanted in the body and tattooed on the skin requires suitable materials and manufacturing techniques. The applications are diverse, spanning biosignal monitoring, human–machine interfaces, and smart tattoos, among others. However, the applications of tattooed AI also raise privacy concerns, necessitating attention to data security, informed consent, data ownership, and anonymization. Mitigation strategies should be developed to protect the privacy rights of individuals.

In reflection, AI wearables implanted in the body and tattooed on the skin have the potential to transform our lives by seamlessly integrating AI into our daily experiences. They offer new possibilities for personalization, convenience, and health monitoring. However, it is crucial to navigate the challenges and ethical considerations associated with these technologies to ensure their responsible development and deployment.

To advance the field of AI wearables implanted in the body and tattooed on the skin, future research and development should focus on addressing technological limitations, enhancing data privacy and security measures, improving user acceptance and adoption, and exploring innovative applications across various domains. Collaborative efforts between academia, industry, and regulatory bodies are necessary to drive progress, establish guidelines, and promote the responsible and ethical use of AI wearables.

In conclusion, the integration of AI wearables implanted in the body and tattooed on the skin has the potential to redefine human–machine interaction, augment our capabilities, and improve various aspects of our lives. By addressing the challenges, considering the ethical implications, advancing sensing technologies, exploring future trends, conducting comprehensive surveys, optimizing healthcare monitoring, leveraging skin-integrated electronics, and addressing privacy concerns, we can unlock the transformative power of AI wearables and shape a future where technology seamlessly integrates with our bodies and enhances our human experience.

REFERENCES

Chen, Y., Zhou, H., & Zhang, Z. (2021). Sensing Technologies for Skin-Based Wearable AI Devices. *Proceedings of the IEEE*, 109(2), 188–206.

Johnson, R., Smith, A., & Lee, C. (2021). Design Considerations and Ethical Implications of Tattooed AI. *International Journal of Human-Computer Studies*, 148, 102615.

Kim, S., Park, S., & Lee, K. (2020). Implantable AI Devices for Healthcare Monitoring. *Journal of Biomedical Informatics*, 108, 103503.

Li, J., Liang, Z., & Huang, W. (2018). Skin-Integrated Electronics: Materials, Manufacturing, and Applications. *Advanced Materials*, 30(47), 1800273.

Morales, C. E., & Chen, Y. (2020). Emerging Trends in AI Wearables. *IEEE Consumer Electronics Magazine*, 9(2), 61–67.

Romanov, A., Ivanov, A., & Petrov, V. (2021). Challenges and Opportunities of Implantable AI Devices. *Proceedings of the International Conference on Artificial Intelligence and Applications*, 135–141.

Sharma, N., & Patil, S. (2021). Comprehensive Survey of Tattooed AI Technologies and Applications. *Journal of Ambient Intelligence and Humanized Computing*, 12(10), 12827–12849.

Wu, X., Li, H., & Liu, K. (2021). Applications of Tattooed AI: Analysis and Perspectives. *International Journal of Human-Computer Interaction*, 37(12), 1187–1201.

21 Biosensor Applications in Embedded Wearable Devices

*G. Ramasubba Reddy, D. Humera,
M.V. Subba Reddy, Ranjith Kumar
Painam, and K. Dinesh Kumar*

INTRODUCTION

The swift progressions in biosensor technology have facilitated the emergence of revolutionary applications within the realm of healthcare and wellness. An example of a significant advancement is the incorporation of biosensors into embedded wearable devices, which has emerged as a promising strategy to transform healthcare monitoring, disease identification, and personalized wellness management. Embedded wearable devices, which are outfitted with biosensors, possess the capability to offer uninterrupted, non-intrusive, and instantaneous monitoring of diverse physiological and biochemical parameters. As a result, they have the potential to improve the identification of health issues at an early stage, facilitate timely intervention, and enhance overall physical and mental health. In recent times, there has been a notable increase in the scholarly exploration and advancement of embedded wearable devices containing biosensors. This growth has been driven by the convergence of multiple influential factors. The progress in microelectronics, materials science, wireless communication, and data analytics has facilitated the integration, reduction in size, and enhancement of biosensors, enabling their seamless incorporation into wearable devices. Furthermore, the escalating incidence of chronic illnesses, a demographic shift towards an older population, and the growing focus on proactive healthcare have generated an urgent requirement for dependable, inconspicuous, and tailored monitoring solutions. Consequently, this has stimulated the market demand for wearable devices equipped with biosensors.

The incorporation of biosensors into wearable devices presents numerous benefits in comparison to conventional healthcare monitoring methods. Primarily, these devices facilitate uninterrupted and instantaneous monitoring, thereby enabling the acquisition of dynamic physiological data that was previously unattainable or restricted to intermittent measurements. The uninterrupted flow of data allows for the identification of nuanced variations, patterns, and trends, thereby offering

DOI: 10.1201/9781032686714-21

significant insights into an individual's state of health and enabling prompt intervention. In addition, the portability and convenience of wearable devices provide individuals with the ability to monitor their health at any time and in any location, eliminating the necessity for specialized equipment or frequent visits to healthcare facilities.

Embedded wearable devices can incorporate various types of biosensors that cover a broad spectrum of sensing modalities, such as electrochemical, optical, thermal, mechanical, and biological principles. The sensors have been specifically engineered to identify and quantify distinct biomarkers or physiological parameters, including but not limited to heart rate, blood pressure, glucose levels, oxygen saturation, body temperature, electrocardiogram (ECG) signals, and sweat composition. The data obtained from these biosensors can undergo additional processing, analysis, and visualization in order to offer valuable feedback, practical insights, and customized recommendations to both users and healthcare professionals. The potential applications of biosensors in embedded wearable devices exhibit a wide range of possibilities and significant implications. In the field of healthcare, these devices possess the capacity to significantly transform the approach to managing chronic illnesses, facilitating timely detection, evaluating the effectiveness of treatments, and enabling the monitoring of patients from a distance. For example, wearable devices equipped with biosensors have the capability to consistently monitor glucose levels in individuals diagnosed with diabetes, providing timely notifications regarding fluctuations and offering guidance for the administration of insulin. In a similar vein, wearable devices equipped with biosensors have the capability to monitor cardiac activity, thereby detecting any abnormalities in heart rhythm and facilitating timely intervention for individuals who are susceptible to cardiovascular diseases. In addition, the integration of biosensors into wearable devices has the potential to facilitate personalized fitness monitoring, sleep analysis, stress mitigation, and overall well-being improvement. This empowers individuals to proactively adopt healthier lifestyle practices.

The field of biosensor applications in embedded wearable devices is experiencing rapid evolution in the research landscape, characterized by the emergence of multidisciplinary collaborations and technological advancements. These factors are instrumental in driving innovation within this domain. Therefore, it is imperative to acquire a thorough comprehension of the present state of research, discern emerging patterns, and examine the prospects and obstacles linked to wearable devices equipped with biosensors. The objective of this study is to offer valuable insights by conducting a bibliometric analysis of the current body of literature. This analysis will illuminate patterns in publication, identify influential researchers, explore collaboration networks, and uncover emerging research themes. This study seeks to provide a comprehensive overview of the research landscape in biosensor applications in embedded wearable devices by conducting a systematic search of major scientific databases and utilizing bibliometric analysis techniques. The present analysis aims to examine the trends in publications, including the temporal increase in published works and the identification of prominent countries and institutions that have made significant contributions to the field. In addition, citation networks serve the purpose of identifying significant works and researchers who have had a significant impact on the field. On the other hand, collaboration networks provide insights into the central knowledge hubs and collaborative partnerships that are instrumental in driving advancements in research. In general, the utilization of biosensor-enabled embedded wearable devices has significant potential to revolutionize the field of healthcare and the management of wellness. Through the utilization of biosensors, these devices have the capability to offer uninterrupted, individualized, and practical observations, thereby enabling individuals to actively oversee and enhance their state of well-being. This chapter presents a comprehensive bibliometric analysis with the objective of contributing to the existing body of knowledge and promoting further advancements in this rapidly evolving field.

ARTIFICIAL INTELLIGENCE (AI) AND BIOSENSING METHODS AND MODELS

The application of artificial intelligence in biosensing techniques and models has attracted significant attention in recent years due to its potential to facilitate profound advancements in the domains of healthcare, diagnostics, and environmental monitoring. AI-based biosensing refers to the amalgamation of AI algorithms with traditional biosensing techniques. The primary objective of this amalgamation is to enhance the sensitivity, accuracy, and efficiency of data analysis. This chapter elucidates various fundamental constituents of biosensing methodologies and models that have been amalgamated with AI.

The data acquisition process encompasses the utilization of biosensors, which are specialized devices engineered to detect and quantify biological or chemical analytes. The utilization of artificial intelligence holds promise in making substantial contributions to the improvement of data acquisition procedures. This can be achieved by leveraging AI techniques to analyze signal quality, optimize sensor placement, and enhance sensor design. For example, algorithms in the field of artificial intelligence have the ability to analyze real-time data acquired by wearable biosensors to detect anomalies or patterns that could potentially indicate different health conditions.

In the domain of biosensors, it is customary to engage in data preprocessing and feature extraction on unprocessed data to enhance signal quality and extract pertinent information. The application of artificial intelligence methodologies, such as signal processing algorithms and deep learning models, can be employed to achieve noise reduction, correction of artifacts, and extraction of relevant features from a given dataset.

The application of AI models, particularly machine learning algorithms, facilitates the training of these models to proficiently classify biosensor data or detect patterns that signify specific conditions or analytes. Artificial intelligence models can develop the capacity to identify disease biomarkers, detect pathogens, and recognize environmental pollutants by undergoing training on labeled datasets.

Predictive modeling encompasses the application of AI techniques to develop models that can effectively forecast and project forthcoming patterns or results through the analysis of biosensor data. These models possess the capacity to be employed in a diverse array of applications, encompassing but not restricted to the prediction of disease progression, estimation of drug response, and evaluation of environmental risks.

AI-driven biosensing systems possess the capacity to provide instantaneous monitoring and decision-making assistance by means of ongoing analysis of biosensor data. These systems have the capability to promptly alert healthcare professionals or users of any potential abnormalities or critical events. The utilization of this technology has the capacity to enable timely interventions and personalized healthcare management.

The utilization of data fusion and integration is a prevalent practice in biosensing applications, where multiple data sources are incorporated, encompassing various sensor types or data acquired from other healthcare systems. Artificial intelligence possesses the capacity to facilitate the integration and fusion of diverse data sources, thereby yielding a comprehensive viewpoint of the entity under surveillance or enabling the extraction of more informative attributes.

The application of reinforcement learning techniques, specifically artificial intelligence, can facilitate the optimization and enhancement of biosensing models. This methodology facilitates the iterative enhancement of these models, resulting in enhanced performance over a period of time. The models possess the capability to adapt and enhance their accuracy and efficiency by receiving feedback from users or engaging in continuous learning processes. The incorporation of AI into biosensing techniques and frameworks has significant potential for enhancing diagnostics, monitoring, and decision-making in various fields. These methodologies possess the capacity to yield increased accuracy and customization in healthcare, prompt detection of diseases, enhanced monitoring of the environment, and overall improvement in well-being.

BIOSENSING WEARABLES – APPLICATIONS AND DEVELOPMENT METHODS

Biosensing wearables encompass the integration of biosensors with wearable technologies, facilitating the continuous monitoring of various physiological and biochemical parameters. Wearable devices offer a diverse range of applications across various domains, including healthcare, sports and fitness, wellness, and related fields. This discourse provides a comprehensive overview of notable applications and development methodologies related to biosensing wearables.

APPLICATIONS

Health monitoring encompasses the utilization of biosensing wearables for the purpose of monitoring crucial physiological parameters, such as heart rate, blood pressure, body temperature, and respiratory rate. These parameters enable the continuous monitoring of diverse health indicators, thereby providing valuable insights into an individual's present state of health. This technology has the potential to expedite the prompt recognition of medical ailments, facilitate the remote tracking of patients, and bolster the execution of individualized healthcare management approaches.

Biosensing wearables possess the capacity to monitor a range of fitness and performance parameters, encompassing, though not limited to, step count, calorie expenditure, sleep quality, and exercise intensity. These devices provide prompt feedback and data analysis, aiding individuals in optimizing their fitness routines, setting goals, and tracking their progress. Advancements in the utilization of wearable devices equipped with biosensors have been observed within the domain of stress management. These devices possess the capacity to identify diverse physiological indicators linked to stress, such as heart rate variability and galvanic skin response. These applications provide individuals with valuable information pertaining to their stress levels and offer a range of strategies for alleviating stress, including engaging in breathing exercises or utilizing mindfulness prompts.

Wearable devices possess the capacity to monitor multiple facets of sleep, encompassing the duration, quality, and distinct stages of sleep. The aforementioned data can be employed for the purpose of identifying sleep disorders, improving sleep hygiene practices, and enhancing overall sleep health. The utilization of biosensing wearables plays a crucial role in the monitoring and management of chronic diseases, including but not limited to diabetes, asthma, and hypertension. These devices greatly assist in the effective control and treatment of such conditions. These devices possess the capacity to monitor diverse physiological parameters, including blood glucose levels, lung function, and blood pressure. This technology allows patients to access real-time data and notifications, thereby enhancing their ability to self-manage their health.

The application of wearable devices integrated with biosensors offers a feasible strategy for the surveillance of diverse environmental parameters, encompassing air quality, UV radiation, and pollution levels, among others. The data mentioned above has the capacity to aid individuals in making informed decisions regarding their level of exposure and subsequently implementing suitable precautionary measures. The subject matter under consideration pertains to methodologies employed in the process of development. The achievement of successful integration of biosensing wearables requires the effective incorporation of appropriate biosensors that possess the capability to accurately measure the desired physiological or biochemical parameters. The careful consideration of sensor selection, calibration, and miniaturization is imperative when designing wearable devices. Figure 21.1 shows various body sensing sensors (Vashistha et al., 2018).

The data acquisition and processing procedure encompasses the retrieval of a substantial amount of data from the integrated sensors of biosensing wearables. To attain accurate measurements and minimize the influence of disturbances and irregularities, it is crucial to utilize effective data acquisition methodologies and implement signal processing techniques. This involves the design and construction of appropriate analog front-end circuitry, application of noise filtering methods, and utilization of signal conditioning algorithms. The management of power is an essential component of wearable technology, given that these devices depend on limited

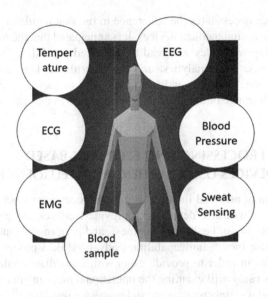

FIGURE 21.1 Various biosensors. Source: Wearable biosensors and recent advancements – Krazytech

power resources, such as batteries or energy harvesting mechanisms. The utilization of power management strategies holds significance in the context of enhancing the energy efficiency of biosensing wearables and extending the longevity of their battery life. The scope of this can involve the advancement of sensor designs that consume minimal power, the integration of intelligent algorithms for power management, and the enhancement of data transmission techniques to maximize energy efficiency.

The transmission and secure storage of data are crucial considerations in the context of biosensing wearables, as they generate continuous data streams that require both secure transmission and storage mechanisms. Wireless communication protocols, such as Bluetooth or Wi-Fi, enable the rapid transmission of data to smartphones or cloud-based platforms. The application of data compression and encryption techniques ensures the enhancement of data transmission and storage procedures, while also guaranteeing the preservation of data integrity and confidentiality (Cui et al., 2020).

The design of the user interface and data visualization for biosensing wearables should prioritize an intuitive and user-friendly experience, facilitating the presentation of real-time data and actionable insights. Development methods involve the systematic creation of user interfaces that are specifically designed for compact form factors, the creation of visually appealing and informative data visualizations, and the integration of interactive elements to enhance user engagement.

The validation and regulatory compliance of biosensing wearables require a thorough validation and testing process to ensure the accuracy, reliability, and safety of these devices. The paramount importance of guaranteeing user safety and achieving market acceptance necessitates the adherence to relevant regulatory standards, particularly those governing medical devices. It is anticipated that the ongoing advancement of biosensing wearables will lead to enhanced capabilities by incorporating sensor technologies, data analytics, and AI algorithms. This advancement will enable more accurate and personalized monitoring of various health parameters, thus expanding the range of their potential applications in the domains of healthcare and wellness.

BIOSENSING PROCESSING AND FEEDBACK-BASED WEARABLE DEVICES USING ARTIFICIAL INTELLIGENCE

The incorporation of artificial intelligence in the processing of biosensing data and the development of wearable devices that provide feedback has gained significant traction in the domains of healthcare, wellness, and performance monitoring. These devices incorporate biosensing capabilities, advanced data processing techniques, and AI algorithms in order to provide users with immediate feedback and practical insights. This essay will examine the operational mechanisms of these devices, including the fundamental components and processes involved.

Biosensors play a crucial role in wearable devices specifically designed for biosensing purposes, as they facilitate the acquisition of physiological and biochemical data from the user. Biosensors encompass a diverse array of devices, such as heart rate monitors, ECG sensors, PPG sensors, temperature sensors, EDA sensors, and other analogous instruments. The incorporation of these sensors into wearable devices is intended to enhance the precision of data acquisition while giving priority to user comfort and convenience. Figure 21.2 shows recent innovation of the biosensing model (Lim et al., 2020).

The data acquisition process encompasses the ongoing measurement of pertinent physiological or biochemical parameters from the user, facilitated by the integration of biosensors within the wearable device. The acquired data is typically expressed in the form of electrical signals, levels of light intensity, or measurements of temperature. Following this, the gathered data is sent to the processing unit of the wearable device for further examination.

The process of data preprocessing plays a vital role in the examination of unprocessed biosensor data, as it aims to enhance the signal's quality and remove any undesirable noise or artifacts. Preprocessing techniques frequently utilized in data analysis encompass a range of procedures, including noise filtering, baseline correction, artifact removal, and normalization. The objective of these procedures is to improve the accuracy and reliability of the data before proceeding with further analysis.

The process of feature extraction plays a vital role in data analysis. It involves extracting relevant features from preprocessed data to accurately capture and represent important information present in the raw data. The choice of feature extraction methods is dependent on the specific application and the parameters being observed. These features may include heart rate variability (HRV), pulse transit

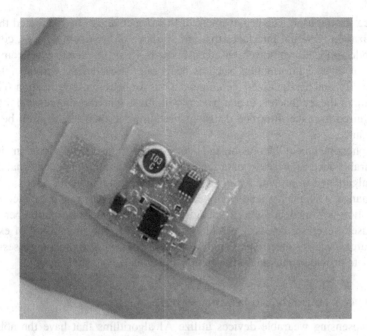

FIGURE 21.2 Recent innovation of the biosensing model

time, respiratory rate, skin conductance, or any other parameter that is considered to be pertinent. The extracted features are employed as input for subsequent artificial intelligence algorithms.

The extracted features are utilized as input for artificial intelligence algorithms in order to facilitate analysis and interpretation. The incorporation of AI algorithms in biosensing wearable devices encompasses a range of methodologies, spanning from traditional machine learning techniques to advanced deep learning models. The algorithms are trained using comprehensive datasets that include labeled instances of various physiological states or conditions. The training process enables algorithms to acquire knowledge, enabling them to identify patterns, correlations, and anomalies related to specific health conditions, wellness metrics, or performance indicators (Jin et al., 2023; Qureshi et al., 2023; Soudier et al., 2022).

Pattern recognition and classification play pivotal roles in the functioning of artificial intelligence algorithms. The aforementioned algorithms examine the extracted features and employ diverse techniques to discern pertinent patterns or states. For instance, these algorithms have the ability to detect abnormal heart rhythms that indicate cardiac arrhythmias, as well as recognize patterns associated with stress levels, sleep stages, or levels of physical activity. Classification models are purposefully designed to assign input data to discrete categories, utilizing the patterns learned during the training phase.

The utilization of processed data enables the facilitation of real-time monitoring and feedback, which in turn allows for timely and continuous evaluation and guidance for the user. The achievement of this objective can be facilitated through the utilization of visual interfaces that are available on wearable devices, companion mobile applications, or web-based dashboards. Real-time feedback encompasses a range of diverse information, such as heart rate, stress levels, activity intensity, sleep quality, and personalized recommendations for behavior modification. The primary aim of the feedback is to equip users with the essential resources and information required to make informed decisions pertaining to their health, well-being, or performance.

The phenomenon of data fusion and integration is frequently observed in biosensing wearable devices, whereby data acquired from diverse sensors or external origins are amalgamated to provide a comprehensive understanding of the user's health or performance. This phenomenon can be observed when heart rate data is combined with motion sensors, leading to the acquisition of significant information pertaining to the user's exercise intensity or calorie expenditure. The integration of external data sources, such as electronic health records or environmental data, possesses the capacity to provide contextual information and enhance the analysis.

Continuous Learning and Personalization

Some biosensing wearable devices utilize AI algorithms that have the ability to continuously acquire knowledge and adapt based on the user's behavior and preferences. Through continuous learning, algorithms improve their accuracy over time and provide more customized feedback. The refinement of the analysis and recommendations offered by the wearable device is achieved through the integration of user feedback and the incorporation of new data.

Ensuring the safeguarding of data privacy and security holds paramount significance in the context of biosensing wearable devices, given their involvement in the processing of highly sensitive personal information. The protection of user information requires the utilization of strong encryption methods, the establishment of secure storage systems, and strict compliance with privacy regulations. During the development and deployment of biosensing wearable devices, it is imperative to consider the inclusion of user consent and the adoption of transparent data handling practices (Kumar et al., 2023). In brief, the incorporation of biosensors, data processing, and AI algorithms in biosensing processing and feedback-based wearable devices facilitates the delivery of instantaneous monitoring, analysis, and tailored feedback to individuals. These devices have the potential to revolutionize the healthcare, wellness, and performance monitoring industries by offering individuals valuable insights into their physiological and biochemical parameters. The ongoing progress in AI technology is expected to lead to further improvements in biosensing wearables, thereby enhancing their capabilities and effectiveness in addressing a wide range of health and well-being issues.

REFERENCES

Cui, F., Yue, Y., Zhang, Y., Zhang, Z., & Zhou, H. S. (2020). Advancing biosensors with machine learning. *ACS Sensors*, *5*(11), 3346–3364.

Jin, X., Cai, A., Xu, T., & Zhang, X. (2023). Artificial intelligence biosensors for continuous glucose monitoring. *Interdisciplinary Materials*, *2*(2), 290–307.

Kumar, M. R., Devi, B. R., Rangaswamy, K., Sangeetha, M., & Kumar, K. V. R. (2023). IoT-edge computing for efficient and effective information process on industrial automation. 2023 International Conference on Networking and Communications (ICNWC), Chennai, India, 1–6.

Lim, S. A., Lim, T. H., & Ahmad, A. N. (2020). The applications of biosensing and artificial intelligence technologies for rapid detection and diagnosis of COVID-19 in remote setting. *Diagnostic Strategies for COVID-19 and Other Coronaviruses* 1, 109–134.

Qureshi, R., Irfan, M., Ali, H., Khan, A., Nittala, A. S., Ali, S., Shah, A., Gondal, T. M., Sadak, F., Shah, Z., & Others. (2023). Artificial intelligence and biosensors in healthcare and its clinical relevance: A review. *IEEE Access* 11, 61600–61620.

Soudier, P., Faure, L., Kushwaha, M., & Faulon, J.-L. (2022). Cell-free biosensors and AI integration. *Cell-Free Gene Expression: Methods and Protocols*, 2433, 303–323.

Vashistha, R., Dangi, A. K., Kumar, A., Chhabra, D., & Shukla, P. (2018). Futuristic biosensors for cardiac health care: An artificial intelligence approach. *3 Biotech*, *8*, 1–11.

22 Real Impacts of AI on Sports Wearables

Panuganti Naveen, R. Shruthi, V. Pranavitha,
Dumpa Sree Harsha Reddy, and Kashika Parmar

INTRODUCTION

Sports wearables refer to electronic gadgets that are intended to be worn during physical exercise to monitor and evaluate information associated with the user's performance [3]. The gadgets encompass smartwatches, fitness trackers, heart rate monitors, global positioning system (GPS) devices, and various other related technologies (Yasar, 2022).

The prevalence of wearable technology in the realm of sports has experienced an upward trend in recent times. As per the findings of a report published by Grand View Research, the worldwide market for sports wearable devices was assessed to be worth USD 13.2 billion in 2020 and is projected to exhibit a compound annual growth rate (CAGR) of 15.9% between 2021 and 2028 (Fortune Business Insights, 2021). The growth of this market can be attributed to several factors, including heightened consciousness regarding health and fitness, augmented disposable incomes, and technological advancements (Henriksen et al., 2018).

Sports wearables are being increasingly utilized by sports teams and coaches to track the performance of their athletes during training and games, in addition to their personal use. The aforementioned data has the potential to facilitate the identification of areas that require improvement, mitigate the occurrence of injuries, and enhance the efficacy of training programs (Sharma et al., 2022).

The usage of sports wearables has gained significant traction among athletes, fitness enthusiasts, and individuals who engage in occasional exercise. Wearable devices provide a diverse array of functionalities, including the ability to monitor physical activity metrics such as step count, distance traveled, calories expended, heart rate, sleep patterns, and stress levels. In addition, personalized feedback and coaching are offered to users to assist them in attaining their fitness objectives (Aroganam, 2021).

The utilization of artificial intelligence (AI) in the sports wearable sector has the capacity to bring about a significant transformation by facilitating more sophisticated data analysis and customized guidance for its users (Sieron, 2023). The utilization of artificial intelligence in sports wearables is advantageous due to its capacity to analyze extensive data sets and furnish users with significant findings.

 DOI: 10.1201/9781032686714-22

AI algorithms have the capability to examine information obtained from diverse sensors embedded in wearable devices, including heart rate monitors and accelerometers. This enables the monitoring of user activity and the provision of individualized feedback. Artificial intelligence has the capability to recognize recurring trends in an individual's workout regimen and propose alterations to facilitate the attainment of their objectives. (Phaneuf, 2022)

Artificial intelligence AI has the potential to improve the precision of tracking capabilities in sports wearables. Artificial Intelligence (AI) algorithms have the capability to consider extraneous variables, such as environmental conditions and topography, in order to furnish more precise GPS tracking information. Accurate location tracking can prove to be particularly advantageous for athletes and sports teams, especially during training and competitive events (Shore, 2022).

The utilization of AI-enabled sports wearables can provide athletes and sports teams with significant insights into their performance, enabling them to make informed decisions based on data to enhance their outcomes. Through the examination of data obtained from wearable sports technology, artificial intelligence algorithms have the capability to detect patterns that may signify an elevated probability of sustaining an injury. The aforementioned data may be utilized to modify training regimens and mitigate the likelihood of physical harm. Artificial intelligence has the potential to offer tailored guidance and evaluation to individuals utilizing athletic wearables, thereby enhancing their ability to attain their fitness objectives with greater efficiency (Roger, 2022).

THEORETICAL BACKGROUND

Machine learning is a specialized area within the broader field of AI that focuses on the development of algorithms capable of learning from data and subsequently utilizing that knowledge to make predictions or decisions. Within the realm of sports wearables, the implementation of machine learning techniques enables the examination of sensor data, thereby facilitating the delivery of tailored coaching to individual users (Sanyal, 2022).

The field of data analytics pertains to the systematic examination of data in order to reveal underlying patterns and insights. Data analytics can be employed in the domain of sports wearables to detect potential areas for enhancement in an athlete's performance or to identify factors that could be contributing to an escalated risk of injury (Chen and Yang, 2023).

The field of psychology encompasses a sub-discipline known as motivation theory, which delves into the various factors that impact human motivation. Motivation theory can be applied to the development of coaching programs for sports wearables, with the aim of customizing them to suit the unique needs and preferences of individual users (Matijevich, 2022).

Human–computer interaction (HCI) is an academic discipline that investigates the ways in which humans interact with technology. HCI can be employed in the realm of sports wearables to develop user interfaces that are user-friendly and furnish users with pertinent data to enhance their athletic abilities (Scott, 2022).

The field of AI ethics is an expanding domain of inquiry that investigates the ethical ramifications of employing AI in diverse settings. When examining sports wearables, it is crucial to take into account concerns pertaining to data privacy and security, algorithmic bias, and the possible effects of artificial intelligence on the independence of athletes and coaches (Pazzanese, 2022).

BENEFITS

The data that can be gathered from sports wearables can be analyzed by artificial intelligence, and the metrics that can be analyzed include, but are not limited to, heart rate, stride length, and a variety of other metrics. The results of this study can be put to use to provide athletes with individualized advice. This can be of assistance to individuals in refining their training routine, reducing the risk that they will develop injuries, and improving their overall performance. Wearable sports technology that is equipped with AI functions has the ability to provide athletes with real-time assessments of how well they are performing their sporting activities. When individuals use this information, it can be helpful in adjusting their approach, speed, and other variables when they are competing or participating in training in order to get greater results. Wearable technology that contains artificial intelligence has the capacity to monitor the physical motions of athletes and identify potential dangers that could result in injury. Consequently, coaches and trainers may make use of this data in order to devise injury prevention routines that specifically target these threats and reduce the likelihood of injuries occurring. AI has the capacity to assess huge quantities of data obtained through sports wearables. This paves the way for the detection of detailed patterns and trends that may be difficult to analyze using traditional methods. This can assist coaches and athletes in making decisions regarding training, strategy for games, and tactics with a higher level of informed accuracy.

DISADVANTAGES

It is possible that the incorporation of AI into sports wearables will result in a greater cost when compared to the cost of conventional wearables. This will reduce the accessibility of the wearables for athletes who may be limited financially. Because the effectiveness of AI is dependent on the availability of considerable data, concerns have been raised over the protection of data privacy and security. The use of sports wearables that are coupled with artificial intelligence technology raises the risk of data theft or the malicious exploitation of users in some way. Athletes who place an excessive amount of dependence on sports wearables that are equipped with AI run the risk of becoming disconnected from their natural instincts and the talents that they were born with. It is possible that as a result of this, the level of inventiveness and spontaneity that was displayed during the performance would drop. In the event of improper calibration or failing sensors, the precision of the data produced by wearables integrated with AI may be jeopardized. As a consequence of this, it is possible that incorrect inferences will be drawn regarding an athlete's performance or their risk of being injured. When the data used for training AI algorithms is not representative of the entire population, there is a risk that the algorithms would produce biased results. There is a risk that certain athletes or teams will receive unequal benefits or disadvantages as a result of the competition.

STATEMENT OF PROBLEMS

The dearth of comprehensive research on the actual effects of AI on sports wearables is notable, despite the increasing prevalence of these wearables and the incorporation of AI technology into them. The utilization of AI-powered sports wearables exhibits encouraging indications of potential benefits (Hilete, 2022). However, additional research is required to comprehensively understand the practical implications of these devices on athletes' performance, injury prevention, and motivation (Trevor, 2023). Furthermore, it is imperative to consider the ethical implications associated with AI implementation, including but not limited to safeguarding data privacy and protection, mitigating algorithmic bias, and assessing the potential ramifications of AI on the autonomy of athletes and coaches. The purpose of this report is to examine the actual effects of AI on sports wearables, specifically analyzing the advantages, obstacles, and ethical implications linked to these gadgets (Bedi et al., 2023).

RESEARCH QUESTIONS

1. In what ways does the implementation of AI technology augment the precision and utility of sports wearables in monitoring athletic performance and health metrics?
2. What are the potential applications of AI-enabled sports wearables for injury prevention and enhancing the general health of athletes?
3. The utilization of AI in sports wearables raises ethical concerns. These concerns pertain to the protection of user privacy and security. Is it imperative to address these ethical implications to ensure that users are safeguarded?
4. What are the existing constraints of AI-driven sports wearables, and what measures can be taken to enhance their functionality for the benefit of athletes and sports organizations?
5. What is the perception of athletes and coaches regarding the utilization of artificial intelligence in sports wearables, and what are the primary determinants that impact their acceptance of these gadgets?
6. What is the impact of incorporating AI technology into sports wearables on user motivation, and what are the optimal strategies for creating tailored coaching programs that utilize this technology?

RESEARCH OBJECTIVES

1. To understand what advantages AI-enabled sports wearables offer in terms of monitoring fitness and performance
2. Examining how coaching programs aided by artificial intelligence affect user engagement and productivity

3. To compare and contrast the efficacy of AI-enhanced sports wearables with more conventional forms of training in improving the performance and health of athletes
4. To investigate how athletes and coaches feel about AI in sports wearables, and to determine what elements contribute to their acceptance
5. To assess how well AI-enabled sportswear performs at reducing injuries and improving athletes' health

RESEARCH METHODOLOGY

A combination of qualitative and quantitative methodologies will be used in the research to examine the actual effects of AI on sports wearables. Athletes' experiences with wearable AI technology may be better understood through qualitative research techniques including in-depth interviews with coaches, athletes, and sports scientists. To acquire a wider variety of viewpoints, surveys and focus groups could also be held. To find patterns and correlations, quantitative research techniques like statistical analysis of performance data from wearables may be applied. An integrated understanding of the true effects of AI on sports wearables could be achieved by triangulating the results from both qualitative and quantitative methodologies using a mixed-methods methodology.

HYPOTHESES

The hypotheses would be as follows:

1. The first hypothesis is that by enhancing performance, AI will benefit sports wearables. Wearables could be developed using AI technology to give athletes immediate feedback on their performance, enabling them to make modifications and thus advance their abilities more quickly.
2. The second hypothesis is that sports wearables driven by AI will aid in injury prevention. AI algorithms might spot trends that suggest a higher risk of injury by examining data from the wearables' sensors, and they could then notify athletes or coaches to take precautionary action.
3. A third hypothesis is that AI will make it possible to create training plans that are better tailored to each individual. AI algorithms could develop specialized training plans that are suited to a person's strengths and weaknesses by gathering and analyzing data on an athlete's performance. This can result in more effective training and improved performance for the athletes.

SCOPE AND LIMITATIONS

The subject of "Real impacts of AI on Sports Wearables" is vast in scope and covers a variety of changes brought about by artificial intelligence in the sports sector. On the plus side, AI-powered sports devices are helping athletes train more effectively, monitor their results, and avoid injuries, among other advantages. These wearables have sensors and algorithms that gather and analyze data in real time, giving athletes insightful information about their bodies and abilities (Salehi et al., 2022).

The application of AI in sports wearables is not without its drawbacks, though. The precision and dependability of the data collected present a considerable issue. The algorithms and sensors might not always be calibrated properly, which could produce false results. Aside from these challenges, there are worries regarding the security and privacy of the data gathered as well as possible moral dilemmas related to the application of AI in sports (Amorim et al., 2019). While the genuine effects of AI on sports wearables are encouraging, there are still significant issues that must be resolved to guarantee the use of AI in sports is both effective and safe.

DISCUSSION

ADVANCEMENTS IN SPORTS WEARABLES WITH AI

With the incorporation of artificial intelligence, sports wearable technology has improved dramatically in recent years, offering even greater potential to improve athletic performance. In order to gather information on an athlete's activity, vital signs, and other physical indicators, current sports wearables frequently include sensors, such as accelerometers and heart rate monitors. The performance of an athlete, their training requirements, and their general health are all revealed by the analysis of this data (Dao, 2022).

Sports wearables are incorporating AI to enhance the precision and accuracy of data processing. Large amounts of data may be processed fast by machine learning algorithms, which can then spot patterns and trends that would be challenging or impossible for a human to spot. In turn, this makes it possible for wearables to offer more exact training, dietary, and recuperation advice based on each athlete's particular requirements and objectives (Baca et al., 2022).

The use of computer vision to track an athlete's motions and technique is one instance of AI in sports wearables. Artificial intelligence and camera-equipped wearables can evaluate an athlete's form and give suggestions for development. Another illustration is the application of AI in injury prediction and avoidance. Machine learning algorithms can spot potential injury indications by examining an athlete's physical data and can then notify coaches and trainers so they can take preventative action (Nadikattu, 2020).

There are many advantages to using AI in sports wearables. Athletes can improve their training and performance by using wearables driven by AI to give more precise and accurate data. They can also speed up the healing process and aid in injury prevention, which will ultimately result in better general health outcomes. AI can also assist players and coaches in making judgments that are based on factual information as opposed to their subjective observations (Ghosh et al., 2023).

The WHOOP Strap, which employs machine learning algorithms to analyze data on a sportsperson's sleep, recovery, and training, is one real-life example of a wearable driven by artificial intelligence. The Catapult Sports wearable, which is employed by professional sports teams, tracks athletes' movements, and provides immediate performance feedback. Another example is the Lumo Run, which analyses a runner's stride using computer vision, and the Zephyr BioHarness, which tracks a competitor's vitals while they are competing (Dai and Li, 2021).

AI in Data Collection and Analysis

Sports wearables must capture data because they can reveal information about an athlete's performance and general health. In the past, data collecting included manually entering information or utilizing sensors to gather data in the present. The use of AI in sports wearables is, however, altering how data is gathered and examined (Ho et al., 2022).

Wearables are now able to gather data more quickly and at higher volumes than ever before thanks to AI. Massive volumes of data may be analyzed in real-time by machine learning algorithms, giving coaches and athletes immediate feedback. As a result, athletes may be able to improve their preparation and performance, which will benefit their general health (Whitford et al., 2019).

The capacity of AI to offer more individualized insights is one of the major advantages of data collecting and analysis for sports wearables. Machine learning algorithms may examine a specific athlete's data, including body composition and biomechanics, and offer personalized suggestions for training and recovery. Athletes may be able to avoid injuries and more effectively meet their performance objectives thanks to this (Gaur et al., 2019).

Wearable sensors are used to track an athlete's heart rate, breathing rate, and other vital signs throughout practice and competition. This is an actual example of AI-powered data collecting and analysis in sports. Coaches can make better choices by analyzing this data using machine learning algorithms, which can also offer feedback on an athlete's performance and recovery (Nahavandi et al., 2022).

Another illustration is the use of artificial intelligence to review athlete video footage and pinpoint areas for technique and form improvement. Athletes may measure their motions and receive real-time feedback on how they are performing thanks to wearables with cameras and machine learning algorithms. This allows athletes to improve their movements and reduce injury risk (Aresta et al., 2023).

Impact of AI on Athlete Performance and Injury Prevention

AI is revolutionizing athlete performance and injury prevention in the sports business. Athletes can now optimize their training and lower their risk of injury thanks to AI technology, which improves their overall health outcomes. Here are some instances of how AI is being applied to enhance athletic performance and prevent injuries (Li and Xu, 2021):

1. **Performance improvement:** By offering individualized insights based on data gathered from wearables, artificial intelligence is being utilized to enhance athletic performance. Machine learning algorithms can examine

a specific athlete's data, including body composition, biomechanics, and heart rate variability to offer individualized training and recuperation suggestions. This can aid athletes in avoiding injuries and effectively achieving their performance objectives. Computer vision is used to track an athlete's motions and technique as one example of AI-powered performance optimization. The form of an athlete can be examined using wearables with cameras and AI, which can then give suggestions for how to improve. Injuries can be prevented and athletic performance improved (Rössler et al., 2015).

2. **Predicting and preventing injuries:** Using AI, injuries can also be anticipated and prevented. With the help of wearables with sensors, it is possible to gather information on an athlete's physiological markers, such as heart rate, breathing rate, and muscle activation, in order to determine the likelihood of injuries. This data can be analyzed by machine learning algorithms, and the alarms generated will allow coaches and trainers to take appropriate preventative action (Hanlon et al., 2019).

AI is revolutionizing athletic performance and injury prevention across various sports including Performance Analysis, Biomechanics and Motion Analysis, Injury Prevention, Game Strategy and Analysis, etc

Professional sports teams employ the Catapult Sports wearable to measure an athlete's movement and deliver real-time performance feedback. The wearable contains sensors that gather information on an athlete's physiological signs. Machine learning algorithms then analyze this data to produce individualized training and recuperation recommendations (Claudino et al., 2019). The WHOOP Strap analyses data on an athlete's sleep, recovery, and training using machine learning algorithms. The wearable gathers information about an athlete's activity level, heart rate, and respiration rate among other things, and offers suggestions for enhancing performance and recovery (Qermane and Mancha, 2020). While training and competing, an athlete's vital signs are tracked by the Zephyr BioHarness. This wearable contains sensors that gather information about an athlete's heart rate, breathing rate, and muscle activation. This information is then processed by machine learning algorithms to identify potential injury risks (Iliadis et al., 2021).

AI has various advantages for improving athletic performance and reducing injuries. Athletes may improve their training and performance with the help of AI-powered wearables, which can deliver more precise and accurate data. Additionally, they can speed up recovery times and assist in injury prevention, which will ultimately result in better overall health results. AI can also assist athletes and coaches in making judgments that are based on facts rather than impressions, which will lead to better outcomes (Khayyam et al., 2019). AI is becoming more crucial to athletic performance and injury avoidance. Athletes may reach their performance goals while reducing the risk of injury with the aid of AI-powered wearables since they offer personalized insights and enable speedier data processing. The potential for additional innovation in this area is enormous because of ongoing breakthroughs in AI technology, and the future is promising for both athletes and sports organizations (Lee and Lee, 2021).

ETHICAL AND SOCIAL IMPLICATIONS OF AI IN SPORTS WEARABLES

The use of AI in sports wearables has the potential to revolutionize how athletes prepare, compete, and recover. To ensure responsible and ethical practices, there are a number of ethical and social issues related to its use that need to be addressed (Phatak et al., 2021).

One of the biggest drawbacks of AI for sports wearables is privacy issues. It's essential to make sure that this data is secure and inaccessible to unauthorized parties given that these devices collect enormous amounts of data regarding an athlete's physical and physiological condition. Additionally, there is a chance of data breaches, which could have detrimental effects on the athlete's safety and privacy (Sui et al., 2023). Wearable sports technology that uses AI raises ethical questions about the reliability of the data collected and how it will be used. An athlete's performance or health may be incorrectly predicted by AI systems, for instance, because of prejudice. Additionally, the application of AI in sports wearables may prompt debate over how far technology should be allowed to track and evaluate athletes' performance (Brown and Brown, 2021).

Affordability and accessibility are two social implications of AI in sports wearables. There may be an unfair playing field for some athletes because AI technology is still in its infancy and is costly. Athletes from affluent backgrounds may have access to more sophisticated technologies, which could further exacerbate already existing inequities (Wu and Luo, 2019).

Several efforts can be taken to resolve these problems and ensure the proper usage of artificial intelligence in sports wearables. First, more stringent regulation of the collecting, use, and sharing of athlete data is required. Second, AI algorithms must be designed with fairness and accuracy in mind, and the training and testing of these algorithms must be transparent. Third, efforts should be made to make AI technology available to all athletes, regardless of financial background (Kah Phooi Seng et al., 2023).

CONFLICT OF INTEREST

When discussing the actual effects of AI on sports wearables, a conflict of interest may occur if the speaker or writer has financial ties to or a personal stake in the firms or products that are being addressed. For instance, if the speaker works as a consultant or investor for a business that manufactures sports wearables driven by AI, their opinions and analysis may be skewed toward pushing the product or downplaying any negative effects (Bargh, 2019).

It's critical to be transparent about any financial relationships or personal interests in order to avoid such conflicts of interest, and it's also crucial to work toward objectivity and balance when discussing the subject. To get a more objective understanding of the true effects of AI on sports wearables, it may also be useful to confer with experts or researchers who are not affiliated with any of the firms or products being addressed (Dimitropoulos et al., 2021).

CONCLUSION AND RESULTS

In conclusion, the practical effects of AI on sports wearables are considerable and far-reaching, with both positive and negative repercussions. On the plus side, AI-powered sports wearables can offer athletes real-time feedback on their performance, assist in injury prevention, and improve overall training and performance. However, there are worries about data privacy, accuracy, and the possible amplification of existing disparities in sports (Raman and Aashish, 2021).

The analysis of the subject includes a number of privacy issues, ethical issues, and social consequences of AI in sports wearables. Due to the enormous amounts of data gathered and the possibility of data breaches, privacy issues are raised. The accuracy of the data gathered and the extent to which technology can track and study athlete behavior are ethical issues. Accessibility and affordability have social repercussions because some athletes have access to more sophisticated technology than others (Nayak et al., 2021).

To address these challenges and ensure the appropriate use of AI in sports wearables, strict data collecting, use, and sharing regulations are required. Furthermore, AI algorithms must be designed with fairness and accuracy in mind, and the process by which these algorithms are taught and tested must be transparent. Finally, efforts should be made to make AI technology available to all athletes, regardless of financial status (Pandey et al., 2021).

The research indicates that, while AI-powered sports wearables provide tremendous benefits, there are significant ethical and societal consequences that must be examined in order to guarantee that the technology is used responsibly and ethically.

REFERENCES

Amorim, J. P. V., Silva, C. M. and Oliveira, A. R. R. (2019). Software and Hardware Requirements and Trade-Offs in Operating Systems for Wearables: A Tool to Improve Devices' Performance. *Sensors*, 19(8), p. 1904.

Aresta, S., Musci, M., Bottiglione, F., Moretti, L., Moretti, B. and Bortone, I. (2023). Motion Technologies in Support of Fence Athletes: A Systematic Review. *Applied Sciences*, 13(3), p. 1654.

Aroganam, G. (2021). Review on Wearable Technology Sensors Used in Consumer Sport Applications. *Sensors*, 19(9), p. 1983. https://doi.org/10.3390/s19091983.

Baca, A., Dabnichki, P., Hu, C.-W., Kornfeind, P. and Exel, J. (2022). Ubiquitous Computing in Sports and Physical Activity—Recent Trends and Developments. *Sensors*, 22(21), p. 8370.

Bargh, M. (2019). Digital Health Software and Sensors: Internet of Things-based Healthcare Services, Wearable Medical Devices, and Real-Time Data Analytics. *American Journal of Medical Research*, 6(2), pp. 61–66. [online] Available at: https://www.ceeol.com/search/article-detail?id=804108 (Accessed 25 Apr. 2023).

Bedi, A., Qadir, S., Ottman, P., Bedi, A., Masri, M. K. A., Hennawi, H. A., Qadir, S. and Ottman, P. (2023). The Integration of Artificial Intelligence Into Patient Care: A Case of Atrial Fibrillation Caught by a Smartwatch. *Cureus*, 15(3). https://doi.org/10.7759/cureus.35941.

Breker. (2022). 74 Technology Trends to Expect in 2023. www.abiresearch.com. [online] Available at: https://www.abiresearch.com/blogs/2023/01/01/74-technology-trends-for -2023/ (Accessed 25 Apr. 2023).

Brown, S. M. and Brown, K. M. (2021). Should Your Wearables Be Shareable? The Ethics of Wearable Technology in Collegiate Athletics. *Marquette Sports Law Review*, 32, p. 97. [online] Available at: https://heinonline.org/HOL/LandingPage?handle=hein.journals/ mqslr32&div=9&id=&page= (Accessed 25 Apr. 2023).

Chen, L. and Yang, S. (2023). Impact of Sports Wearable Testing Equipment Based on Vision Sensors on the Sports Industry. *Journal of Sensors*, 2021(2), p. e2598721. https://doi.org /10.1155/2021/2598721.

Claudino, J. G., Capanema, D. de O., de Souza, T. V., Serrão, J. C., Machado Pereira, A. C. and Nassis, G. P. (2019). Current Approaches to the Use of Artificial Intelligence for Injury Risk Assessment and Performance Prediction in Team Sports: A Systematic Review. *Sports Medicine - Open*, 5(1).

Dai, X. and Li, S. (2021). Application Analysis of Wearable Technology and Equipment Based on Artificial Intelligence in Volleyball. *Mathematical Problems in Engineering*, 2021, pp. 1–10.

Dao, N.-N. (2022). Internet of Wearable Things: Advancements and Benefits from 6G Technologies. *Future Generation Computer Systems*, 138, 172–184.

Dimitropoulos, N., Togias, T., Zacharaki, N., Michalos, G. and Makris, S. (2021). Seamless Human–Robot Collaborative Assembly Using Artificial Intelligence and Wearable Devices. *Applied Sciences*, 11(12), p. 5699.

Fortune Business Insights. (2021). Fitness Tracker Market Size, Share, Trends | Global Analysis, 2020–2027. www.fortunebusinessinsights.com. [online] Available at: https:// www.fortunebusinessinsights.com/fitness-tracker-market-103358.

Gaur, B., Shukla, V. K. and Verma, A. (2019). Strengthening People Analytics through Wearable IOT Device for Real-Time Data Collection. 2019 International Conference on Automation, Computational and Technology Management (ICACTM), London, UK.

Ghosh, I., Ramamurthy, S. R., Chakma, A. and Roy, N. (2023). Sports Analytics Review: Artificial Intelligence Applications, Emerging Technologies, and Algorithmic Perspective. In Mehmed Kantardzic, Witold Pedrycz (Eds.) *Wiley Interdisciplinary Reviews-Data Mining and Knowledge Discovery*. Wiley-Blackwell.

Hanlon, C., Krzak, J. J., Prodoehl, J. and Hall, K. D. (2019). Effect of Injury Prevention Programs on Lower Extremity Performance in Youth Athletes: A Systematic Review. *Sports Health: A Multidisciplinary Approach*, 12.1, p. 12–22.

Henriksen, A., Haugen Mikalsen, M., Woldaregay, A. Z., Muzny, M., Hartvigsen, G., Hopstock, L. A. and Grimsgaard, S. (2018). Using Fitness Trackers and Smartwatches to Measure Physical Activity in Research: Analysis of Consumer Wrist-Worn Wearables. *Journal of Medical Internet Research*, 20(3), p. e110. https://doi.org/10.2196/jmir.9157.

Hilete. (2022). Wearables AI Global Market Report 2022: Rapid Adoption of New Technologies Such as Artificial Intelligence and 5G Smartphones Drives Growth. *GlobeNewswire News Room*. [online] Available at: https://www.globenewswire.com/ en/news-release/2022/12/29/2580610/28124/en/Wearables-AI-Global-Market-Report -2022-Rapid-Adoption-of-New-Technologies-Such-as-Artificial-Intelligence-and-5G -Smartphones-Drives-Growth.html (Accessed 25 Apr. 2023).

Ho, M.-T., Mantello, P., Ghotbi, N., Nguyen, M.-H., Nguyen, H.-K. T. and Vuong, Q.-H. (2022). Rethinking Technological Acceptance in the Age of Emotional AI: Surveying Gen Z (Zoomer) Attitudes Toward Non-Conscious Data Collection. *Technology in Society*, 70, p. 102011.

Iliadis, A., Tomovic, M., Dervas, D., Psymarnou, M., Christoulas, K., Kouidi, E. J. and Deligiannis, A. P. (2021). A Novel mHealth Monitoring System during Cycling in Elite Athletes. *International Journal of Environmental Research and Public Health*, 18(9), p. 4788.

Kah Phooi, Seng, Ang, L.-M., Peter, E. and Mmonyi, A. (2023). Machine Learning and AI Technologies for Smart Wearables. *Electronics*, 12(7), pp. 1509–1509. https://doi.org /10.3390/electronics12071509.

Khayyam, H., Javadi, B., Jalili, M. and Jazar, R. N. (2019). Artificial Intelligence and Internet of Things for Autonomous Vehicles. *Nonlinear Approaches in Engineering Applications*, pp. 39–68.

Lee, H. S. and Lee, J. (2021). Applying Artificial Intelligence in Physical Education and Future Perspectives. *Sustainability*, 13(1), p. 351.

Li, B. and Xu, X. (2021). Application of Artificial Intelligence in Basketball Sport. *Journal of Education, Health and Sport*, 11(7), pp. 54–67.

Matijevich. (2022). Three Wearable That Could Impact the World of Sport. *Sportsmith*. [online] Available at: https://www.sportsmith.co/articles/the-next-big-thing-three-new -wearable-features-that-could-impact-the-world-of-sports-performance/ (Accessed 25 Apr. 2023).

Nadikattu, R. R. (2020). Implementation of New Ways of Artificial Intelligence in Sports. papers.ssrn.com, Rochester, NY. [online] Available at: https://papers.ssrn.com/sol3/ papers.cfm?abstract_id=3620017.

Nahavandi, D., Alizadehsani, R., Khosravi, A. and Acharya, U. R. (2022). Application of Artificial Intelligence in Wearable Devices: Opportunities and Challenges. *Computer Methods and Programs in Biomedicine*, 213, p. 106541.

Nayak, B., Bhattacharyya, S. S., Kumar, S. and Jumnani, R. K. (2021). Exploring the Factors Influencing Adoption of Health-Care Wearables among Generation Z Consumers in India. *Journal of Information, Communication and Ethics in Society*. 20.1, 150–174

Pandey, S., Chawla, D., Puri, S. and Jeong, L. S. (2021). Acceptance of Wearable Fitness Devices in Developing Countries: Exploring the Country and Gender-Specific Differences. *Journal of Asia Business Studies*, 16.4, 676–692

Pazzanese, C. (2022). Ethical Concerns Mount as AI Takes Bigger Decision-Making Role. *Harvard Gazette*. [online] Available at: https://news.harvard.edu/gazette/story/2020/10 /ethical-concerns-mount-as-ai-takes-bigger-decision-making-role/.

Phaneuf, A. (2022). Latest Trends in Medical Monitoring Devices and Wearable Health Technology. *Insider Intelligence*. [online] Available at: https://www.insiderintelligence .com/insights/wearable-technology-healthcare-medical-devices/.

Phatak, A. A., Wieland, F.-G., Vempala, K., Volkmar, F. and Memmert, D. (2021). Artificial Intelligence Based Body Sensor Network Framework—Narrative Review: Proposing an End-to-End Framework using Wearable Sensors, Real-Time Location Systems and Artificial Intelligence/Machine Learning Algorithms for Data Collection, Data Mining and Knowledge Discovery in Sports and Healthcare. *Sports Medicine - Open*, 7(1).

Qermane, K. and Mancha, R. (2020). WHOOP, Inc.: Digital Entrepreneurship During the Covid-19 Pandemic. *Entrepreneurship Education and Pedagogy*, p. 251512742097518.

Raman, P. and Aashish, K. (2021). Gym Users: An Enabler in Creating an Acceptance of Sports and Fitness Wearable Devices in India. *International Journal of Sports Marketing and Sponsorship*, 23.4, 707–726

Rössler, R., Donath, L., Bizzini, M. and Faude, O. (2015). A New Injury Prevention Programme for Children's Football – FIFA 11+ Kids – Can Improve Motor Performance: A Cluster-Randomised Controlled Trial. *Journal of Sports Sciences*, 34(6), pp. 549–556.

Roger, B. (2022). How Wearable Technology Can Benefit Your Workout. butlertechnologies .com. [online] Available at: https://butlertechnologies.com/blog/wearable-technology -can-benefit-workout.

Salehi, W., Gupta, G., Bhatia, S., Koundal, D., Mashat, A. and Belay, A. (2022). IoT-Based Wearable Devices for Patients Suffering from Alzheimer Disease. *Contrast Media & Molecular Imaging*, 2022, pp. 1–15.

Sanyal, S. (2022). How Are Wearables Changing Athlete Performance Monitoring? *Forbes*. [online] Available at: https://www.forbes.com/sites/shourjyasanyal/2018/11/30/how-are -wearables-changing-athlete-performance-monitoring/?sh=10db23fcae09   (Accessed 25 Apr. 2023).

Scott. (2022). What Is Human-Computer Interaction (HCI)? *The Interaction Design Foundation*. [online] Available at: https://www.interaction-design.org/literature/topics /human-computer-interaction.

Sharma, A., Singh, A., Gupta, V. and Arya, S. (2022). Advancements and Future Perspectives of Wearable Sensing Technology for Healthcare Applications. *Sensors & Diagnostics*, 44(2). https://doi.org/10.1039/d2sd00005a.

Shore, V. (2022). How Do Wearable Fitness Trackers Measure Steps? *News-Medical.net*. [online] Available at: https://www.news-medical.net/health/How-do-wearable-fitness -trackers-measure-steps.aspx.

Sieron, T. (2023). Design Wearables That People Want to Use. *punktum*. [online] Available at: https://punktum.net/insights/design-wearables-people-want-to-use/ (Accessed 25 Apr. 2023).

Sui, A., Sui, W., Liu, S. and Rhodes, R. (2023). Ethical Considerations for the use of Consumer Wearables in Health Research. *Digital Health*, 9, p. 205520762311537.

Trevor. (2023). Wearable Devices in Veterinary Medicine: The Big Data or Digital Revolution in Animal Health. *Pashudhan Praharee*. [online] Available at: https://www.pashud-hanpraharee.com/wearable-devices-in-veterinary-medicine-the-big-data-or-digital -revolution-in-animal-health/.

Whitford, D. K., Gage, N. A., Katsiyannis, A., Counts, J., Rapa, L. J. and McWhorter, A. (2019). The Exclusionary Discipline of American Indian and Alaska Native (AI/ AN) Students with and Without Disabilities: A Civil Rights Data Collection (CRDC) National Analysis. *Journal of Child and Family Studies*, 28(12), pp. 3327–3337.

Wu, M. and Luo, J. (2019). Wearable Technology Applications in Healthcare: A Literature Review. www.himss.org. [online] Available at: https://www.himss.org/resources/wear-able-technology-applications-healthcare-literature-review.

Yasar, K. (2022). What Is Wearable Technology? - Definition from WhatIs.com. *SearchMobileComputing*. [online] Available at: https://www.techtarget.com/searchm obilecomputing/definition/wearable-technology.

23 Recommendation Systems for Customers

Krishnachaitanya Pabbu and Rizwan Zhad

INTRODUCTION

The way we get information has changed significantly because of technological improvements and online services. Users may now rate, comment on, and exchange ideas online, which has produced an enormous amount of data. Recommender systems, which try to cut down on the time and effort needed for searches, have evolved as a solution to this problem. These systems offer customers personalized recommendations in a variety of businesses, including web services, e-commerce, tourism, and entertainment [1]. However, there is still potential for improvement in the efficacy and adaptability of recommender systems across several areas, notwithstanding their advancement. It is necessary to conduct additional study and research, paying particular attention to various applications, algorithmic categorization, and approaches. Consideration of simulation platforms is essential and also offers thorough explanations of datasets, which are frequently ignored features in this subject.

In conclusion, it is critical to carry on with research and innovation to enhance the capabilities of recommender systems and overcome the challenges they encounter. It is crucial to continually improve the efficacy of these systems because they are crucial for providing customized recommendations.

The three primary categories of recommender systems are content-based recommenders, collaborative recommenders, and hybrid recommenders. These divisions show the many approaches used in recommendation algorithms [2].

THE SIGNIFICANCE OF RECOMMENDATION SYSTEMS ACROSS DISCIPLINES AND INDUSTRIES

- **E-commerce**: By taking into consideration customer preferences, browsing habits, and previous transactions, online shopping platforms' recommendation systems can produce personalized product recommendations. This strategy encourages customer interaction, improves the whole shopping experience, and boosts revenue.
- **Entertainment:** Recommendation systems are generally used by streaming services for films, TV episodes, and music to make content suggestions that suit the preferences and tastes of their users. To provide individualized

DOI: 10.1201/9781032686714-23

recommendations, these systems examine users' viewing or listening habits, preferred genres, and social connections.

- **Online advertising:** Through the delivery of relevant ad suggestions to users while taking into consideration their demographic data, surfing habits, and previous interactions, recommendation systems optimize online advertising campaigns. The efficacy of advertising and user engagement is increased, thanks to this tailored strategy.

- **Curation of news and content:** Recommendation systems are crucial in generating personalized news stories, blog posts, and other sorts of information based on users' tastes and reading habits because there is so much content available online. For users, these technologies make it easier to find trustworthy and pertinent information.

- **Social media:** Recommendation engines are used by social media platforms to connect users with organizations, people, and relationships that are relevant to their interests, social networks, and activities. Users of these platforms can widen their social networks and find fresh information.

- **Travel and tourism:** Recommender systems are essential in the travel and tourism sector because they make recommendations for locations, lodgings, restaurants, and activities based on user preferences, historical travel behavior, and comments from other users. These programs improve the procedure for making travel arrangements and provide tailored advice in line with user preferences.

- **Education and e-learning:** In the field of e-learning, recommendation systems are increasingly used to suggest appropriate courses, learning materials, and educational resources while taking into account students' learning goals, academic backgrounds, and performance. This encourages personalized learning paths and improves academic results.

In conclusion, recommendation systems are now crucial in a wide range of industries since they provide consumers with individualized and pertinent ideas. They improve user experiences, encourage engagement, and support the expansion and prosperity of companies across various industries [3].

THE POWER OF RECOMMENDATION SYSTEMS: IMPROVING USER ENGAGEMENT, SALES, AND CUSTOMER EXPERIENCE

By employing diverse approaches, recommendation systems greatly enhance user engagement, foster sales growth, and elevate the overall customer experience.

1. **Personalized suggestions:** By examining unique tastes, browsing habits, and prior purchases, recommendation systems improve the consumer experience and offer personalized recommendations. The total customer experience is eventually improved by this personalization feature, which makes sure that suggestions are in line with unique interests and preferences.

2. **Finding new and useful products/content:** Recommendation systems help users find new and valuable products or information by introducing

them to products that are related to or complementary to their current selections. Users are exposed to more options, which boosts engagement and satisfaction.

3. **Decision-making that is simplified:** Recommendation systems make decision-making easier by relieving the load of having to consider a large number of possibilities. They reduce information overload by giving users a short list of options that are catered to their tastes. Users may then make informed decisions, which improves the user experience.

4. **Upselling and cross-selling:** Systems that provide recommendations are excellent at promoting cross-selling and upselling tactics. By proposing related or more expensive products that go well with users' current selections, these systems encourage users to look into new options and may enhance revenue. These systems may encourage higher-value purchases by suggesting goods that match users' interests, thereby increasing sales income.

5. **Improved user involvement:** The level of user engagement can be significantly increased by using personalized and pertinent suggestion systems. These systems successfully catch and hold users' attention by continually providing interesting and appealing recommendations, which results in prolonged interactions and higher user satisfaction.

6. **Increased customer loyalty:** It fosters loyalty and pleasure by continually providing personalized recommendations that take users' preferences and needs into account. As a result, this promotes return business, raises client retention rates, and fosters good word-of-mouth, all of which help to boost sales and revenue.

7. **Targeted advertising:** Systems that make recommendations are essential for enabling customized advertising efforts. These systems can recommend suitable adverts that are in line with the tastes and behaviors of users. By increasing the likelihood of user engagement and conversion, this improved targeting makes advertising more effective.

In conclusion, recommendation systems significantly affect user engagement, sales growth, and the overall customer experience. By offering personalized suggestions, streamlining decision-making and the discovery process, cross-selling and upselling, increasing user engagement, boosting customer loyalty, and using targeted advertising, these systems improve customer satisfaction, increase sales revenue, and create an interactive and engaging user journey.

TYPES OF RECOMMENDATION SYSTEMS

COLLABORATIVE FILTERING

Collaborative filtering is a popular technique used in recommender systems that relies on the idea that past agreements among people can predict future agreements and preferences. It can be categorized into memory-based and model-based methods, with user-based algorithms and matrix factorization being prominent examples [4].

The strength of collaborative filtering lies in its ability to provide accurate recommendations without requiring an in-depth understanding of individual items. Various algorithms, such as k-nearest neighbor (KNN) and Pearson correlation, assess user or item similarity.

Data collection methods in collaborative filtering can be explicit or implicit. Explicit methods involve direct input from users, while implicit methods analyze user actions to uncover preferences.

Collaborative filtering faces challenges such as the cold-start problem, scalability issues, and the sparsity problem arising from a large number of available items.

Amazon.com's item-to-item collaborative filtering serves as an excellent example of its effectiveness. Collaborative filtering continues to play a crucial role in hybrid systems that combine multiple approaches to deliver exceptional recommendations.

CONTENT-BASED FILTERING

Content-based filtering is an intriguing approach used in recommender systems that focuses on item characteristics rather than relying solely on user data. It creates item profiles based on attributes and keywords, enabling algorithms to recommend items similar to those the user has shown interest in. Unlike other methods, content-based filtering doesn't require user sign-in, resulting in a more dynamic profile. The system compares candidate items with previously rated ones to find the best matches, drawing on research from information retrieval and filtering.

To construct a user profile, the system takes into account two types of information: the user's preference model and their interaction history with the recommender system. An item presentation algorithm abstracts features using techniques like tf-idf (term frequency–inverse document frequency) or vector space representation. This generates a content-based user profile based on a weighted vector of item features, indicating the user's preferences. Various techniques, from simple averaging to advanced methods like Bayesian classifiers and artificial neural networks, can be employed to achieve this.

One challenge in content-based filtering is learning user preferences from one content source and applying them across different types of content. Recommender systems become more valuable when they can suggest diverse content, such as music, videos, products, and discussions, based on a user's browsing behavior. To address this, hybridization techniques are commonly used in content-based recommender systems.

Opinion-based recommendations can also be incorporated into content-based filtering. Users can provide reviews or feedback on items, which serve as implicit data for the system. Extracting features from these reviews enhances the item's metadata and captures aspects that interest users. Techniques such as text mining, information retrieval, sentiment analysis, and deep learning are employed to analyze and leverage this valuable feedback.

By combining content-based filtering with other approaches, recommender systems can provide more accurate and diverse recommendations by considering both item attributes and user preferences.

Hybrid Recommendation Approaches

Recommender systems have evolved by blending collaborative filtering, content-based filtering, and other techniques, resulting in hybridization. This fusion of methods offers unlimited possibilities, allowing for separate predictions, merging capabilities, enrichment, and the creation of unified models.

Comparative studies have demonstrated that hybrid methods outperform pure collaborative and content-based approaches, delivering highly accurate recommendations. Hybridization addresses challenges like the cold-start problem and sparsity, while also overcoming bottlenecks in knowledge engineering.

Netflix serves as an example of the power of hybrid recommender systems. By analyzing user behavior, it suggests movies based on similar viewers (collaborative filtering) and shared attributes with highly-rated films (content-based filtering), providing engaging and personalized recommendations.

Hybridization techniques, such as weighted combinations, switching, mixed recommendations, feature combination and augmentation, cascade, and meta-level chaining, bring flexibility and innovation to recommendation systems.

Through the use of hybrid approaches and their diverse techniques, recommender systems are transforming the way users discover and enjoy personalized content.

Exciting ways to improve the accuracy of recommendations are provided by hybridization approaches in recommendation systems. Here are a few inventive strategies:

1. **Weighted fusion:** Numerical methods are used to combine the scores from different suggestion component scores to produce a pleasing synthesis of insights.
2. **Switching strategy:** The system carefully chooses the best suggestion component out of the available ones, bringing a sense of surprise to the recommendation process.
3. **Mixed ensemble:** Imagine a group of recommenders pooling their talents to create a single recommendation that embodies the best of all possible worlds.
4. **Feature melting pot:** One recommendation algorithm is powered by a potent combination of features drawn from various information sets.
5. **Feature boost:** Imagine a feature acquiring superhuman abilities! It combines forces with the following technique after being computed and strengthened, giving the suggestion process an additional advantage.
6. **Cascade hierarchy:** Top contenders are given priority in a properly crafted hierarchy of recommenders that compete for attention. The lower priority ones eagerly await their opportunity to sway the ultimate recommendation, just as in a royal court.
7. **Meta magic:** A chain of fascinating suggestions is created when one recommendation technique takes the initiative and creates a distinctive model that serves as the trigger for the following technique.

A captivating user experience is ensured by these cutting-edge hybridization approaches, which enable recommender systems to provide even more compelling and personalized choices.

PERSONALIZED RECOMMENDATIONS FOR USERS' CURRENT NEEDS: DEMOGRAPHIC-BASED RECOMMENDATION SYSTEMS

The development of utility models for specific items by demographic-based recommendation systems takes into account several user-relevant variables. These systems evaluate each item's utility and recommend the one with the highest utility. They evaluate suggestions depending on the user's current needs and the options available, providing individualized and context-specific advice.

Utility-based systems, however, are constrained by the absence of detailed descriptive features in the items. Even if user recommendations match their preferences, insufficient information may mask important advice.

Additionally, to personalize recommendations, demographic-based recommendation systems employ user demographic data. Age, gender, and geography allow for targeted and personalized recommendations, which improves the user experience overall.

UTILITY-BASED RECOMMENDATION SYSTEMS

A utility-based recommender system generates recommendations by building a model of utility for each item based on user preferences. These systems create a multi-attribute utility function for the user and explicitly suggest the item with the highest calculated utility. The advantage of a supply-based system is that non-product-related attributes such as availability, and supplier reliability can be incorporated into the supply function. It can also provide up-to-date information on item availability and characteristics, allowing users to assess the current status of an item. Unlike other systems, utility-based systems prioritize users' immediate needs and available choices rather than long-term generalizations. However, usage-based system limitations arise when a product lacks sufficient descriptive capabilities. In such cases, the user will not see recommendations that may be relevant, even if they match the user's preferences [5].

It's important to emphasize that user-based recommender systems offer a dynamic and personalized approach to recommendations. These systems aim to meet the specific needs of individual users by incorporating real-time data and considering various attributes of utility models.

KNOWLEDGE-BASED RECOMMENDATION SYSTEMS

Systems that employ explicit knowledge about users and items to make suggestions based on predetermined standards are known as knowledge-based recommendation systems. They don't rely too heavily on initial statistics or user reviews. Instead, they consider the preferences and wants of the user to make suggestions for products that meet those demands. These systems have benefits such as not requiring as many user ratings as machine learning-based techniques do to learn. Furthermore, since recommendations are not tailored to specific user preferences, they don't need a lot of user information. Knowledge-based systems may work alone or in conjunction with other recommendation systems.

However, a noteworthy downside of knowledge-based recommendation systems is the possibility of a knowledge acquisition bottleneck. It can be time-consuming and difficult to define rules, requirements, and ontologies while obtaining the explicit knowledge required for these systems. In these systems, learning, and knowledge acquisition are governed by Batesonian theories.

The use of natural language processing (NLP) techniques to extract knowledge from unstructured text sources like product descriptions, customer reviews, and online forums is an intriguing trend in knowledge-based recommendation systems. These systems can expand their knowledge base and offer consumers more precise and pertinent recommendations by analyzing and digesting textual data.

DATA COLLECTION FOR RECOMMENDATION SYSTEMS

1. **User interactions:** User interactions are typically the primary source of data for recommendation systems. These interactions include explicit feedback such as clicks, purchases, or likes and implicit input such as ratings, reviews, and likes. User interactions are gathered and recorded to provide useful data on user preferences and behavior.
2. **Item attributes:** Along with user interactions, gathering item attributes is essential for creating recommendation systems. Title, description, category, genre, release date, and other pertinent information that describes the recommended things are all included in item attributes. These characteristics provide context for the objects, assisting in understanding user preferences and producing exact suggestions.
3. **Contextual data:** Additional information that is pertinent to the recommendation process is included in contextual data. It includes the demographics of the user, location, time of day, type of device, and other contextual elements that may affect the user's preferences. The effectiveness and personalization of the recommendation system can be increased by using contextual data.

PREPROCESSING STEPS FOR RECOMMENDATION DATA

1. **Data cleaning:** Data cleaning includes addressing missing values, correcting discrepancies within the collection, and removing irrelevant or noisy data. This procedure ensures the data's excellent quality, enabling its efficient use for modeling purposes.
2. **Data transformation:** Data transformation entails transforming data into a format that recommendation systems can use. This can entail actions such as normalizing numerical features, translating categorical variables into

numerical representations, or using dimensionality reduction methods like matrix factorization.

3. **Data sampling and balancing:** The dataset may occasionally show imbalance, indicating a significant difference in the volume of interactions among various objects or people. This issue can be resolved by using sampling strategies to balance the dataset, ensuring fair representation and avoiding bias in the recommendation process.

4. **Feature engineering:** In recommendation systems, feature engineering comprises creating new features or representations from already-existing data to collect more pertinent information. Building user or object profiles, calculating similarity scores between individuals or groups, and incorporating contextual variables are all tasks included in this process.

5. **Data splitting:** In the end, the dataset is usually divided into three sets. These include training, validation, and test sets. The training set is utilized to construct the recommendation model, the validation set aids in adjusting the model's parameters, and the test set is employed to assess the ultimate performance of the model.

The data can be cleaned, transformed, and prepared by following these pretreatment steps to build recommendation algorithms that can provide accurate and individualized suggestions.

EVALUATION METRICS

Understanding a recommendation system's fundamental characteristics and creating quantitative standards to gauge its quality is necessary for assessing its efficacy. Recommendation systems' performance is evaluated using widely used measures such as recall, precision, accuracy, receiver operating characteristic (ROC) curves, and F-score. These measurements enable us to demonstrate the true allure of a superb recommendation system [6].

RECALL AND PRECISION

Information retrieval (IR) and recommender systems both aim to help users find what they need. Imagine a treasure hunt where IR finds relevant documents and RS uncovers exciting treasures from vast collections. In this adventure, IR uses two of its powerful metrics as gauges: precision and recall. Accuracy measures the accuracy of recommender systems by determining the percentage of truly valuable items out of all recommendations. Recall, on the other hand, determines the number of notable recommendations about the total number of gems that should have been found. It's an exciting journey and the success of the referral depends on finding the perfect treasure to captivate users. To convey this sense of excitement, we refer to a mystical matrix that resembles a treasure map outlining the four possible outcomes of a recommendation. If the recommended treasure captivates the user, that's a glorious win. Otherwise, this is a bold attempt and needs improvement.

CONFUSION MATRIX FOR A RECOMMENDER SYSTEM

We commonly assess the effectiveness of recommendation systems using metrics such as precision and coverage. While coverage evaluates the depth and diversity of recommendations, precision evaluates the advice's accuracy. Consider a situation where your recommendation system suggests items to users as a way to demonstrate this. Let's now examine the specifics of this scenario.

Precision can be thought of as a spotlight that shows how precisely a system reaches its intended goal. It displays the number of things that were successfully recommended and retrieved (also known as true positives; denoted by the letter "a"). But precision can occasionally stray since it is not perfect. Some items (often referred to as "b" items) might have been marked as suggested yet the system may have overlooked them. Also possible are items that were mistakenly added to the recommended pool (also known as disqualified items; denoted by "c"). We also take into account the items that were correctly retrieved as "not recommended" and correctly classified as "true negatives" (designated with the letter "d") to get a complete picture of accuracy.

A top-notch recommendation engine aims to strike a positive balance between precision and coverage. It aims to give users a wide variety of recommendations, providing comprehensive coverage. However, concentrating only on general recommendations could perhaps hurt the system's precision as not all items in the pool might be beneficial to the user.

$$Precision = \frac{a}{a + b}$$

$$Recall = \frac{a}{a + c}$$

ACCURACY

Selecting the best recommendation system is a difficult undertaking, comparable to looking for a needle while blinded in a haystack. Predicted ratings are frequently used by organizations as an evaluation metric, but precisely determining how well a recommendation system is working is a challenging task in and of itself. There isn't a quick, miraculous way to determine precision. Instead, we take the system through a split-validation process where we expose it to 80% of the dataset of a user's purchasing history and ask it to forecast the remaining 20%. Between the system's recommendations and the decisions made, a cerebral conflict results. The system's accuracy and the hidden insights buried in the data can only be discovered at that point. It turns into a search for the most accurate advice, where each prediction error serves as a clue that brings us closer to our ultimate objective.

$$Accuracy = \frac{Number\ of\ successful\ recommendations}{Total\ Number\ of\ recommendations}$$

watched by both users

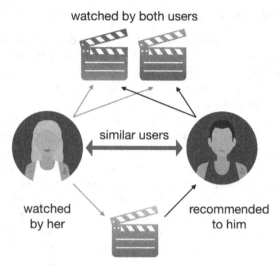

Image source : vitalflux.com

FIGURE 23.1   Architecture of the proposed system Source: vitalflux.com

FIGURE 23.2   Costumer buying preferences Source: medium.com

A key consideration in rating recommendation systems is accuracy, and the root mean square error (RMSE) is a popular metric for this task. RMSE is widely used to assess algorithms since it reflects the overall accuracy of ratings, whether they are positive or negative. RMSE is frequently favored despite the existence of alternative metrics such as mean average error and normalized mean average error that provide various evaluation viewpoints. This is because RMSE treats all rating errors equally, making sure that even minute differences in ratings are taken into account. This is particularly helpful when anticipating rating discrepancies between stars that are close to one another because correct recommendations depend on the capacity to recognize even the smallest variations.

ROC Curve

ROC analysis offers a different strategy than precision/recall evaluation when comparing performance evaluation methods. When precision is functioning extraordinarily well in the context of recommendations, the recall tends to drop.

The relationship between fallout (false positive rate) and recall (true positive rate) is shown by the ROC curve. Its objective is to minimize fallout or false positives while maximizing recall, which represents the retrieval of pertinent items. The ROC curve gives us the ability to control the threshold for classifying objects as recommended or not recommended and to understand the balance between recall and precision. Figure 23.3 demonstrates that to maximize memory and precision, the peak of the curves must be moved closer to the point at which recall and precision are equal to 1.

The ROC curve would include all important data before including any irrelevant data in an ideal forecasting system. The ROC metric implies binary relevance, classifying items as successful or unsuccessful suggestions, hence the order in which relevant items are presented has no bearing on the ROC meter. The ability of the system to discern between these two categories is what is stressed. The area under the curve must be looked at when evaluating performance using the ROC curve. The likelihood that the system will correctly distinguish between a randomly selected item from the set of acceptable things and one from the set of unsatisfactory items is shown in this area.

Content-based filtering

Read by user

Similar articles

Recommended to user

FIGURE 23.3 Content based filtering

F-SCORE

A metric called the F-score combines precision and recall to provide a thorough assessment of system performance. In comparison to using precision and recall separately, it provides a more thorough evaluation. The F-score can emphasize either precision or recall by modifying the value. The F1 score, which creates a balanced assessment of recall and precision, is the most widely used variation. When the greatest F-score value is 1, all forecasts are considered reliable recommendations.

Precision represents the percentage of relevant items in the top-k recommendations for ranked retrieval, whereas recall measures the percentage of relevant items found in the top-k recommendations. Because a wider set of recommendations results in a stronger recall, the F-score tends to rise as the k's value rises. Below are the equations for calculating F and F1.

$$F_\beta = \frac{precision \times recall}{(1 - \beta) \times precision + \beta \times recall}$$

$$F_1 = \frac{2 \times precision \times recall}{precision + recall}$$

CHALLENGES AND FUTURE DIRECTIONS

Analyzing the performance of recommendation systems is similar to navigating a dynamic environment. Although there is no surefire way to gauge user satisfaction, we may assess the system's capabilities by looking at how it performs when faced with typical problems. The cold-start problem, accuracy restrictions, worries about limited data, scalability problems, and diversity issues are just a few of the formidable recommendation-related roadblocks we examine in this investigation.

Cold Start

The "cold-start" issue in recommendation systems is comparable to starting a car's engine on a chilly day. Similar to how engines need time to warm up and function

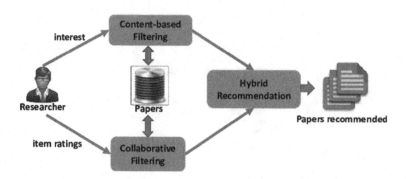

FIGURE 23.4 Hybrid model for recommendation Source: Analytics India Magazine

properly, recommendation algorithms struggle when there is not enough data. There are two sorts of cold starts: user cold starts and product cold starts.

When a new product debuts on an e-commerce platform without any reviews or user interactions, this is known as a "cold start." Without user feedback, it is challenging for the recommendation algorithm to deliver appropriate adverts. When new users join without preferences or browsing history, this is known as a "user cold start," which makes personalized recommendations challenging. Existing users may also experience cold starts if their settings abruptly change and the system is left in the dark about their preferences.

The cold-start problem is addressed using metrics, approaches, and techniques including projection in weighted alternating least squares (WALS) and Bayesian classifiers. WALS enables the system to determine user embeddings for new objects, and Bayesian models provide reasoning in recommendation systems. These methods successfully address the issue of cold starts.

Recommendation systems can deliver worthwhile recommendations even with little data by overcoming the cold-start issue, resulting in a better user experience.

$$min_{ui0 \in Rd} \|Ai0 - ui0VT\|$$

A fascinating aspect of the WALS method, which successfully maintains the user's embeddings, is presented by the equation. The system takes on the task of creating embeddings for new objects during each iteration to keep the model up to date. Heuristic techniques are used to estimate embeddings for these additional items when interactions are insufficient. One intriguing method encourages adaptation and adds creativity to the suggestion process by averaging the embeddings of things that fall into the same category.

DATA SPARSITY

When users have only given ratings to a small number of things, data sparsity poses a problem for recommendation algorithms and can result in erroneous

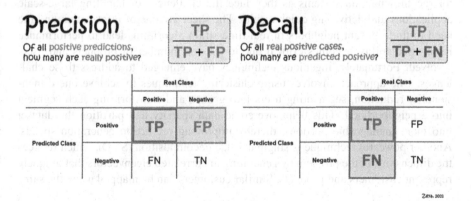

FIGURE 23.5 Confusion Matrix Source: Medium.com

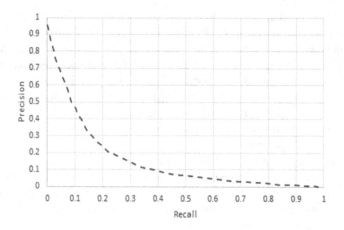

FIGURE 23.6   AUC-ROC Curve

recommendations. By utilizing user behaviors and reliable social ties, trust-based solutions have arisen to overcome this problem. By combining trust assertions, creating nodes for users, and directing edges for trust connections, trust networks are created. Users' distances from one other are calculated using their relationships of trust to determine trustworthiness. Through a web of trust, these trust-aware systems determine user trust levels. By including reviews from reliable neighbors, taking similarity into account, and averaging ratings on frequently rated products, trust-based strategies like the merge approach increase the accuracy of recommendations. These creative methods get beyond the drawbacks of sparse data by combining trust networks and users' confidence in one another, producing more precise and reliable results.

SCALABILITY

With the rapid growth of e-commerce, scalability becomes a pressing concern for recommendation systems as they face the challenges of handling large-scale applications and delivering quick results. The sheer volume of information and the need to find relevant neighbors in real-time strain algorithms lead to performance issues. Scalability becomes critical due to the vast number of products and users involved. Fortunately, ingenious techniques have emerged to address these challenges. One approach involves using clustering techniques to achieve one-dimensionality reduction, segmenting users into groups and transforming each segment into a neighborhood. This helps overcome data sparsity and partition the dataset into more manageable portions, thereby improving prediction generation speeds. Another powerful technique is singular value decomposition (SVD), which reduces the dimensions of the problem by generating uncorrelated eigenvectors that uniquely represent customers and products. Similar customers can be mapped using the same

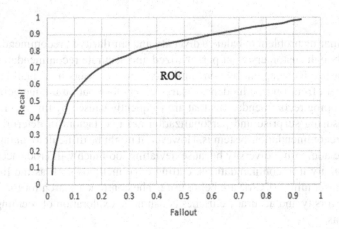

FIGURE 23.7 ROC Curve

eigenvectors based on their ratings of similar but not identical products. By decomposing the rating matrix into SVD component matrices, predictions can be generated by calculating cosine similarities between pseudo-customers and pseudo-products. These innovative methods enable recommendation systems to overcome scalability challenges, enhancing both efficiency and accuracy. It's like embarking on a fascinating data-driven journey to uncover perfect recommendations in the vast realms of e-commerce.

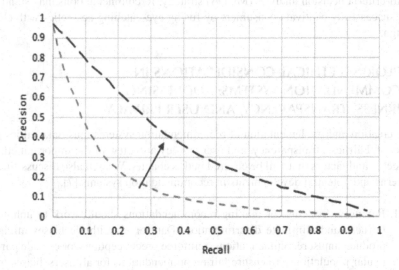

FIGURE 23.8 AUC-ROC Curve 1

DIVERSITY

It's a fascinating problem to balance diversity and similarity in recommendation systems. Although customers want personalized and accurate recommendations based on their tastes, focusing entirely on similarity can keep users in a small bubble and prevent them from finding hidden treasures. Users lose out on unique goods due to the overlapping recommendations that are frequently caused by this diversity problem. Measuring surprise and personalization are two factors in determining how diverse a recommendation system is. However, it might be difficult to maintain accuracy while addressing diversity because deviating too much will erode accuracy. To boost diversity in recommendations, cutting-edge methods such as the linear time closed itemset miner (LCM) are being used. The continued search for the ideal balance of diversity and accuracy enriches consumers' exploration of seemingly limitless options.

HABITUATION EFFECT

Recommendation interfaces are essential for enhancing marketing tactics and delivering persuasive content. Examining components such as the volume of recommendations, aesthetically pleasing photos, intriguing explanations, and appealing layouts will help you increase their effectiveness. It is vital to deliver advice in visually appealing ways because consumers frequently experience information overload and have the propensity to ignore banners. To grab attention, marketers use strategies such as animations and flashing effects. Marketers assess visual intensity, the capacity to captivate attention, and the amount of time needed to do so by using a multi-criteria decision analysis (MCDA) strategy. Recommendations may stand out and successfully captivate consumers by fusing eye-catching aesthetics with clever design.

EXPLORING ETHICAL CONSIDERATIONS IN RECOMMENDATION SYSTEMS: ADDRESSING FAIRNESS, TRANSPARENCY, AND USER PRIVACY

The creation and implementation of recommendation systems must take ethics into account. Fairness, transparency, and user privacy are a few of the important ethical concerns and factors that will be covered in this chapter. We will also discuss current patterns and potential developments in recommendation systems [7].

1. **Fairness:** Systems for making recommendations should work to uphold fairness and eliminate discrimination. Data or algorithmic biases might produce unjust recommendations, reinforce preconceptions, or exclude particular populations. To ensure fair recommendations for all users, biases in data collecting, algorithm design, and decision-making processes must be identified and addressed.

2. **Transparency:** Users should be informed clearly about the process used to create suggestions. It is important to explain the inner workings of recommendation algorithms and give justifications for the decisions that are made. Users can understand and trust the advice they receive when using transparent systems, which promotes a sense of control and allays worries about algorithmic "black boxes."

3. **User privacy:** Respecting user privacy is required when gathering and using user data for recommendation purposes. Users should have control over the collection, storage, and use of their data, and it should be managed securely. To safeguard user information and uphold user confidence, privacy laws, and practices should be adhered to.

EMERGING TRENDS AND FUTURE DIRECTIONS

1. **Explainable AI:** It is becoming more and more important for recommendation systems to provide justifications for their proposals. Explainable AI techniques aim to provide concise and understandable reasoning for the recommendations made. This not only boosts user confidence but also makes it possible for consumers to see how the recommendations are influenced, allowing them to make more educated choices.

2. **Personalized recommendations:** One significant development in recommendation systems is the continued emphasis on personalization. These systems can now gather and use ever-more complex user preferences thanks to ongoing developments in machine learning and data analytics. They are therefore able to offer highly customized advice that is tailored to certain requirements and interests. This individualized strategy greatly raises user satisfaction and system engagement.

3. **Multi-stakeholder recommendation approaches:** In recommendation systems, taking into account the interests and perspectives of different stakeholders is becoming more important. The preferences of users, content providers, marketers, and other pertinent entities must be taken into consideration to achieve this. Recommendation systems can produce outcomes that are advantageous to all parties by balancing the demands and goals of many stakeholders.

4. **Ethical guidelines and standards:** Establishing moral principles and criteria that are particular to recommendation systems are gaining ground. To define the best procedures and guidelines for the responsible design, implementation, and application of these systems, numerous organizations and business groups are working together. These rules are meant to encourage fairness, transparency, privacy, and responsibility in the recommendation-making process.

5. **Context-aware recommendations:** Context-aware recommendation systems are becoming more significant as a result of their ability to take into account the unique conditions of each user. To give recommendations that

are both relevant and timely, these systems analyze variables including time, location, and user context. This improves the entire user experience.

6. **Cross-domain recommendations:** The transmission of knowledge and recommendations between several domains is now being studied within the topic of cross-domain recommendation systems. Utilizing user preferences and behaviors from one domain to make suggestions in another is an example of this. This new trend gives consumers the ability to find interesting and relevant goods or materials outside of their main area of interest.

Fairness, openness, and user privacy are all included in the ethical considerations for recommendation systems. Making the systems understandable, individualized, and inclusive of various stakeholders is the current trend. Furthermore, cross-domain suggestions, context-aware recommendations, and ethical principles are becoming more popular. The goal of these developments is to create recommendation systems that demonstrate responsibility, accountability, and concern for user needs and societal norms in addition to efficacy and accuracy.

CONCLUSION

In conclusion, recommendation systems are crucial for organizations and customers, providing advantageous benefits to both. They improve interaction with customers, increase revenue, and foster client loyalty for businesses. Customers, on the other hand, benefit from convenient and personalized shopping experiences, learning about new products that suit their tastes.

Future developments in recommendation systems will be significant as more sophisticated algorithms make use of technologies like deep learning and reinforcement learning. Along with preserving privacy and establishing trust, ethical issues will be essential. Recommendation systems will continue to spur innovation, improve consumer experiences, and cement the bond between companies and their clients. The ability for personalized recommendations to become even more impactful and seamless in the future will revolutionize how businesses interact with their customers.

REFERENCES

1. https://vitalflux.com/wp-content/uploads/2022/08/collaborative-filtering-recommender-system.png
2. https://www.appier.com/en/blog/what-is-a-recommendation-engine-and-how-does-it-work#:~:text=There%20are%20three%20main%20types,a%20hybrid%20of%20the%20two.
3. https://indatalabs.com/blog/big-data-behind-recommender-systems
4. https://www.mdpi.com/2076-3417/10/21/7748
5. https://www.analyticsvidhya.com/blog/2021/07/recommendation-system-understanding-the-basic-concepts/#:~:text=A%20recommendation%20system%20is%20a,suggests%20relevant%20items%20to%20users.
6. https://en.wikipedia.org/wiki/Recommender_system
7. https://analyticsindiamag.com/a-guide-to-building-hybrid-recommendation-systems-for-beginners/

24 Analysis of Medical Wearable Device Experience
A Machine Learning Model Using Structural Topic Modeling

Vinay Chittiprolu and Pradeep Sharma

INTRODUCTION

Since the creation of the first sensors, the wearable technology industry has grown rapidly. The shrinking of components has allowed for this evolution to extend beyond hardware, technology, the creation of new and more effective batteries, as well as where and how sensors are used (Patel et al. (2022); Chidambaram et al. (2022)). Wearable technology can be used in a variety of fields, from entertainment to medicine and life-saving systems (Trauth & Browning, 2022). Wearable technology can monitor a patient's vital signs, improve human abilities, replace and enhance the senses, and even warn of and respond to medical situations thanks to the sensors that are already accessible (Ronghe et al. (2022); Webster, Scheeren, and Wan (2022)).

Let us take for example, one of the newest innovations in the development of information technology, smartwatches – mini-computers – have several uses in addition to telling the time. This illustrates the idea that because important information will be comfortably presented on the user's wrist, smartwatches could cause customers to give less attentiveness to other smart devices, like smartphones. However, due to their access to emails, texts, and various other features, smartwatches could potentially attract a user's attention more frequently (Rafl et al., 2022). A classic watch's functions as well as visibility could also elevate it to the status of a luxury item. To put it another way, rather than being hidden, technology and fashion converge to become important aspects of a user's identity.

DOI: 10.1201/9781032686714-24

THE RISE OF WEARABLE DEVICES IN INDIA

Wearable technology use in India has increased as a result of a number of factors, including:

1. Technological advancements
2. Evolving lifestyles
3. Rising consumer awareness

The popularity of wearable devices could increase over time. Thanks to the range of attributes and capabilities, including smartwatches, fitness trackers, smart bands, and augmented reality (AR) glasses, that can enhance users' everyday experience.

INDIA'S ADOPTION OF WEARABLE DEVICES AND ITS CAUSES

1. **Rise of health and fitness consciousness:** India has experienced a surge in health and fitness consciousness among its population. Wearable devices, such as smartwatches, provide relevant tools for users to monitor their health and lead an active and healthy lifestyle (Times of India, 2023).
2. **Technological advancements:** The use and popularity of these gadgets among Indian consumers have expanded as a result of advancements in wearable technology. Component shrinking, longer battery life, and improved connectivity have made wearable technology more practical and user friendly. The development of technology has allowed manufacturers to offer a wide range of reasonable cost solutions that are targeted to specific Indian market sectors. Due to the availability of affordable smartphones with desirable features and functionalities, their use has increased.
3. **Integration of smartphones:** The widespread use of smartphones in India has significantly influenced the uptake of wearable technology. In order to improve the user experience, several wearable devices effortlessly interact with smartphones. These gadgets enable users to stay connected and get immediate access to critical information from their wrists by providing functions like phone notifications, message alerts, and music control. For Indian customers, the use of wearables has become more convenient and alluring because of the interaction between wearable technology and smartphones.
4. **Fashion and style:** Customer tastes in India are influenced by the factors of fashion and style. Producers of wearable devices have taken note of this and focused on creating products that are not only technically cutting-edge but also visibly pleasing. Users of wearable devices can express their individual styles by choosing from a pool of beautiful designs, colors, as well as watch straps. Wearable technology has become more convincing to Indian customers who are interested in fashion, thanks to the combination of technology and style.

5. **Impact of e-commerce and online retail chains:** The wide spread of India's e-commerce market has boosted wearable devices' availability, and the use of wearable devices has been greatly facilitated by e-commerce platforms. It was observed that the Indian wearable devices market has seen a 47% growth year-on-year in 2022 with shipments touching 100 million units, led by smart watches (30.7 million Units) and emerged as the largest market in Q3 2022 as it touched 171% (Business Insider, (2022); Amazon Sales Data (2022)).

LITERATURE REVIEW

Any electronic device that is made to be worn on the user's body is considered a wearable device. Gadgets of this kind come in a variety of sizes and shapes, which include accessories, jewelry, medicative features or medicative equipment, and clothing-related items. While the term "wearable devices" implies communication, certain specific devices can indeed be quite advanced or sophisticated in their capabilities. The high-end wearable devices cover holographic computers which are in the form of virtual-reality headsets, smart glasses, HoloLens, smartwatches, and artificial intelligence-enabled hearing aids. One example of a simpler kind of wearable technology is a disposable skin patch which includes sensors that wirelessly transmit a patient's data to a control unit at a healthcare establishment.

With respect to the evolving lifestyle, acceptance of new technology, increased health consciousness, and also fashion constraints, the demand for wearable devices particularly smartwatches has been increasing since COVID-19. Wearable devices, initially are expected not to support only the ascending demand for medical services with no geographical restrictions or restrictions on time but are also expected to support the increasing demand for medicative services of senior citizens who experience decreased physical movements and a lack of access to specialized medicative care.

India has the highest adolescent youth population in the world (United Nations Population Fund). The youth population is highly cautious, aware of technology, and interested in electronic gadgets. Therefore, it stands as a potential market, particularly for wearable devices in the coming years (Times of India, 2023). When it comes to particular wearable devices like smartwatches, it was witnessed that the demand and trend surged in the last 2 years. Also, it was observed that online shopping has become the new norm among the Indian youth. The advent of the internet has made customers more aware of the products available and made them hard to convince (Consumer Behavior, Schiffman). Online reviews of these wearable products would have a significant impact on purchase decisions.

Many theories put forth on exploring how people are accepting new technology by continuing to develop in the ever-changing digital environment. Ajzen (1991) developed the theory of planned behavior, and Venkatesh (2003) had his unified theory of acceptance and use of technology are a couple of examples. The innovation diffusion model (Rogers, 1962) is another.

The technology acceptance model (TAM) is one that is most frequently used to assess the individual acceptability of developing information and communication technologies (ICTs). The theory of reasoned action, which is a well-researched hypothesis with roots in behavioral psychology, is where TAM gets its start. The earliest version of TAM describes in which ways the users would adopt computer information systems at the workplace (Davis, 1989). The validity, practicality, and capacity of TAM account for the variation in terms of intention to use, and behaviors that are to be supported by the empirical data. For instance, Venkatesh (2000) claims that "significant theoretical and empirical support has accumulated in Favor of TAM," while King (2006) claims that "TAM is a powerful and robust predictive model." TAM is therefore frequently chosen over competing models, like the theory of planned behavior and the theory of reasoned action, across a wide range of user situations.

METHODOLOGY

DATA

This study collected the smartwatch user experiences from the Amazon India website. We collected data manually and stored it in Excel format. We chose three premium brands and three budget brands' user data. The data has review, reviewer, and brand information. The review information contains the numerical ratings of the product and text reviews of the product. The reviewer information has the reviewer's name and geographical location. The brand name and features were part of the brand information. A total of 500 reviews were collected for further analysis.

TEXT MINING

Analyzing the text data has numerous advantages, by using text data reviewers can identify the new or hidden dimensions from the data and compare the dimensions between premium and budget brands. However, analyzing text data has some challenges. First, text data doesn't have any proper format. The style of writing and wording may be different. Second, the text data needs to be cleaned before analysis because the data may contain stop words, brand names, and some unnecessary information. Finally, the data needs to be normalized by transforming the words to their root words (Tirunillai and Tellis (2014)).

TEXT CLEANING

Text cleaning involves transforming raw text data into a standardized and clean format that is appropriate for analysis. The procedure can be summarized as the

1. Removal of special characters, numbers, symbols, emojis, punctuations, numbers, stop words, and special words
2. Lemmatization of words

3. Conversion text to lowercase
4. Removal of words with fewer characters

STRUCTURAL TOPIC MODEL

Structural topic modeling (STM) is a powerful machine learning algorithm for uncovering latent thematic topics in text corpora. It is a more advanced model than latent Dirichlet allocation (LDA) to provide insights into the relationships between metadata and textual content. Each document is a mixture of latent topics, and each topic is a combination of words. However, STM advances the uses of LDA by incorporating covariates that capture the metadata associated with each document, such as brand reputation, and brand value characteristics. Also, STM calculates the correlation between topics which gives a thorough understanding of the topics' relations. Finally, the interaction effect is also possible using structural topic modeling. By considering these covariates, STM enables the investigation of how topics vary across different groups or conditions.

THE PROCESS OF STM ANALYSIS

1. Creation of document feature matrix (dfm)
2. Identify number of topics (K)
3. Run the stm model
4. Calculate the coherence score and FREX (Frequently inclusive and exclusive) score
5. Label the topics
6. Do the co-variate analysis (Here brand value)
7. Do the correlation analysis
8. Compare with literature

The entire analysis was performed using R studio (R Studio is tool https://posit.co/download/rstudio-desktop/) and sentiment, stm, ggplot, dplyr, tidytext, wordcloud2, and igraph packages.

RESULTS

After cleaning data, the technical features and sentiment score were calculated using sentiment R package. Table 24.2 shows the technical features of the reviews by rating.

DATA EXPLORATION

As part of our data exploration, we conducted a comprehensive analysis of unigrams and bigrams. Figure 24.1 and Figure 24.2 display the results of this analyses. By conducting both unigram and bigram analyses, we can gain a comprehensive view of the textual data. These analyses help us identify significant terms, discover meaningful

TABLE 24.1
Previous studies

Authors	Methodology	Results
Adapa et al. (2018)	The approach was based on rich data collected in the form of interviews. Sample Size 25 in Midwestern Technological State University, Texas USA	Factors like functions, functionality, compatibility, look and feel, battery life, GPS, battery heat, technology, data privacy, interface, weight, hands-free interaction, waterproofing, and notifications.
Sergueeva, Shaw, and Lee (2020)	Data is processed through the partial least squares(PLS) method. (Canada)	The findings of this study demonstrated a favorable correlation between the intention of consumers to use wearable devices and the dimensions of performance expectancy, facilitation circumstances, social influence, habit, hedonic incentive, and personalization.
Pathania, Dixit, and Rasool (2022)	Data was analyzed using partial least square structural equation modeling (PLS-SEM) with smart PLS on a survey of 434 wearable technology users.	The structure path analysis reveals that watching homophilous users' online evaluations as well as ambiguities regarding one's own tendency and a company's reputation favorably impacts people's herding in the adoption of wearable technology for personal healthcare.
Chuah et al. (2016).	Students studying business at a Malaysian university were asked to complete a survey. 226 complete paper-pencil surveys were gathered.	Findings support previous TAM research by confirming that attitudes about using smartwatches are influenced by perceived utility and visibility, which results in adoption intention.
Lee and Lee (2020)	The constructs were evaluated using a five-point Likert scale in the questionnaire. To fit this research, measurement items from earlier studies were adjusted. SPSS (Statistical Package for the Social Sciences) 23.0 and AMOS (Analysis of Moment Structures) 23.0 were used in this investigation. The method of choice was structural equation modeling (SEM), which offers the instruments required to evaluate the assumptions.	The study's findings supported the favorable impacts of internal variables' knowledge (H1), attitudes (H2), and beliefs (H3) on the actual usage of medical wearables and applications.

(Continued)

TABLE 24.1
(Continued)

Authors	Methodology	Results
Vijayalakshmi et al. (2018).	In this work, the done market research to analyze several aspects that affect the demand for wearable technology	In 2019, the number of wearable devices reached a record high of 68.7 million, which had a huge impact on businesses and people's way of life. The domains of medicine and safety are other areas where wearables are useful. By 2020, 20 million units are anticipated to be sold in the E-Textile market. Additionally, Gartner predicts a 17% increase in smartwatch sales by 2021 compared to now. The sale of smart eyewear is anticipated to increase by 40% by 2020.
Channa, Popescu, Skibinska, and Burget (2021)	This particular research was carried out entirely by PRISMA guidelines	In order to remedy the issue, this research advises that governments enact security legislation.
How online reviews affect purchase intention: a new model based on the stimulus-organism-response (SOR) framework	The SOR framework of how online reviews affect purchasing behavior served as the sole theoretical foundation for the research.	According to the findings, there is a direct link between online reviews and online purchasing behavior. The more favorable internet evaluations are, the more likely people are to make a purchase.

TABLE 24.2
Technical features of the reviews

Rating	Mean Word Length	Mean Sentiment Score	cCunt	Negative	Positive	Sentiment
1	59.88	-0.15	17	46	17	-29
2	66.57	-0.14	14	35	19	-16
3	58.94	-0.07	82	139	152	13
4	44.35	0.34	225	184	410	226
5	53.91	0.44	139	105	428	323

FIGURE 24.1 Unigram analysis

associations between words, and extract valuable insights that contribute to a more thorough exploration of the dataset.

Figure 24.1 shows the top 50 frequently repeated words. Words with higher frequency have larger font compared to words with lower font. Consumers talk about device features, charging issues, battery back-up, tracking time, app features, display features, etc.

Figure 24.2 shows the bigrams. Bigrams display the relationship between two associated words. Based on the word's association, some clusters have been formed. Consumers spoke more about battery issues, call features, product quality, brand value, service centers, heart rate monitors, display features, and sports modes.

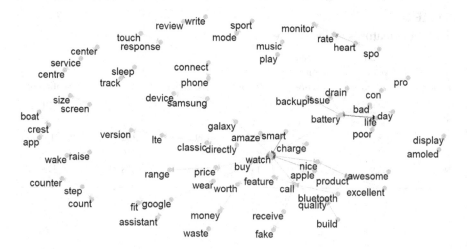

FIGURE 24.2 Bigram analysis

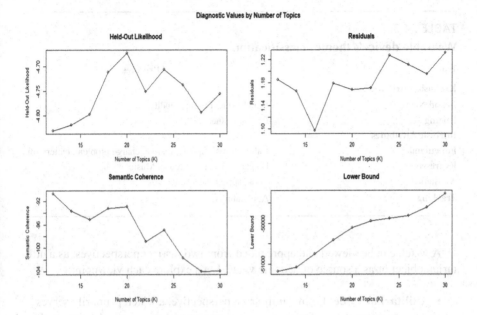

IMAGE 24.3 Finding the number of topics

Researchers get the initial or basic idea about user experience by doing unigram and bigram analysis. To get a full idea, researchers have to do thematic or topic model analysis.

STM Results

By using the *quanteda* package, the dfm matrix was created. Identifying a number of topics involves scientific procedure. We followed the below studies to identify the number of topics (for example, Ding et al. (2020); Park, Chae, and Kwon (2020); Delgosha, Hajiheydari, and Talafidaryani (2022)). Image 24.3 shows the diagnostic values by number of topics. The lower the residual values, the better the model. After 17 topics there is not much increase in the lower bound values and residuals. Hence, we choose k=17 topics and run the model.

A total of 17 topics were identified and broadly categorized into two categories: intrinsic and extrinsic attributes. Extrinsic attributes refer to the tangible or external characteristics of a product or service that persuade consumers' perceptions and purchase decisions. Brand name and pricing are part of extrinsic attributes. Whereas intrinsic attributes refer to the intangible or internal characteristics of a product or service that are intrinsic to the offering itself. These attributes are associated with the product's qualities, features, or characteristics that exist regardless of external factors. Functional, expressive, aesthetic, and tracking attributes were part of the intrinsic attributes of wearable devices. All 17 topics were mapped to intrinsic and extrinsic attributes and shown in Table 24.3.

TABLE 24.3

Wearable device's themes classification

Type	Attributes
Extrinsic Attributes	
Brand Name	Brand value, Product quality
Pricing	Product value
Intrinsic Attributes	
Functional	Call features, support, service, charge problem, battery life
Expressive	Display, Face, Experience, Music
Aesthetic	Look, appearance, watch size
Tracking	App features

A watch can be viewed and appreciated from two distinct perspectives: as a utilitarian object or as a source of hedonic value. Let's explore each viewpoint:

- **Utilitarian value:** From a utilitarian perspective, a watch primarily serves as a functional tool for timekeeping and providing practical features. It is seen as a practical accessory that helps individuals manage their daily activities and maintain punctuality. Utilitarian watch features may include accurate timekeeping, alarms, timers, calendars, and other practical functions that aid in time management and organization. In this context, the value of the watch lies in its ability to fulfill specific functional needs efficiently and reliably.
- **Hedonic value:** From a hedonic perspective, a watch holds value beyond its mere functionality and serves as a source of pleasure, aesthetics, self-expression, and personal satisfaction. A watch with hedonic value is chosen for its design, craftsmanship, brand reputation, exclusivity, and emotional appeal. It reflects the wearer's individuality, style, and taste. Hedonic watches often emphasize aesthetics, materials, intricate detailing, and unique features that enhance the overall appeal and create an emotional connection with the wearer. The pleasure derived from owning and wearing such a watch goes beyond pure functionality and can provide a sense of luxury, status, or personal enjoyment. It's important to note that these perspectives are not mutually exclusive, and many watches combine both utilitarian and hedonic aspects. The specific balance between utilitarian and hedonic attributes varies among individuals based on their preferences, needs, and motivations. Some individuals prioritize the utilitarian aspects and seek watches that excel in functionality and reliability, while others prioritize the hedonic aspects and value watches for their design, brand, or emotional connection. Ultimately, the perceived value of a watch is subjective and can

Topic wise terms

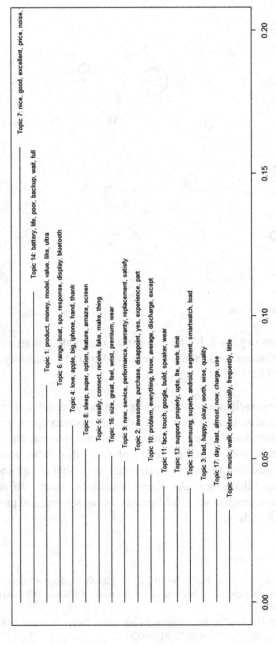

Topic 7: nice, good, excellent, price, noise,

Topic 14: battery, life, poor, backup, wait, full

Topic 1: product, money, model, value, like, ultra

Topic 6: range, boat, spo, response, display, bluetooth

Topic 4: love, apple, big, iphone, hand, thank

Topic 8: sleep, super, option, feature, amaze, screen

Topic 5: really, correct, receive, fake, make, thing

Topic 16: size, great, feel, wrist, premium, wear

Topic 9: new, service, performance, warranty, replacement, satisfy

Topic 2: awesome, purchase, disappoint, yes, experience, part

Topic 10: problem, everything, know, average, discharge, except

Topic 11: face, touch, google, build, speaker, wear

Topic 13: support, properly, upto, lte, work, limit

Topic 15: samsung, superb, android, segment, smartwatch, load

Topic 3: bad, happy, okay, worth, wise, quality

Topic 17: day, last, almost, now, charge, use

Topic 12: music, walk, detect, actually, frequently, little

0.00   0.05   0.10   0.15   0.20

Expected Topic Proportions

IMAGE 24.4 Topic summary

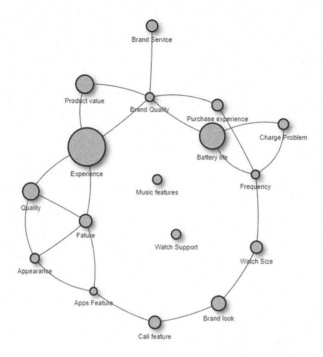

IMAGE 24.5   Correlation between topics

vary greatly from person to person. It's the combination of both utilitarian and hedonic attributes that contributes to the overall appeal and desirability of a watch in the eyes of the wearer.

Finally, Image 24.5 shows the correlation between topics. Appearance, app features, and call features were correlated with each other. First, when a consumer appreciates appearance the same consumer praises the app and call features. Second, battery life, charging issues, and brand quality were correlated with each other. When a consumer complained or appreciated about battery life, the same consumer complained or appreciated about battery and charging issues. Third, brand look, watch size, and call features were correlated with each other. Brand quality, brand service, and product value were correlated with each other.

REFERENCES

Adapa, A., Nah, F. F. H., Hall, R. H., Siau, K., & Smith, S. N. (2018). Factors influencing the adoption of smart wearable devices. *International Journal of Human–Computer Interaction*, *34*(5), 399–409.

Ajzen. (1991). "The theory of planned behavior," *Organ Behav Hum Decis Process*, *50*(2), 179–211. doi: https://doi.org/10.1016/0749-5978(91)90020-T.

Business Insider. (2023). India dominates smartwatch market accounts for over a quarter of global sales in q1. Available at: https://www.businessinsider.com/tech/news/india -dominates-smartwatch-market-accounts-for-over-a-quarter-of-global-sales-in-q1 -2023/articleshow/100522354?IR=T

Channa, A., Popescu, N., Skibinska, J., & Burget, R. (2021). The rise of wearable devices during the COVID-19 pandemic: A systematic review. *Sensors, 21*(17), 5787.

Chidambaram, S., Maheswaran, Y., Patel, K., Sounderajah, V., Hashimoto, D. A., Seastedt, K. P., ... Darzi, A. (2022). Using artificial intelligence-enhanced sensing and wearable technology in sports medicine and performance optimisation. *Sensors, 22*(18), 6920.

Chuah, S. H. W., Rauschnabel, P. A., Krey, N., Nguyen, B., Ramayah, T., & Lade, S. (2016). Wearable technologies: The role of usefulness and visibility in smartwatch adoption. *Computers in Human Behavior, 65*, 276–284.

Davis, F. D. (1989). Perceived usefulness, perceived ease of use, and user acceptance of information technology. *MIS Quarterly*, 13.3, 319–340.

Delgosha, M. S., Hajiheydari, N., & Talafidaryani, M. (2022). Discovering IoT implications in business and management: A computational thematic analysis. *Technovation, 118*, 102236.

Ding, K., Choo, W. C., Ng, K. Y., & Ng, S. I. (2020). Employing structural topic modelling to explore perceived service quality attributes in Airbnb accommodation. *International Journal of Hospitality Management, 91*, 102676.

King, W. R., & He, J. (2006). A meta-analysis of the technology acceptance model. *Information & Management, 43*(6), 740–755.

Lee, S. M., & Lee, D. (2020). Healthcare wearable devices: An analysis of key factors for continuous use intention. *Service Business, 14*(4), 503–531.

Park, E., Chae, B., & Kwon, J. (2020). The structural topic model for online review analysis: Comparison between green and non-green restaurants. *Journal of Hospitality and Tourism Technology, 11*(1), 1–17.

Patel, V., Chesmore, A., Legner, C. M., & Pandey, S. (2022). Trends in workplace wearable technologies and connected-worker solutions for next-generation occupational safety, health, and productivity. *Advanced Intelligent Systems, 4*(1), 2100099.

Pathania, A., Dixit, S., & Rasool, G. (2022). 'Are online reviews the new shepherd?'–examining herd behaviour in wearable technology adoption for personal healthcare. *Journal of Marketing Communications*, 1–27.

Rafl, J., Bachman, T. E., Rafl-Huttova, V., Walzel, S., & Rozanek, M. (2022). Commercial smartwatch with pulse oximeter detects short-time hypoxemia as well as standard medical-grade device: Validation study. *Digital Health, 8*, 20552076221132127.

Ronghe, V., Mahakalkar, M., Thool, B., Ankar, R., Gomase, K., & Teltumbde, A. (2022, August). A wearable device for remote vital signs monitoring: A review article. In *2022 3rd international conference on electronics and sustainable communication systems (ICESC)* (pp. 966–969). IEEE.

Sergueeva, K., Shaw, N., & Lee, S. H. (2020). Understanding the barriers and factors associated with consumer adoption of wearable technology devices in managing personal health. *Canadian Journal of Administrative Sciences/Revue Canadienne des Sciences de l'Administration, 37*(1), 45–60.

Times of India. (2023). India wearable market grows 47% YoY in 2022 led by smartwatches. Available at: https://timesofindia.indiatimes.com/gadgets-news/india-wearable-market -grows-47-yoy-in-2022-led-by-smartwatches/articleshow/97737160.cms

Tirunillai, S., & Tellis, G. J. (2014). Mining marketing meaning from online chatter: Strategic brand analysis of big data using latent dirichlet allocation. *Journal of Marketing Research, 51*(4), 463–479.

Trauth, E., & Browning, E. R. (2022). Technologized talk: Wearable technologies, patient agency, and medical communication in healthcare settings. In E. Chocolate Ave., Hershey, PA (ed.) *Research anthology on improving health literacy through patient communication and mass media* (pp. 558–587). IGI Global.

Venkatesh, V., Morris, M. G., Davis, G. B., & Davis, F. D. (2003). User acceptance of information technology: Toward a unified view. *MIS Quarterly*, 27.3, 425–478.

Venkatesh, V., & Davis, F. D. (2000). "Theoretical extension of the Technology Acceptance Model: Four longitudinal field studies," *Manage Sci*, *46*(2), 186–4204. doi: 10.1287/mnsc.46.2.186.11926.

Vijayalakshmi, K., Uma, S., Bhuvanya, R., & Suresh, A. (2018). A demand for wearable devices in health care. *International Journal of Engineering & Technology*, 7(1), 1–4.

Webster, C. S., Scheeren, T. W., & Wan, Y. I. (2022). Patient monitoring, wearable devices, and the healthcare information ecosystem. *British Journal of Anaesthesia*, *128*(5), 756–758.

25 Case Study

Hemachandran K, Someshwar, Raul V. Rodriguez,
Channabasava Chola, and Kyung Hee

INTRODUCTION

Ever since its birth in 1956, when John McCarthy first coined the term, artificial intelligence (AI) has been seeing an upsurge. In the past decade, (since 2010), the number of publications in the field of AI has increased by 205%.[1] Ever since, AI has been adopted in multiple sectors such as gaming, finance, social media, marketing, education, healthcare, space, robotics, and other sectors. This rapid growth in AI across sectors has been possible due to its diverse and versatile nature.

Companies have been deploying AI models and techniques in their businesses to either build new products and services or improve their existing products and services to gain a competitive advantage. Be it the voice assistants of Apple, Google, and Amazon (Siri, Google Assistant, and Alexa respectively) or the recommendation systems of TikTok, Google, and Netflix, or the recent strides of OpenAI and Google in the field of natural language programming (NLP) (GPT-4 and BARD respectively), AI has been slowly creeping its way into our day to day lives. Companies and researchers have been constantly looking at new sectors, trends, and applications of AI to enhance our quality of life.

One such recent trend is in wearables. Wearables or wearable devices refers to all kinds of electronic devices or computers that have been integrated into clothing or accessories and are worn by people or are attached to their bodies to provide personalized functionalities (such as step tracking, calories burned, blood oxygen, heart rate, time asleep, etc.) These wearable devices include glasses, watches, headbands, helmets, and jewelry, or any other electronic device, software, or sensors, like electronic skin patches as well as smart chest straps, contact lenses, electronic skin products, or smart patches that can be attached to clothing or other accessories, and can be worn on the body.[2]

Wearables were first introduced when Edward Thorp and Claude Shannon wanted to help gamblers beat the casinos in the game of roulette. To this end, they both invented a wearable device in the year 1961, with 4 buttons, that could fit inside a shoe or could be worn around the wrist. This device would be used to predict where the ball would land.[3] In the year 2004, after Apple released its latest product, the "Apple Watch," the wearable devices gained massive stardom.[4] Subsequently, after wearables were enabled with AI, the wearable devices industry has evolved rapidly and hasn't turned back since.

DOI: 10.1201/9781032686714-25

As of 2023, the global wearable AI market is valued at $20 Billion USD and is expected to reach $230 Billion USD in the next 10 years with an astonishing compound annual growth rate (CAGR) of 27.6%.[5] This surge in the global wearable AI market is fuelled by both technological advancements (in the form of increasing AI assistant usage, growing penetration of AI in business, and the development of Internet of Things (IoT) and wireless technology) and consumer demands (such as rapid urbanization, rising disposable income, and increasing concerns about chronic diseases by using wearable devices such as fitness trackers, body monitors, and smartwatches).[6]

The rest of the chapter is organized as follows. The second section covers the first case study that uses wearable devices to track the activity of workers in car manufacturing. The third section covers the second case study that uses wearable devices to monitor worker fatigue. The fourth section covers the third case study that uses wearable devices to monitor children's learning. I the final section, we conclude the chapter and highlight future scope.

CASE STUDY 1 – WEARABLE DEVICES IN CAR PRODUCTION

OVERVIEW

The wearIT@work is a 4.5-year-long project that is funded by the European Union to promote real-life industrial development of wearable devices. This project has €23.7 Million Euros in funding and 42 partners. This project seeks to promote wearable device applications in four sectors – aircraft maintenance, car production, healthcare, and emergency response.[7]

This study was carried out in cooperation with the European car manufacturer, Skoda[8]. The researchers found that there was insufficient research available on the complex activity recognition for industrial applications; hence, the purpose of this study was to investigate the real-life applications of wearable devices in the context of car production. Through this study, the researchers wanted to develop a context-aware wearable device that would track and assist a production or maintenance worker by identifying the tasks that the worker is performing and delivering accurate and timely information about the tasks to be performed. The objective of this study was to 1) offer interactive assistance to the workers 2) allow the supervisor to track and monitor the progress of multiple workers at once, and 3) create a log sheet that could be used to evaluate the progress of the workers/trainees.

CASE DESCRIPTION

The researchers decided to go ahead with using wearable devices for tracking complex activity recognition in car production because they saw the following benefits of wearable devices.

- **Overt benefits:** Wearables are physically unobtrusive which leaves the hands of the workers free, giving them full and free range of motion.

- **Covert benefits:** Using wearables to assist and track the tasks of the workers ensures that they can avoid distractions and the minimize cognitive load taken by the worker.

The researchers were from Spain's Tekniker Foundation which was supported by the German consulting company Unity. To carry out this study, the researchers used three of the facilities of Skoda which were located in the cities of Mlada, Boleslav, and Vrchlabi,[9] and recreated the Skoda facilities at the wearable computing lab at ETH in Zurich, Switzerland. Moving forward, this study has been split into two parts so that readers can easily follow: Part I focuses on the use of wearable devices to train context-aware workers for the assembly line, Part II focuses on the use of wearable devices to provide assistance to workers for quality control.

Part I

The training process in the training center of Skoda for an assembly line worker involved two steps. In the first step, the workers were taught all the theoretical concepts related to the assembly of the car in a classroom. After this, the workers were then shifted to a "learning island," where they were given practical and hands-on training. To put it quite simply, a learning island is nothing but a car body that has been stripped down completely, where the workers repeatedly practice assembling and removing the components of the car.

As it wouldn't be feasible to carry out the study in the actual production facilities, the researchers, along with the help of Skoda, were able to recreate the entire setup of the learning island on their own land. For the purpose of this study, the researchers decided to focus on training the workers with the installation of only headlights. The primary reason for the researchers to choose this was that the installation of headlights was representative of a wide range of tasks in terms of complexity as it involves handheld and hand-operated tools, maneuvring gestures, and multiple other interactions between the worker, assembly parts, and the car body. The secondary reason for this was that the researchers observed that there was a considerable skill gap between the workers when it came to installing headlights.

Setup

The installation of the headlights consisted of four steps:

1. Inserting the headlight
2. Using three screws and a cordless screwdriver to mount a supportive plastic bar
3. Fixing the headlight on the supportive plastic bar with the help of two screws and a cordless screwdriver
4. Alignment and adjustment check of the headlight.

In order to track these steps, the researchers used the sensors discussed below which were placed on both the body of the worker and the car body.

Radio Frequency Identification (RFID) Sensors

The researchers used RFID tags and sensors.

- The researchers placed RFID tags on the tools used in the setup.
- The researchers placed RFID readers between the thumb and index finger of the worker to identify which tool was being used by the worker.

Force-Sensitive Resistors (FSRs)

The researchers used FSRs on both the car body and on the body of the worker.

- The researchers placed FSRs on the car body, placed next to the screw joints. These FSRs are able to convey the attachment of the headlights and the supportive bar on the car body by indicating the tightening of the corresponding screw.
- The researchers have also placed FSRs on the straps that are worn by the workers around their forearms. These FSRs are able to detect the muscle contractions in the forearms as and when the worker grabs the tools (such as the screwdriver or the screws) and the assembly parts (the headlight or the supportive bar).

Inertial Measurement Unit (IMU)

To determine whether the worker has completely or partially tightened the screw, the researchers used an IMU which was mounted at the back of the palm of the worker. Once the screw was completely tightened, the screwdriver reaches the torque limiter. This causes the screwdriver and the hand of the worker to vibrate and shake. The IMU captures this information. These IMUs also help track the hand movements of the workers.

Magnetic Switches

- Four magnetic sensors are placed in the headlight cavity to detect the insertion of the headlight. These switches help detect subtitle movements when the worker tries to verify the headlight adjustment using a special adjustment tool.

Apart from these four primary sensors, the researchers have also used a data acquisition module (that has been attached to the car body) and a Bluetooth link (that has been integrated into the strap worn by the workers).

Process

To track the setup process of the headlight, the researchers used a task modelling scheme which is similar to a finite state machine (FSM). With the FSM model, the researchers have renamed the four steps involved in the installation of the headlight into "states." The researchers have used the word "transitions" to refer to the possible sequences of the states and they indicate the progression from 1 assembly state to another.

Figure 25.1 gives a detailed description of the FSM model and how the researchers have used the sensors to track the different states. The study done by T. Stiefmeier et al, on the paper "Event-Based Activity Tracking in Work Environments" provides a better and in-depth explanation of the entire process.[10]

In Figure 25.1, "A" stands for screws and A1, A2, and A3 refer to the three screws required to mount the supportive plastic bar, and A4 and A5 refer to the two screws needed to fix the headlight on the supportive plastic bar. The FSRs that are attached to the car body are used to detect this.

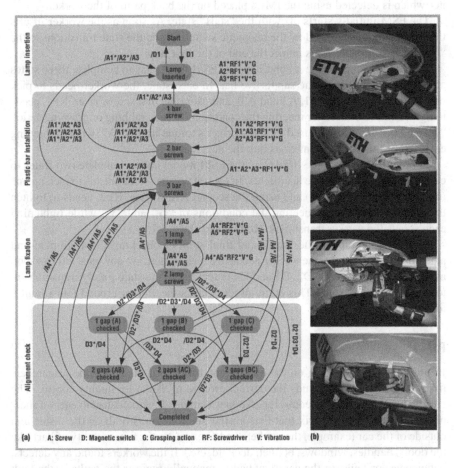

FIGURE 25.1 The "/" represents the logical operator "not" and the "*" represents the logical operator "and." **Source:** Stiefmeier, Thomas; Roggen, Daniel; Ogris, Georg; Lukowicz, Paul; Tr, Gerhard (2008). Wearable Activity Tracking in Car Manufacturing. *IEEE Pervasive Computing*, 7(2), 42–50. doi: 10.1109/MPRV.2008.40

"D" stands for magnetic switch. D1 refers to the first magnetic switch that is used to detect the insertion of the headlight. D2, D3, D4 are used to check the alignment of the headlight. The magnetic switches that are installed in the car body are used to detect this.

"G" stands for the grabbing, or the grasping action made by the worker to grab tools (screwdriver or the screws). The FSRs that are placed on the strap that is worn around the forearm of the worker is used to detect this.

"RF" stands for screwdriver. RF1 refers to the first cordless screwdriver that is used to insert the headlight and RF2 refers to the second cordless screwdriver that is used to fix the headlight onto the supportive bar. The RFID tag and sensors are used to detect this.

"V" stands for the vibration caused by the onset of the screwdriver's torque limiter which is detected using the IMUs placed on the back palm of the worker.

The FSM initially starts off with the "start" state. As soon as the worker inserts the headlight, it is detected by the magnetic switch D1and the state transitions to the "lamp inserted" state. This marks the end of the first step.

In order to complete the second step, the FSM model has to complete three states after the worker mounts the supportive bar. In each state, the worker needs to pick one of the three screws (A1, A2 and A3), and using the first screwdriver, RF1, needs to tighten the screws. After meeting the conditions of the grabbing or grasping action "G," and the vibrations caused by screwdriver's torque limiter "V," while tightening each screw, the worker transitions into the next state.

In order to complete the third step, the FSM model has to complete two states to complete the fixation of the headlights. In each state, the worker needs to pick up one of the two screws (A4, A5) and using the second screwdriver, RF2, needs to tighten the screws. Again, after meeting the conditions of "G," and "V," while tightening each screw, the worker transitions into the next state.

In order to complete the fourth and final step, of checking the alignment and adjustment of headlight, the worker needs to check if there are any gaps left. As soon as the magnetic switches D2, D3 and D4 detect them, the state finally transitions in to the "completed" state. This entire process takes the worker around four minutes to complete.

Part II

In order to begin working on the quality control part of the study, the researchers firstly had to assess and study the current process in Skoda for quality control. The current process of quality control in Skoda was manual; workers had to carry a sheet of paper with a fault matrix that lists all the possible faults both on the inside and outside of the car (example, the workers had to examine the functionality of part like the doors, handles, windows, bonnet, deck lid, etc.). If the workers found any defects, deficiencies or faults in the car, they had to manually register the faults in the fault matrix sheet and submit it to the supervisor.

The primary functions of wearable devices in this study are to:

1. Raise warning signs every time a worker misses a step in the quality control process
2. Enable workers to directly enter any defects or faults found into an electronic database

Setup

To track the workers in this study, the researchers have decided to go ahead with using only body-worn sensors. The researchers chose not to mount any sensors on the car bodies as in the previous study due to the lack of feasibility of replicating it in car production. To this end, the researchers have built a "motion jacket" to track the motion of the worker and their location. The motion jacket was built with the following sensors:

Inertial measurement unit: The researchers have used seven IMUs in total to track the motion in the upper body of the worker and their activities. The IMUs are placed on the torso and on either side of the upper arm, lower arm, and hands.

- **Force-sensitive resistors (FSRs):** The researchers have attached FSRs near the forearms of the sleeves of the "motion jacket" to track the muscle movements.
- **Other tags and sensors:** The researchers have used ultrawideband sensors from Ubisense to determine the workers' location with respect to the car. The researchers attached these tags on the sleeves of the worker's shoulders to better track their position with the help of the four transmitters positioned around the car. Apart from these, the researchers have also placed two data acquisition and transfer units on the motion jacket.

Process

For this quality control procedure, the researchers were able to identify 46 activities that could be tracked using the wearable devices. A worker would require seven minutes to complete all of these activities. The researchers asked eight workers to use the motion jacket and asked each worker to perform the quality control procedure ten times. Figure 25.2 shows the activity tracking of workers while performing six activities. The stick figures above each picture shows the upper body motion of the worker while performing these activities.

In order to accurately collect the data stream coming from all the various sensors, the researchers put one person in charge of annotating the starting and ending points of each activity to provide ground truths, and one person in charge of annotating the worker's location. Both used the context recognition network toolbox and they had to work simultaneously so that the annotated streams synchronized with the data streams[11]. With this, the researchers were able to collect data for nearly 3,680 activities for 560 minutes of data. Out of the 560 minutes of data, the researchers identified only 320 minutes of actual data (the remaining minutes contained unnecessary noise such as activities that are not directly related to the study, transitions in between activities, etc.)

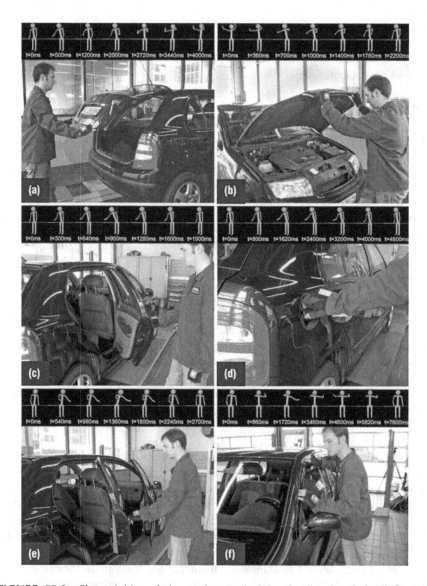

FIGURE 25.2 Six activities of the worker tracked by the "motion jacket." **Source:** Stiefmeier, Thomas; Roggen, Daniel; Ogris, Georg; Lukowicz, Paul; Tr, Gerhard (2008). Wearable Activity Tracking in Car Manufacturing. *IEEE Pervasive Computing, 7*(2), 42–50. doi: 10.1109/MPRV.2008.40

Generally speaking, there are two steps involved when building any activity tracking algorithm. In the first step, we need to segment the continuous stream of data from the sensors by identifying and highlighting only sections that contain relevant information (hence, the need for two people to constantly and simultaneously annotate the data at ground level). The second and final step is to classify the identified sections (hence why the researchers narrowed down the data stream from 520 to 320 minutes).

For this study, the researchers were able to build a new activity tracking algorithm that can perform both of the steps involved in the activity tracking algorithm mentioned above in just one step. This new algorithm was inspired by the approximate string-matching technique and relies only on simple arithmetic operations[12]. With the help of this new algorithm, the researchers were able to achieve an accuracy of 74% in the data stream and were able to narrow down the 3,680 activities, to just 480 precise activities.

CONCLUSION

Let us revisit the objectives of this study, which where to: 1) offer interactive assistance to the workers 2) allow the supervisor to track and monitor the progress of multiple workers at once, and 3) create a log sheet that could be used to evaluate the progress of the workers/trainees.

Through this study, the researchers were able to achieve the first objective with the help of the context aware wearable device that tracks the worker and provides assistance to them in real time. This helped the workers to receive timely information and instructions about their activities, reduced their learning time, and improved their efficiency. The researchers were able to achieve the second and third objectives by developing algorithms and using sensors and the motion jacket to track the progress of activities and evaluate the worker in real time.

When deploying such futuristic technologies (such as context recognition, activity tracking, etc.), it is common for the workers to show resistance and raise concerns about their privacy and autonomy. However, this study found that none of the workers had raised any such concerns. On the contrary, the workers had a positive response towards them. The workers said that task tracking feature and the ability of the wearable device to display error messages whenever they either skipped a step or made a mistake was of great value to them.

In summary, this study shows the benefits of utilizing wearable AI devices in the industrial context and how easy it is to implement this technology. It also provides insight into how the adopters (or the end users) show zero resistance to such technologies if they find them to truly aid, add value and improve the efficiency of their work rather than replace them.

CASE STUDY 2 – WEARABLE DEVICES TO MONITOR FATIGUE

OVERVIEW

Fatigue is generally defined as "a state of feeling tired, weary, or sleepy that results from prolonged mental and physical work, extended periods of anxiety, exposure to harsh environment, or loss of sleep."[13]. Some of the reasons for fatigue include long working hours, harsh work environments, medical reasons, lengthy and complex production process, poor management, working with inconvenient and awkward postures, etc.[14]

Prolonged exposure to fatigue has severe health implications on a worker. The short-term health implications include discomfort, reduced strength and stamina, and reduced motor control. Some of the long-term health implications include development of musculoskeletal disorders, development of chronic fatigue syndrome and diminished immune function.[15] Worker fatigue also has some negative implications on the companies such as an overall reduction in performance, productivity, and quality of work and an increase in the number of worker accidents and human errors.[16]

Due to the above-mentioned negative implications that fatigue has on both the workers and the companies, it is important for us to track the onset of fatigue in workers so that we can negate these implications, and this precisely is the purpose of the researchers in this study.

CASE DESCRIPTION

The objective of this study is to:

1. Detect the onset of fatigue in the workers
2. Classify the different stages of onset and development of fatigue
3. Learn how fatigue varies from person to person[17]

For the purpose of this study, the researchers have limited themselves to detecting only physical fatigue in the workers as it is relatively the easiest one to track.

Currently, there are three ways to track and measure physical fatigue.

- **Blood sampling:** As the name suggests, in this method, we need to take the blood samples of the workers to measure levels of hormones (e.g., cortisol and adrenaline) and other metabolic changes to detect fatigue. Out of all the three methods, this is the most accurate and reliable method. However, it is very intrusive and expensive and cannot be practically implemented.
- **Electromyography (EMG):** In this method, we need measure the electrical activity of the muscles by placing electrodes on the skin over the muscles. This method is highly accurate and relatively less intrusive and expensive than blood sampling. However, EMG is suitable in scenarios where there are only stationary tasks, hence, it is unsuitable for our setting.
- **Feature-based methods:** In this method, we use sensors to track changes in physical parameters of the body (such as posture, walking patterns, and eye movements) to detect fatigue. These methods are the least intrusive and expensive, but they may not be as accurate as the other methods.

After weighing out the pros and cons of the three methods, the researchers have decided to go ahead with using the feature-based methods to detect fatigue. The researchers have also limited themselves to only consider the gait parameters (stride length, height, and duration) to detect fatigue as it is shown that walking is ubiquitous in numerous manufacturing sectors such as construction, mining, nursing, distribution centres, etc.

The research was carried out by gathering data from 15 participants who had sensors attached to them and were asked to preform various physical activities (similar to the ones they perform in their job) for 3 hours. They then combined exploratory data analysis (EDA) methods, multivariate nonparametric change point methodology, and time series clustering to analyse the data and to answer the research questions.

SETUP

For this study, the researchers used an inertial measurement unit sensor and attached it to the right ankle of each participant. The researchers used these IMU sensors as these sensors included three more sensors, namely, an accelerometer (to measure the force exerted by the body), a gyroscope (to measure the angular rate of change in the foot movement), and a magnetometer (to measure the magnetic field).

Prior to attaching the IMUs and starting the study, the participants were briefed about the experimental setup and were given time to practice the procedure and warm up their bodies before the actual study started. The participants were then asked to perform high level manual material handling tasks for three hours. Firstly, the participants were asked to randomly pick up one of 18 cartons; half of the cartons were painted a blue colour and the half ared and were divided into 3 lots based on weights – 10 kg, 18 kg and 26 kg with 6 cartons in each weight. The participants were then asked to deliver the cartons selected by them to the destination by loading the cartons on a two wheeled dolly and pushing it.

The distance between the loading point and the destination was around 80 meters. The participants were expected to complete all the delivers within 30 minutes. Every 10 minutes, the participants provided a rating of their perceived exertion (RPE) on a scale of 6-20 and every 30 minutes, the participants provided a rating of subjective fatigue level (SFL).[18]

Figure 25.3 shows a visual summary of the experimental setup.

PROCESS

The researchers divided the process of the extracting data from the IMU sensors into four steps.[19]

1. **Stride segmentation:**The first step involves the identification of the beginning and end points of each stride from the data series. The researchers were able to identify these by looking at readings from the accelerometer that were close to zero. The researchers also identified instances when the readings from the gyroscope were close to zero. If readings from both were close to zero, the researchers were able to deduce that the foot was stationary on the ground and used these data points to demarcate strides.

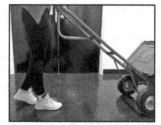

Picking up the box   Loading the box on dolly   Pushing the dolly on the path

Pushing the dolly on side 1 of   Pushing the dolly on side 2 of   The type of footwear and
the walkway (front view)   the walkway (back view)   placement of the IMU at the ankle

FIGURE 25.3 A detailed visual summary of the experimental setup. **Source** Baghdadi, Amir; Megahed, Fadel M.; Esfahani, Ehsan T.; Cavuoto, Lora A. (2018). A Machine Learning Approach to Detect Changes in Gait Parameters Following a Fatiguing Occupational Task. *Ergonomics*, 1–21. doi: 10.1080/00140139.2018.1442936

2. **Rotational orientation estimation:**In the second step, the researchers use data from the gyroscope to estimate the orientation of the foot over time. However, the readings from the gyroscope are generally susceptible to drift (which means that it slowly accumulated errors over time). To rectify this, the researchers have used a Kalman filter (KF). KF is a statistical method that is used to estimate the state of a system from noisy measurements. In this case, the state of the system is the orientation of the foot.

3. **Translational velocity estimation:**In the third step, the researchers use data from the accelerometer to record the velocity of foot in air over time. However, data for the accelerometer is also susceptible to drift. To correct this issue, the researchers made adjustments so that the velocities are set to zero when the foot is both stationary and in contact with the ground.

4. **Trajectory formation:**The final step involves building a trajectory of the foot's movement over time. The researchers have accomplished this by integrating all the corrected velocities (adjusted for drift and errors) from the previous three steps.

Figure 25.4 shows a visual overview of the entire process.

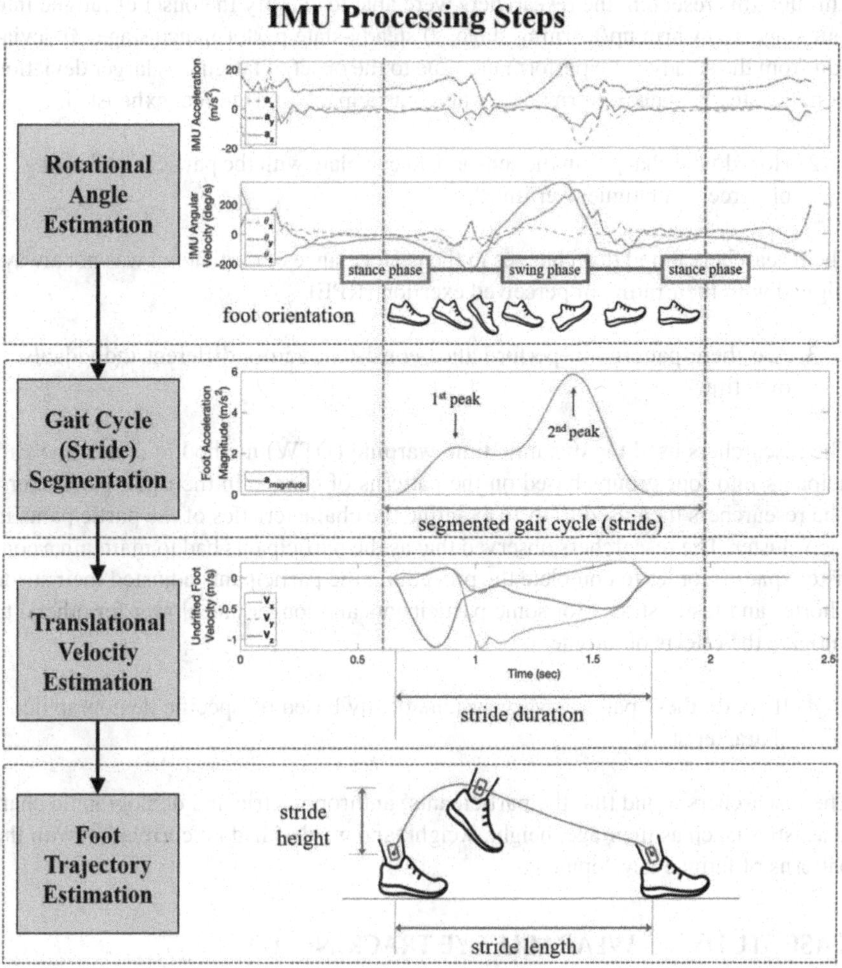

IMU Processing Steps

FIGURE 25.4 Visual overview of IMU processing steps. **Source:** Amir Baghdadi, Lora A. Cavuoto, Allison Jones-Farmer, Steven E. Rigdon, Ehsan T. Esfahani & Fadel M. Megahed (2021) Monitoring Worker Fatigue Using Wearable Devices: A Case Study to Detect Changes in Gait Parameters, *Journal of Quality Technology*, 53(1), 47–71, DOI: 10.1080/00224065.2 019.1640097 Rebula, John R.; Ojeda, Lauro V.; Adamczyk, Peter G.; Kuo, Arthur D. (2013). Measurement of Foot Placement and its Variability with Inertial Sensors. *Gait & Posture*, 38(4), 974–980. doi: 10.1016/j.gaitpost.2013.05.012

CONCLUSION

The key findings from this study can be summed up by answering the four research questions that were asked by the researchers:

1. How do important gait parameters (such as stride length, height, and duration) change over time?

Through this research, the researchers were able to classify the onset of fatigue into four stages: 1) warm up/learning stage, 2) steady-state performance stage, 3) deviation from the steady-state performance due to the onset of fatigue, 4) larger deviation from the steady-state performance, where participants start to feel exhausted.

  2. How do the changes in the sensor data correlate with the participants' rating of perceived fatigue over time?

The researchers found that changes in the performance of participant was not always aligned with their rating of perceived exertion (RPE).

  3. Are there patterns in performance consistent across different individuals over time?

The researchers used the dynamic time warping (DTW) method to cluster the participants into four groups based on the patterns of change in their gait parameters. The researchers then used EDA to examine the characteristics of the participants in each cluster. The researchers observed that as the participants had to maintain a consistent pace in order to complete the procedure, the participants adjusted their stride (shorter and faster strides for some participants and longer and slower for others) to mitigate the effects of fatigue.

  4. If so, do these patterns vary systematically based on specific demographic characteristics?

The researchers found that the participants' anthropometric and demographic characteristics (such as their age, height, weight, and gender) had no correlation with the patterns of fatigue development.

CASE STUDY 3 – WEARABLE EYE TRACKING TO MONITOR CHILDREN'S LEARNING GAINS

OVERVIEW

Traditional methods of assessing student learning have relied upon quantitative methods of evaluation (such as standardized tests, assignments, and examinations) rather than qualitative methods of evaluation (such as case studies, artifacts, interaction analysis, etc). These quantitative methods of evaluation are useful for evaluating the knowledge retention and comprehensive skills of a student; however, they fail to provide an exhaustive understanding of a student's learning gains.

In today's rapidly evolving world, where all the knowledge in the world sits right inside our pockets, there in a dire need for students to be able to think critically and solve problems. This requires more than just knowledge retention. It requires the students to be able to apply their knowledge in creative new ways.

As a result, educators and researchers have been exploring new avenues to assess the learning gains of students. One such avenue is the use of wearable eye-tracking devices to monitor a students' learning gains. These wearable eye-tracking devices collect data by capturing the eye movements (gaze) of student and provide insights into a students' visual attention, cognitive processes, and emotional engagement. This data can then be used to paint a better picture of a students' learning gains and identify areas in which the student is lacking.

Although there have been studies about the use of wearable eye-tracking devices to understand and predict learning gains in students, these studies usually focus on university students or learning adults and are carried out in controlled environments. However, there are very few studies that look at children. Hence, to bridge this gap, the researchers have carried out this study with the main objective being to predict the learning gains of children by tracking their gaze.

CASE DESCRIPTION

To address the objective of the study, the researchers have come up with tworesearch questions

1. Can wearable eye-tracking devices be utilized to predict children's learning during a making-based coding activity workshop?
2. What are the most important gaze-based predictors of children's learning?[20]

This study was supported in part by the European Commission's Horizon 2020 COMnPLAY-Science project and in part by the Norwegian Research Council under the projects FUTURE LEARNING and was carried out in the Norwegian University of Science and Technology.

This study involved 44 child participants between the ages of 8 and 17. This study was carried out over the course of five workshops totalling to around four hours which taught students coding activities and were completed within the duration of two weeks. The researchers conducted the study only after taking consent from the teachers, parents, and the children themselves. Another notable observation was that when the researchers explained the procedure to the children, they became inquisitive about the eye-tracking glasses they were about to wear and provided their full cooperation during the course of the study.

SETUP

To track the children's gaze, the researchers used SensoMotoric Instruments (SMI) and TOBII mobile eye-trackers the recorded data at a sampling rate of 60 Hz.

The workshop was based on a constructionist approach, which is a learning-by-doing method. The workshop was conducted in an informal setting (i.e., outside the premises of schools) and was an out of school activity. This workshop was designed for children with little to no previous experience with coding. The children were divided into groups of three and student volunteers were assigned to one or two groups to assist and guide the children.

The workshop was divided into two parts –

- **Part 1 – Interacting with the robots:** A series of activities were conducted in order to acquaint the children with the robots. The children were taught how to perform simple programming tasks and loops to make the robots react with visual effects such as moving a specific part of the robot when light coming from a sensor was dim. To create this, the researchers used the hardware platform Arduino (Arduino is an open-source electronics platform that allows users to create interactive projects) and Scratch (Scratch is a block-based programming language that allows users to create interactive stories, games, and animations) for Arduino tools. This part of the workshop lasted between 45 minutes to 90 minutes.
- **Part 2 – Developing games with Scratch:**In the second part of the workshop, the children had to develop their own games. The children had to work collaboratively to design and program a game using Scratch. The children were given a step-by-step process of how to do so. Firstly, they had to make a draft storyboard of their idea, then, they had to write code using simple loops and other computational think concepts taught during the workshop, and finally they had to debug their game. This part of the workshop lasted for nearly three hours.

Apart from the activities, the children had to complete a pre and post knowledge test. These tests were used to provide ground truths. These tests were designed to assess the children's computational thinking skills and consisted of nine questions of increasing difficulty. It took the children around ten minutes to complete the tests. The children wore the eye-tracking glasses throughout the course of the workshop to monitor their gaze.

PROCESS

The researchers wanted to assess the effects of the following measures on the children.

- **Effect of coding activity on children's learning:** The researchers used relative learning gains (RLG)[21] to measure the effect of coding on children's learning. The researchers chose RLG instead of simply measuring learning gains because RLGs take into account the difficulty of learning new information when the learner already knows a lot about a subject. In our context, RLG measures how much a child's learning has increased after participating in the workshop, compared to how much their knowledge would have increased if they had not participated in the workshop. It is calculated by comparing a child's pre and post knowledge test scores.

- **Effect of cognitive load on children's learning:** Cognitive load refers to the amount of mental effort required by a person to perform a task. For this study, the researchers used the method proposed by another study done for measuring cognitive load using eye-tracking data.[22] This method involves using three parameters to calculate cognitive load, namely, 1) pupil diameter, 2) number of long fixations, and 3) saccade speed.

Pupil diameter is a measure of how wide a person's pupils are. When people focus intensely on something, their pupils tend to dilate, hence, this is used to measure cognitive load.

Number of long fixations is a measure of how often and for how long a person fixates their eyes on a particular object. When people tend to pay attention and concentrate on something, their eyes seem to focus and fixate on that particular object for a long period of time, hence, this is used to measure attention.

Saccade speed is a measure of how far and quick a person's eyes move (also called a saccade). When a person is familiar with something, their saccade speed tends to be shorter, hence, it can be used to measure anticipation.

It is calculated by combing the mean and standard deviation of pupil diameter along with the number of long fixations and saccade length.

- **Effect of anticipation on children's learning:** Anticipation is the ability to predict what will happen next (in our context, the ability to predict where to find the elements of interface in Scratch). If a person is familiar with the interface of Scratch, they will be able to move their eyes between different parts of the interface more quickly to solve the given problem. Saccade velocity refers to the speed at which the eyes move when they make a saccade. It is calculated using the skewness of the saccade velocity histogram.

CONCLUSION

Through this study, the researchers were able to find two of the most important gaze-based predictors of children's learning.

- **Children's anticipation:** The researchers found that children who either had prior knowledge about coding or children who developed knowledge quickly during the workshop where able to anticipate which task they had to do next. This allowed them to learn quicker than their peers. The researchers found that the children who were able to anticipate their next tasks had shorter saccades. This suggests that these children were able to move their eyes quickly to the parts of the Scratch interface that they needed to focus on.
- **Children's attention, joint attention, and the level of focus** It was found that children with higher numbers of long fixations reported higher levels of focus and attention. This means that children who spent more time with

interface and on-screen, had higher RLG. it was also found that children who experienced moments of joint attention (when two or more children worked together on a same task), they reported higher RLG.

As we can see, cognitive load is not one of the top predictors of learning gains. This means that the amount of mental efforts and energy a child puts into learning does not correlate to how much they learn. This indicated that it is essential for a child to relax their minds during learning and not feel overwhelmed. Meaningful learning occurs when the cognitive load of the children is moderate. Hence, it is important to design learning frameworks that mix mentally heavy activities with playful and social elements. This will make sure that the children feel challenged and engaged, but at the same time do not feel overwhelmed.

CONCLUSION

In this chapter, we have looked at some of the applications of using wearable AI in an industrial and educational setting. Let us briefly summarize each of the case studies.

1. **Case study 1:** Wearable AI is used to provide interactive assistance to the users as well as track and monitor their progress in real time. These devices have boosted the efficiency of the user and reduced their learning time.
2. **Case study 2:** Wearable AI provides an alternative to the current methods to track fatigue, thus, helping mitigate the harmful health effects related to prolonged exposure to fatigue by identifying the different stages of its onset.
3. **Case study 3:** Wearable AI helped continuously monitor the learning gains in children in an unobtrusive and unintrusive way.

Case studies 1 and 2 demonstrate how we can use wearable AI to significantly boost the efficiency, productivity, and safety of its users in an industrial setting. Case studies 1 and 3 demonstrate how we can use wearable AI to assist and monitor learning progress of its users; be it adults or children.

REFERENCES

1. (2022). Artificial Intelligence Index Report 2022. Chapter 1: Research & Development. https://aiindex.stanford.edu/wp-content/uploads/2022/03/2022-AI-Index-Report_Chapter-1.pdf
2. Wright, R.; Keith, L. (2014). Wearable Technology: If the Tech Fits, Wear It. *Journal of Electronic Resources in Medical Libraries*, 11(4), 204–216. doi: 10.1080/15424065.2014.969051
3. Yasar, Kinza (2022, May). Wearable Technology. https://www.techtarget.com/searchmobilecomputing/definition/wearable-technology#:~:text=In%201961%2C%20Edward%20Thorp%20and,strapped%20around%20the%20user's%20waist
4. Singh, Peeyush (2023, June). Wearable AI: What Does the Implementation Mean for the Digital World. https://appinventiv.com/blog/ai-and-wearable-technology/

5. Fact.MR. Wearable AI Market: Wearable AI Market Analysis By Product (Smartwatches, Ear Wear, Eye Wear & Others), By Operation (On-Device AI, Cloud-based AI), By Application & By Region – Global Market Insights 2023 to 2033. https://www.factmr.com/report/wearable-ai-market

6. Patidar, Shubham (2023, February). Increasing Adoption of Wearables and Smart Technologies to Drive Sales Growth of Wearable AI Devices. https://www.globenews-wire.com/news-release/2023/02/27/2616278/0/en/Increasing-Adoption-of-Wearables-and-Smart-Technologies-to-Drive-Sales-Growth-of-Wearable-AI-Devices-Fact-MR-Report.html

7. Lukowicz, Paul; Timm-Giel, Andreas; Lawo, Michael; Herzog, Otthein (2007). WearIT@work: Toward Real-World Industrial Wearable Computing. *IEEE Pervasive Computing, 6*(4), 8–13. doi: 10.1109/MPRV.2007.89

8. Stiefmeier, Thomas; Roggen, Daniel; Ogris, Georg; Lukowicz, Paul; Troester, Gerhard (2008). Wearable Activity Tracking in Car Manufacturing. *IEEE Pervasive Computing, 7*(2), 42–50. doi: 10.1109/MPRV.2008.40

9. Stiefmeier, T. et al. (2006, Spring). Event-Based Activity Tracking in Work Environments. *Proceedings 3rd International Forum Applied Wearable Computing,* 3rd International Forum on Applied Wearable Computing 2006, Bremen, Germany, 91–100.

10. Bannach, D. et al. (2006, Spring). Distributed Modular Toolbox for Multimodal Context Recognition. *Proceedings Architecture of Computing Systems Conference, LNCS, 3894,* 99–113.

11. Stiefmeier, T.; Roggen, D.; Tröster, G. (2007). Fusion of String-Matched Templates for Continuous Activity Recognition. *Proceedings 11th IEEE International Symposium Wearable Computers, IEEE CS Press,* 41–44.

12. Sadeghniiat-Haghighi, K.; Yazdi, Z. (2015). Fatigue Management in the Workplace. *Industrial Psychiatry Journal, 24*(1), 12–17. doi: 10.4103/0972-6748.160915

13. Singh, Dr. Anuradha (2021). Fatigue: Types, Causes and Methods to Reduce Fatigue at Work Place. https://www.mddmcollege.ac.in/wp-content/uploads/2021/06/file_60c89147ab613.pdf

14. Lu, Lin; Megahed, Fadel M.; Sesek, Richard F.; Cavuoto, Lora A. (2017). A Survey of the Prevalence of Fatigue, its Precursors and Individual Coping Mechanisms Among U.S. Manufacturing Workers. *Applied Ergonomics.* doi: 10.1016/j.apergo.2017.06.004

15. Yung, Marcus; Bigelow, Philip L.; Hastings, Darnell M.; Wells, Richard P. (2014). Detecting within- and Between-Day Manifestations of Neuromuscular Fatigue at Work: An Exploratory Study. *Ergonomics, 57*(10), 1562–1573. doi: 10.1080/00140139.2014.934299

16. Baghdadi, Amir; Cavuoto, Lora A.; Jones-Farmer, Allison; Rigdon, Steven E.; Esfahani, Ehsan T.; Megahed, Fadel M. (2021). Monitoring Worker Fatigue Using Wearable Devices: A Case Study to Detect Changes in Gait Parameters. *Journal of Quality Technology, 53*(1), 47–71. doi: 10.1080/00224065.2019.1640097

17. Baghdadi, Amir; Megahed, Fadel M.; Esfahani, Ehsan T.; Cavuoto, Lora A. (2018). A Machine Learning Approach to Detect Changes in Gait Parameters Following a Fatiguing Occupational Task. *Ergonomics,* 1–21. doi: 10.1080/00140139.2018.1442936

18. Rebula, John R.; Ojeda, Lauro V.; Adamczyk, Peter G.; Kuo, Arthur D. (2013). Measurement of foot placement and its variability with inertial sensors. *Gait & Posture, 38*(4), 974–980. doi: 10.1016/j.gaitpost.2013.05.012

19. Giannakos, Michail N.; Papavlasopoulou, Sofia; Sharma, Kshitij (2020). Monitoring Children's Learning Through Wearable Eye-Tracking: The Case of a Making-Based Coding Activity. *IEEE Pervasive Computing,* 1–12. doi: 10.1109/MPRV.2019.2941929

20. Dillenbourg, P.; Lemaignan, S.; Sangin, M. *et al.* (2016). The Symmetry of Partner Modelling. *International Journal of Computer-Supported Collaborative Learning, 11,* 227–253. doi: 10.1007/s11412-016-9235-5

21. Buettner, R. (2013). Cognitive Workload of Humans Using Artificial Intelligence Systems: Towards Objective Measurement Applying Eye-Tracking Technology. In: Timm, I. J.; Thimm, M. (eds.), *KI 2013: Advances in Artificial Intelligence.* Lecture Notes in Computer Science, vol. 8077. Springer. doi: 10.1007/978-3-642-40942-4_4

22. Maurtua, P. T. K.; Stiefmeier, T.; Sbodio, M. L.; Witt, H. (2007). A Wearable Computing Prototype for supporting training activities in Automotive Production. 4th International Forum on Applied Wearable Computing 2007, Tel Aviv, Israel, pp. 1–12.

Index

Printed in the United States
by Baker & Taylor Publisher Services

Printed in the United States
by Baker & Taylor Publisher Services